Mathematical Research

Quaternionic Analysis and Elliptic Boundary Value Problems

K. Gürlebeck
W. Sprößig

Volume 56

AKADEMIE-VERLAG BERLIN

In this series original contributions of mathematical research in all fields are contained, such as

- research monographs
- collections of papers to a single topic
- reports on congresses of exceptional interest for mathematical research.

This series is aimed at promoting quick information and communication between mathematicians of the various special branches.

In diese Reihe werden Originalbeiträge zu allen Gebieten der mathematischen Forschung aufgenommen wie

- Forschungsmonographien
- Sammlungen von Arbeiten zu einem speziellen Thema
- Berichte von Tagungen, die für die mathematische Forschung besonders aktuell sind.

Die Reihe soll die schnelle Information und gute Kommunikation zwischen den Mathematikern der verschiedenen Fachgebiete fördern.

Manuscripts
in English and German comprising at least 100 pages and not more than 500 pages can be admitted to this series. With respect to a quick publication the manuscripts are reproduced photomechanically. Authors who are interested in this series please turn directly to the 'Akademie-Verlag'. Here you will get more detailed information about the form of the manuscripts and the modalities of publication.

Manuskripte
in englischer und deutscher Sprache, die mindestens 100 Seiten und nicht mehr als 500 Seiten umfassen, können in diese Reihe aufgenommen werden. Im Interesse einer schnellen Publikation werden die Manuskripte auf fotomechanischem Weg reproduziert. Autoren, die an der Veröffentlichung entsprechender Arbeiten in dieser Reihe interessiert sind, wenden sich bitte direkt an den Akademie-Verlag. Sie erhalten dort genauere Informationen über die Gestaltung der Manuskripte und die Modalitäten der Veröffentlichung.

K. Gürlebeck · W. Sprößig

Quaternionic Analysis and Elliptic Boundary Value Problems

Mathematical Research · Mathematische Forschung

Wissenschaftliche Beiträge
herausgegeben von der
Akademie der Wissenschaften der DDR
Karl-Weierstraß-Institut für Mathematik

Band 56

Quaternionic Analysis and Elliptic Boundary Value Problems

by
K. Gürlebeck
and
W. Sprößig

Quaternionic Analysis and Elliptic Boundary Value Problems

by

Klaus Gürlebeck

and

Wolfgang Sprößig

Akademie-Verlag Berlin 1989

Autoren:

Dr. Klaus Gürlebeck
Technische Universität Karl-Marx-Stadt
Sektion Mathematik

Prof. Dr. Wolfgang Sprößig
Bergakademie Freiberg
Sektion Mathematik

Die Titel dieser Reihe werden vom Originalmanuskript der Autoren reproduziert.

ISBN 3-05-500756-5
ISSN 0138-3019

Erschienen im Akademie-Verlag Berlin, Leipziger Str. 3-4, Berlin, DDR-1086

Lizenznummer: 202.100/413/89
Printed in the German Democratic Republic
Gesamtherstellung: VEB Kongreß- und Werbedruck, Oberlungwitz, DDR-9273
Lektor: Dr. Reinhard Höppner
LSV 1035
Bestellnummer: 764 072 6 (2182/56)

03400

Prefatory Notes

It is well known that complex methods may be advantageously used for the treatment of boundary value problems of partial differential equations in the plane. Moreover it is very important to transfer results of the classical function theory to function theories over domains in R^n. A comprehensive description of hypercomplex function theories is being made by the research group of R.DELANGHE (Gent) with their book "Clifford Analysis" ([BDS]). The application to solving boundary value problems by the help of hypercomplex function theories is not developed in the same extent.
The main aim of this book consists in the statement of a new strategy for solving linear and nonlinear boundary value problems of partial differential equations of mathematical physics by the help of hypercomplex analysis. In our opinion, it is the first summarizing presentation of a complete hypercomplex solution theory including analytical and numerical investigations in only one closed theory.
Using a special operator calculus and a hypercomplex function theory, the authors study questions of the existence, uniqueness, representation and regularity of solutions of above mentioned problems in a unified form. For the sake of simplicity, the authors restrict their investigations to the case of quaternionic calculus. Sometimes, if it seems to be necessary, it is referred to general results in CLIFFORD algebras. Furthermore suitable numerical approaches which are well-adapted to the quaternionic calculus are included too. The authors not only give an insight into boundary collocation methods but also introduce a new collocation procedure. Occurring for the first time, a discrete model of the quaternionic function theory was developed and applied to constructing and investigating of finite difference methods.
The first chapter makes the reader familiar with a basic knowledge in the field of quaternionic analysis. Most of the results are also valid in more general algebras.
In Chapter 2 the authors have studied algebraic and functional analytical properties of generating operators F_Γ, T_G, and D, which denote a CAUCHY-type operator, a quaternionic analogue to the complex T-operator and a generalized CAUCHY-

RIEMANN operator, respectively.
The third chapter only contains an orthogonal decomposition of the space $L_{2,H}(G)$ of quaternionic-valued functions into the subspaces $\ker D \cap L_{2,H}(G)$ and $D(\overset{\circ}{W}{}^1_{2,H}(G))$. This decomposition is an essential methodological instrument throughout the following explanations.
In Chapter 4 a series of linear and nonlinear elliptic boundary value problems of mathematical physics has been investigated by the help of a unified method in a rather complete manner.
Starting with some fundamental functional analytic theorems, a quaternionic version of the boundary collocation method is treated in Chapter 5.
Finally, in Chapter 6 a discrete quaternionic function theory is introduced. These results are used in order to find a well-adapted numerical approach to the analytical theory given in Chapter 4. The line of action is demonstrated by considering the discrete NAVIER-STOKES problem.
The book finishes with an Appendix. It is intended to give a short survey about other questions in the hypercomplex theory which have been investigated recently. The authors apologize in advance that in this summary not all important ideas and papers can be mentioned.
The monograph is suitable for mathematicians, physicists and engineers in research institutes. It has the character of a textbook. All the necessary mathematical preparations are made available. The structure of the method presented is very simple and makes possible a formal use for practical computations. Suitably chosen examples make the reader familiar with the topics and methods of quaternionic analysis. Other special branches such as approximation theory, theory of right-invertible operators, boundary collocation methods, finite difference methods and equations of mathematical physics will be mentioned in the book. Knowledge in numerical mathematics is desirable and facilitates the understanding.
We have to thank Prof. B.SILBERMANN, Prof. H.JÄCKEL (Karl-Marx-Stadt University of Technology) for suggesting the writing of this book Thanks are also due to Prof. P.LOUNESTO (Helsinki University of Technology), Prof. R.DELANGHE (Gent State University), Dr. V.SOUCEK (Prague Charles University),

Prof. B.GOLDSCHMIDT (Halle University), Doz. Dr. H.MALONEK (Pedagogical University of Halle) for stimulating discussions and useful hints for references.
We also wish to thank Miss. BERNHARDT, who looked for mistakes after typing. Furthermore we should like to thank Mr. M.STRAUCH for giving essential advice concerning the English language of the manuscript. Finally, our thanks go to the Akademie-Verlag, especially to Dr.R.HÖPPNER for the realization of this monograph.

Karl-Marx-Stadt, Freiberg, January 1989, K.GÜRLEBECK, W.SPRÖSSIG

CONTENTS

1. QUATERNIONIC ANALYSIS

It is natural to look for generalizations of the theory of functions over the complex field to higher dimensions. The development of physics at the end of the last century proposed new questions in mathematics. Above all it was necessary to find algebraic possibilities in order to advantageously carry out calculations with vector functions over 3-dimensional domains. An algebraic assumption for such applications was HAMILTON's invention of the quaternions in 1843. This discovery was published in a final form in his **famous paper [Ham] "Elements of Quaternions" in 1866.**
Initially no special class of 'regular' functions among all quaternionic-valued functions was considered, similarly to the 1-dimensional case. This decisive step was made many years later in the important papers by R.FUETER [Fue1] and G.MOISIL/N.TEODORESCU [MT], which may be regarded as the starting point for the function theory of quaternions.
The aim of the present chapter is to give a short survey of a real quaternion theory in such a way as it is needed for our further considerations. In this context it is completely impossible to mention all the things which are known about quaternionic functions. For this purpose we refer to the essential papers by A.SUDBERY [Sud], P.LOUNESTO [Lou], B.GOLDSCHMIDT [Gol1], J.RYAN [Ry], V.SOUCEK [Sou] and J.BURES [Bu] and the lecture notes by R.FUETER [Fue2], [Fue3]. **We will not deal with generalizations of the classical function theory in other abstract algebras, either. In the book of F.BRACKX/R.DELANGHE/F.SOMMEN [BDS] the reader can find a function theory in CLIFFORD algebras, while** the publications of K.HABETHA [Hab1], [Hab2] contain fundamental function-theoretical theorems in more general algebras.

1.1 Algebra of Real Quaternions

In this section we will present all the algebraic properties of quaternions which are used throughout the whole book.

1.1.1 Let R^4 be the 4-dimensional Euclidean vector space.

We choose the orthonormal basis $e_0=(1,0,0,0)$, $e_1=(0,1,0,0)$, $e_2=(0,0,1,0)$, $e_3=(0,0,0,1)$. Hence, a vector $a=(a_0,a_1,a_2,a_3) \in R^4$ can be written as

$$a = \sum_{i=0}^{3} a_i e_i \, . \qquad (1.1)$$

Introducing the abbreviation $\hat{a} = \sum_{i=1}^{3} a_i e_i$ we obtain $a=a_0e_0 + \hat{a}$ .Let $b \in R^4$ be another vector,then a multiplication law is given by

$$ab = (a_0b_0 - \hat{a}\cdot\hat{b})e_0 + a \times b + a_0\hat{b} + \hat{a}b_0 \quad , \qquad (1.2)$$

where $\hat{a}\cdot\hat{b}$ and $\hat{a} \times \hat{b}$ are the scalar product and the vector product in R^3, respectively. Obviously, this product is not commutative in general.In this way the vector space R^4 is furnished with the algebraic structure of a ring, which will be named algebra of real quaternions and denoted by H. This letter is chosen in honour of the discoverer of the quaternions, W.R.HAMILTON.

1.1.2 Quaternions may be identified with a special kind of real 4 x 4 - matrices, which have the form

$$a = \begin{bmatrix} a_0 & -a_1 & -a_2 & -a_3 \\ a_1 & a_0 & -a_3 & a_2 \\ a_2 & a_3 & a_0 & -a_1 \\ a_3 & -a_2 & a_1 & a_0 \end{bmatrix} . \qquad (1.3)$$

Here a_i , $i=0,1,2,3$, denote real numbers. Similarly to the representation (1.1), a basis e_0, e_1, e_2, e_3 is given in the matrix calculus by

$$e_0=\begin{bmatrix} 1 & 0 & 0 & 0 \\ 0 & 1 & 0 & 0 \\ 0 & 0 & 1 & 0 \\ 0 & 0 & 0 & 1 \end{bmatrix} , \; e_1=\begin{bmatrix} 0 & -1 & 0 & 0 \\ 1 & 0 & 0 & 0 \\ 0 & 0 & 0 & -1 \\ 0 & 0 & 1 & 0 \end{bmatrix} , \; e_2=\begin{bmatrix} 0 & 0 & -1 & 0 \\ 0 & 0 & 0 & 1 \\ 1 & 0 & 0 & 0 \\ 0 & -1 & 0 & 0 \end{bmatrix} , \; e_3=\begin{bmatrix} 0 & 0 & 0 & -1 \\ 0 & 0 & -1 & 0 \\ 0 & 1 & 0 & 0 \\ 1 & 0 & 0 & 0 \end{bmatrix} .$$

Sometimes this notation is replaced by 1,i,j,k . In [BDS] it is shown that H is an even subalgebra of the well-known PAULI-algebra of quantum physics.

1.1.3 We define the conjugate quaternion

$\bar{a} = a_0e_0 - \sum_{i=1}^{3} a_i e_i$. From (1.2) it immediately follows

$a\bar{a} = \bar{a}a = \sum_{i=0}^{3} a_i^{\,2}$, and therefore the Euclidean norm $|a| = \sqrt{a\bar{a}}$.

We define the real part and the purely imaginary part of a by

$$\mathrm{Re}\ a = a_0 e_0 = \frac{1}{2}(\ a + \bar{a}\ )\ ,\quad \mathrm{Im}\ a = \sum_{i=1}^{3} a_i e_i = \frac{1}{2}(\ a - \bar{a}\ )\ .$$

The inverse of the quaternion a ($a \neq 0$) is obtained as

$a^{-1} = \bar{a}\ |\ a\ |^{-2}$.

A straightforward computation leads to the identities

$$\begin{array}{ll} \text{(i)} & \overline{ab} = \bar{b}\ \bar{a}, \\ \text{(ii)} & |\ ab\ | = |\ a\ ||\ b\ |, \\ \text{(iii)} & \mathrm{Re}(ab) = \mathrm{Re}(ba)\ , \end{array} \qquad (1.4)$$

where a, b are arbitrary elements of H.

1.1.4 It is easy to show by the use of the multiplication rule (1.2) that the basis quaternions fulfil the following relations:

$$\begin{array}{ll} \text{(i)} & e_0^{\,2} = e_0\ ,\quad e_i^{\,2} = -e_0\ ,\ i = 1,2,3, \\ \text{(ii)} & e_i e_j + e_j e_i = 0\ ,\ i \neq j\ ,\ i,j = 1,2,3, \\ \text{(iii)} & e_0 e_i = e_i e_0 = e_i\ ,\ i = 0,1,2,3, \\ \text{(iv)} & e_1 e_2 = e_3\ ,\ e_2 e_3 = e_1\ ,\ e_3 e_1 = e_2\ . \end{array} \qquad (1.5)$$

1.2 H-regular Functions

1.2.1 Throughout the whole book let G be a bounded domain of R^3, and $\partial G = \Gamma$ a sufficiently smooth LIAPUNOV surface. Then functions u defined in G and on Γ, respectively, with values in H will be considered. The so-called H-valued functions may be written as

$$u(x) = \sum_{i=0}^{3} u_i(x) e_i\ ,\quad x \in G\ , \qquad (1.6)$$

where the functions $u_i(x)$ are real-valued. Properties such as continuity, differentiability, integrability and so on , which are ascribed to u have to be possessed by all the components $u_i(x)$, $i = 0,1,2,3$. In this manner the BANACH spaces of H-valued functions are denoted by $C_H^{m,\alpha}$ ($m=0,1,2,\ldots$, $\alpha \in [0,1]$), $L_{p,H}$ ($p \geq 1$), H_H^s ($s \geq 0$) and

$W^k_{p,H}$ ($p\geq 1$, $k=0,1,2,\ldots$). Moreover, $\overset{\circ}{W}{}^1_{2,H}(G)$ stands for all H-valued functions of the space $W^1_{2,H}(G)$, which are vanishing on the boundary Γ. In case of $p = 2$ we introduce in $L_{2,H}(G)$ the inner product

$$(u,v)=\int_G \bar{u}v \; dG \; . \tag{1.7}$$

1.2.2 Let $\vec{a}=(a_0,a_1,a_2,a_3)$ be an arbitrary vector in R^4. The vector $\vec{a}$ corresponds to the quaternion a, which is given by

$$a = \sum_{i=0}^{3} a_i e_i$$

Replacing $\vec{a}$ by the formal vector $\vec{D}=(0,D_1,D_2,D_3)$, where $D_i = \frac{\partial}{\partial x_i}$, $i = 1,2,3$, we obtain the so-called CAUCHY-RIEMANN operator

$$D = \sum_{i=1}^{3} D_i e_i \quad , \tag{1.8}$$

which is an analogue to the 2-dimensional CAUCHY-RIEMANN operator $\frac{\partial}{\partial \bar{z}}$, $z=x+iy$. Its action on the H-valued function

$$u = \sum_{j=0}^{3} u_j e_j$$

is denoted by Du from the left and by uD from the right.

1.2.3 Definition

An H-valued function u is called H-left regular in G iff $u \in C^1_H(G)$ and $Du=0$ in G. If an H-valued function $u \in C^1_H(G)$ solves the equation $uD=0$, then it is named H-right regular in G. Throughout the whole book we agree upon the fact that H-left regular functions shall be called H-regular.

1.2.4 Using (1.5) it is easy to see that Du may be written as

$$Du = (-\operatorname{div} \hat{u})e_0 + \operatorname{grad} u_0 + \operatorname{rot} \hat{u} \quad , \tag{1.9}$$

where

$$\operatorname{grad} u_0 = (D_1 u_0)e_1 + (D_2 u_0)e_2 + (D_3 u_0)e_3 ,$$

$$\operatorname{div} \hat{u} = D_1 u_1 + D_2 u_2 + D_3 u_3 ,$$

and

$$\operatorname{rot} \hat{u} = \det \begin{vmatrix} e_1 & e_2 & e_3 \\ D_1 & D_2 & D_3 \\ u_1 & u_2 & u_3 \end{vmatrix} .$$

Therefore an H-valued function u is named H-regular if u satisfies the so-called MOISIL-TEODORESCU system

$$\begin{aligned} &\operatorname{div} \hat{u} = 0 , \\ &\operatorname{grad} u_0 + \operatorname{rot} \hat{u} = 0 . \end{aligned} \tag{1.10}$$

In particular, where $u_0 = \text{const}$, this system describes an irrotational fluid without sources or sinks.
The consideration of system (1.10) was the starting point in the development of the hypercomplex function theory. The famous paper of G.MOISIL and N.TEODORESCU [MT] appeared in 1931.

1.2.5 The set of all H-regular functions in G will be denoted by $A_H(G)$. Notice that

$$DD = - \Delta \quad , \tag{1.11}$$

where Δ is the Laplacian in R^3. Obviously, if $u \in C_H^2(G)$, then the equation $\Delta u=0$ follows from $Du=0$ in G.
That means that each component of an H-regular function is a harmonic function. Therefore the theory of H-regular functions considered here refines the theory of harmonic functions.

1.2.6 The following simple example shows that the quaternionic product of two H-regular functions is not necessarily H-regular. We take $u=x_1e_1 - x_2e_2$, then we have
$Du=0$, while

$$D(uu) = Du^2 = [-\operatorname{grad} (x_1^2 + x_2^2)]e_0 = -2x_1e_1 - 2x_2e_2 \neq 0 .$$

Assume that $0 \notin G$, then in G $u(x) \neq 0$. In contrast to the 1-dimensional complex function theory the function

$$u^{-1}(x) = (-x_1e_1 + x_2e_2)(x_1^2 + x_2^2)^{-1}$$

is not H-regular. This is because

$$Du^{-1}(x) = \frac{2(x_1^2 - x_2^2)}{(x_1^2 + x_2^2)^2} e_0 + \frac{4x_1x_2}{(x_1^2 + x_2^2)^2} e_3 \neq 0 .$$

The main trouble, however consists in the non-commutativity of the product of two H-valued functions, which causes most of the other peculiarities. It is also the reason why a theory of conformal mappings has been developed only in a smaller extent, than in the classical complex theory. Some results in this field will be referred in the Appendix.

1.2.7 In order to overcome the above mentioned difficulties, R. DELANGHE introduced in [Del1] the idea of totally analytic variables. In this sense, a variable $z \in H$ is called totally analytic if and only if for each $k \in N$ the k-th power of z is H-regular. Let z be represented by

$$z = \sum_{i=0}^{3} x_i d_i \qquad (1.12)$$

with $d_i \in H$ (not necessarily linearly independent).
In [Del1] it is proved that the identity $d_i d_j = d_j d_i$, $i,j=1,2,3$ is sufficient in order that z be a totally analytic variable. In [Gue1] it is pointed out that the property of commutativity of the products $d_i d_j$ is necessary in case of H-valued functions. In the following we intend to study some structural properties of H-regular functions which are connected with totally analytic variables. R. DELANGHE proved it in CLIFFORD algebras.

1.2.8 Let $d_i \in H$, $i = 0,1,2,3$. Then there exist real numbers a_{ij} which are defined by

$$\begin{pmatrix} d_0 \\ d_1 \\ d_2 \\ d_3 \end{pmatrix} = \begin{pmatrix} a_{00} & a_{01} & a_{02} & a_{03} \\ a_{10} & a_{11} & a_{12} & a_{13} \\ a_{20} & a_{21} & a_{22} & a_{23} \\ a_{30} & a_{31} & a_{32} & a_{33} \end{pmatrix} \begin{pmatrix} e_0 \\ e_1 \\ e_2 \\ e_3 \end{pmatrix} = A \{e_i\}_{i=0}^{3} . \qquad (1.13)$$

All the further considerations in this subsection will be concerned with the rank of the matrix

$$A' = \begin{pmatrix} a_{01} & a_{02} & a_{03} \\ a_{11} & a_{12} & a_{13} \\ a_{21} & a_{22} & a_{23} \\ a_{31} & a_{32} & a_{33} \end{pmatrix} . \qquad (1.14)$$

The exact proofs of the following properties of H-regular functions will be omitted here. The reader can find them in [Gue1].

Property 1

$d_i d_j = d_j d_i$, $i,j = 0,1,2,3$, if and only if $\operatorname{rank} A' < 2$.

Property 2

Let $a = (a_0, a_1, a_2, a_3) \in R^4$. We denote by $M(a)$ the matrix

$$M(a) := \begin{pmatrix} a_0 & a_1 & a_2 & a_3 \\ a_1 & -a_0 & -a_3 & a_2 \\ a_2 & a_3 & -a_0 & -a_1 \\ a_3 & -a_2 & a_1 & -a_0 \end{pmatrix}.$$

Further let $u \in A_H(G)$ with $u(x) \neq 0$ in G, then $u^{-1} \in A_H(G)$ if and only if $(|u|^2 D - M(\mathrm{grad}\,|u|^2))\bar{u} = 0$.

Property 3

If $z \in A_H(G)$, rank $A' < 2$, then $z^{-1} \in A_H(G')$ with $G' = \{x \in R^4 : z(x) \neq 0\}$.

Property 4

Let $u(x) = \sum_{i=0}^{3} u_i(x) d_i$, $v(x) = \sum_{i=0}^{3} v_i d_i$ and rank $A' < 2$.

Then the product rule

$$D(uv) = (Du)v + (Dv)u \tag{1.15}$$

holds.

Property 5

If $u \in A_H(G)$, $u(x) = \sum_{i=0}^{3} u_i(x) d_i$, rank $A' < 2$, then $u^n \in A_H(G)$ for all $n \in N$.

Property 6

If $z \in A_H(G)$, rank $A' < 2$, then $z^{-n} \in A_H(G')$ for all n with $G' = \{x \in R^4 : z(x) \neq 0\}$.

1.2.9 Theorem (LAGRANGE's Interpolation Polynomial)

Let z be a totally analytic variable, rank $A' < 2$, let $u = (u_1, \ldots, u_n) \in H^n$. The H-valued function

$$(L_n u)(x) = \sum_{k=1}^{n} \{ \prod_{j \neq k} [(z(x) - z(a_j))(z(a_k) - z(a_j))^{-1}] \} u_k , \tag{1.16}$$

where $a_i \in R^4$, $i = 1, \ldots, n$; $z(a_k) \neq z(a_j)$ for $k \neq j$, fulfils the conditions

(i) $L_n u \in A_H(R^4)$,

(ii) $(L_n u)(a_j) = u_j$, (1.17)

(iii) $(L_n u)^k \in A_H(R^4)$

independently of the exact values of the entries in matrix A.

__Proof__

By using the properties formulated in the previous section a straightforward calculation leads to the validity of our assertion.

#

__1.2.10 Proposition__

__(Problem of Multipliers of H-regular Functions)__

Let $v \in A_H(G)$ and $uv \in A_H(G)$ for all $u \in A_H(G)$, then we have $v = \text{const}$.

__Proof__

An extensive computation shows that the conditions $u \in A_H(G)$, $v \in A_H(G)$ and $uv \in A_H(G)$ imply the following system of partial differential equations :

$$\begin{pmatrix} -\partial_0 v_1 - \partial_1 v_0 + \partial_2 v_3 - \partial_3 v_2 & -\partial_0 v_2 - \partial_1 v_3 - \partial_2 v_0 + \partial_3 v_1 & -\partial_0 v_3 + \partial_1 v_2 - \partial_2 v_1 - \partial_3 v_0 \\ \partial_0 v_0 - \partial_1 v_1 + \partial_2 v_2 + \partial_3 v_3 & \partial_0 v_3 - \partial_1 v_2 - \partial_2 v_1 - \partial_3 v_0 & -\partial_0 v_2 - \partial_1 v_3 + \partial_2 v_0 - \partial_3 v_1 \\ -\partial_0 v_3 - \partial_1 v_2 - \partial_2 v_1 + \partial_3 v_0 & \partial_0 v_0 + \partial_1 v_1 - \partial_2 v_2 + \partial_3 v_3 & \partial_0 v_1 - \partial_1 v_0 - \partial_2 v_3 - \partial_3 v_2 \\ \partial_0 v_2 - \partial_1 v_3 - \partial_2 v_0 - \partial_3 v_1 & -\partial_0 v_1 + \partial_1 v_0 - \partial_2 v_3 - \partial_3 v_2 & \partial_0 v_0 + \partial_1 v_1 + \partial_2 v_2 - \partial_3 v_3 \end{pmatrix} \begin{pmatrix} u_1 \\ u_2 \\ u_3 \end{pmatrix} = 0 \tag{1.18}$$

where $u = \sum_{i=0}^{3} u_i e_i$, $v = \sum_{i=0}^{3} v_i e_i$.

Setting one by one $u = e_i$, $i = 1,2,3$, and investigating system (1.18) we obtain $v = \text{const}$.

#

__1.2.11 Proposition__

There is an H-regular function v with $uv \in A_H(G)$ for all $u \in A_H(G) \setminus \{ u \mid u = ce_0 , c \in R \}$.

__Proof__

Setting $v = x_0 e_0 + \sum_{i=1}^{3} \frac{x_i}{3} e_i$

and substituting v in system (1.18) we obtain our assertion.

#

__1.2.12 Remark__

Each H-regular function with values in $\text{span}(\{d_i\}_{i=0}^{3})$ is a nontrivial solution of system (1.18) if rank $A' < 2$. That means that totally analytic variables are multipliers from the right with respect to u. Using that, it enables us to clarify the structure of the range of the H-regular function u. More exactly holds the

1.2.13 Proposition

Let $u \in A_H(G)$, $z = \sum_{i=0}^{3} x_i d_i$, rank $A' = 1$, $uz \in A_H(G)$;

then

$$u(x) = \sum_{i=0}^{3} u_i(x) d_i \quad .$$

Proof

Let $u = \sum_{i=0}^{3} u_i(x) e_i$. Replacing in (1.18) v by z we get to the system

$$\begin{bmatrix} -a_{01}-a_{10}+a_{23}-a_{32} & -a_{02}-a_{13}-a_{20}+a_{31} & -a_{03}+a_{12}-a_{21}-a_{30} \\ a_{00}-a_{11}+a_{22}+a_{33} & a_{03}-a_{12}-a_{21}-a_{30} & -a_{02}-a_{13}+a_{20}-a_{31} \\ -a_{03}-a_{12}-a_{21}+a_{30} & a_{00}+a_{11}-a_{22}+a_{33} & a_{01}-a_{10}-a_{23}-a_{32} \\ a_{02}-a_{13}-a_{20}-a_{31} & -a_{01}+a_{10}-a_{23}-a_{32} & a_{00}+a_{11}+a_{22}-a_{33} \end{bmatrix} \begin{bmatrix} u_1 \\ u_2 \\ u_3 \end{bmatrix} = 0. \qquad (1.19)$$

It is easy to see that $z = \sum_{j=0}^{3} \left(\sum_{i=0}^{3} x_i a_{ij} \right) e_j$.

As $z \in A_H(G)$ follows from (1.19)

$$\begin{bmatrix} a_{23} - a_{32} & a_{31} - a_{13} & a_{12} - a_{21} \\ a_{22} + a_{33} & a_{03} - a_{21} & -a_{02} - a_{31} \\ -a_{03} - a_{12} & a_{11} + a_{33} & a_{01} - a_{32} \\ a_{02} - a_{13} & -a_{01} - a_{23} & a_{11} + a_{22} \end{bmatrix} \begin{bmatrix} u_1 \\ u_2 \\ u_3 \end{bmatrix} = 0. \qquad (1.20)$$

We assumed that all subdeterminants of the matrix A' of an order greater than one must be vanished. So we conclude that the vectors

$$V_0 = \begin{bmatrix} 0 \\ a_{23}-a_{32} \\ a_{31}-a_{13} \\ a_{12}-a_{21} \end{bmatrix}, \quad V_1 = \begin{bmatrix} 0 \\ a_{22}+a_{33} \\ a_{03}-a_{21} \\ -a_{02}-a_{31} \end{bmatrix}, \quad V_2 = \begin{bmatrix} 0 \\ -a_{03}-a_{12} \\ a_{11}+a_{33} \\ a_{01}-a_{32} \end{bmatrix}, \quad V_3 = \begin{bmatrix} 0 \\ a_{02}-a_{13} \\ a_{01}-a_{23} \\ a_{11}+a_{22} \end{bmatrix}$$

belong to ker A. Applying $z \in A_H(G)$ and $d_i d_j = d_j d_i$, $i,j = 0,1,2,3$, we gain that the matrix of the system (1.19) has the rank 2. Because rank A = 2 we also have only two linearly independent solutions of $Au = 0$. Therefore it holds dim ker A = 2 .

Further we know that ker A = span { v_0, v_1, v_2, v_3 }. On the other hand , we have $(u_0, u_1, u_2, u_3)^T$ is orthogonal to the kernel of A if we use the scalar product in R^4. By virtue of the validity of FREDHOLM's theorems the system

$A^T h = u$ possesses a solution, whence it follows

$$\sum_{j=0}^{3} u_j e_j = \sum_{j=0}^{3} \sum_{i=0}^{3} h_i a_{ij} e_j = \sum_{i=0}^{3} h_i \left(\sum_{j=0}^{3} a_{ij} e_j \right) = \sum_{i=0}^{3} h_i d_i \, .$$ #

With the next theorem we want to clarify the question about the possibilities of constructing totally analytic functions.

1.2.14 Theorem

$z^k \in A_H(G)$ for all $k \in N$ if and only if $z \in A_H(G)$ and rank $A' \leq 1$.

Proof

The necessity is proved in [Del1].

Putting $u_j = \sum_{i=0}^{3} x_i a_{ij}$, j = 1,2,3, and

$$S = \begin{pmatrix} a_{23}-a_{32} & a_{31}-a_{13} & a_{12}-a_{21} \\ a_{22}+a_{33} & a_{03}-a_{21} & -a_{02}-a_{31} \\ -a_{03}-a_{12} & a_{11}+a_{33} & a_{01}-a_{32} \\ a_{02}-a_{13} & -a_{01}-a_{23} & a_{11}+a_{22} \end{pmatrix} .$$

Since $z^2 \in A_H(G)$, we get $Su=0$ by consideration of (1.17) , where

$$u = \left(\sum_{j=0}^{3} x_j a_{j1} , \sum_{j=0}^{3} x_j a_{j2} , \sum_{j=0}^{3} x_j a_{j3} \right)^T .$$

As it has been true for any $x = (x_0, x_1, x_2, x_3)^T$, we conclude $S(a_{j1}, a_{j2}, a_{j3})^T = 0$ for j = 0,1,2,3 , which means

$(a_{j1}, a_{j2}, a_{j3})^T \in$ ker S.

Now we distinguish different cases with respect to dim ker S:

(i) dim ker S = 0 : That can only be realized if $a_{ji} = 0$ for i = 1,2,3 ; j = 0,1,2,3 , therefore S=0 which would imply dim ker S = 3. This is in contradiction to our assumption.

(ii) dim ker S = 1 : Because all vectors $(a_{j1}, a_{j2}, a_{j3})^T$ belong to ker S, it clearly follows rank $A' \leq 1$.

(iii) dim ker S = 2 : Handling this case is much more complicated. First of all it is clear that rank $A' \le 2$ and rank S = 1, whence

$$\det \begin{bmatrix} a_{23}-a_{32} & a_{31}-a_{13} \\ a_{02}-a_{13} & -a_{01}-a_{23} \end{bmatrix} = 0$$

This leads to

$$a_{23}^2 + a_{13}^2 - a_{32}a_{23} - a_{13}a_{31} - a_{03}a_{12} - a_{03}a_{21} = 0 . \qquad (1.21)$$

The last identity was obtained because of

$(a_{01}, a_{02}, a_{03})^T \in \ker S$. Similarly we obtain the identities

$$\det \begin{bmatrix} a_{23} - a_{32} & a_{12} - a_{21} \\ -a_{03} - a_{12} & a_{01} - a_{32} \end{bmatrix} =$$

$$= a^2_{32} + a^2_{12} - a_{23}a_{32} - a_{12}a_{21} - a_{02}a_{31} + a_{02}a_{13} = 0, \qquad (1.22)$$

$$\det \begin{bmatrix} a_{31} - a_{13} & a_{12} - a_{21} \\ a_{03} - a_{21} & -a_{02} - a_{31} \end{bmatrix} =$$

$$= a^2_{21} + a^2_{31} - a_{01}a_{23} + a_{01}a_{32} - a_{13}a_{31} - a_{12}a_{21} = 0 . \qquad (1.23)$$

Adding (1.21), (1.22), (1.23) and taking into consideration that $(a_{01}, a_{02}, a_{03})^T \in \ker S$ we obtain

$$(a_{23}-a_{32})^2 + (a_{12}-a_{21})^2 + (a_{13}-a_{31})^2 = 0 \quad , \text{ whence}$$

$$a_{ij} = a_{ji} , \quad j,i = 1,2,3 . \qquad (1.24)$$

On account of rank S = 1 we get by the additional use of relation (1.24)

$$\det \begin{bmatrix} a_{22}+a_{33} & a_{03}-a_{21} \\ -a_{03}-a_{12} & a_{11}+a_{33} \end{bmatrix} = \det \begin{bmatrix} a_{22}+a_{33} & -a_{02}-a_{31} \\ a_{02}-a_{13} & a_{11}+a_{22} \end{bmatrix} =$$

$$= \det \begin{bmatrix} a_{11}+a_{33} & a_{01}-a_{32} \\ -a_{01}-a_{23} & a_{11}+a_{22} \end{bmatrix} = 0$$

the relations

$$a^2_{03} + a^2_{13} + a^2_{33} + a_{22}a_{33} + a_{02}a_{13} - a_{03}a_{12} = 0,$$

$$a^2_{02}+a^2_{22}+a^2_{23}+a_{11}a_{22}+a_{01}a_{23}-a_{02}a_{13} = 0,$$

$$a^2_{01}+a^2_{11}+a^2_{12}+a_{11}a_{33}+a_{03}a_{12}-a_{01}a_{23} = 0.$$

The sum of these relations leads to

$$a^2_{01}+a^2_{02}+a^2_{03}+a^2_{13}+a^2_{12}+a^2_{23} + \frac{1}{2}(a_{33}+a_{22})^2 + \frac{1}{2}(a_{33}+a_{11})^2 +$$

$$+ \frac{1}{2}(a_{11}+a_{22})^2 = 0$$

and therefore must be for $i=0,1,2,3$ and $j=1,2,3$ $a_{ij}=0$. Consequently, we have rank $A' = 0 < 2$.

(iv) dim ker $S = 3$: This implies rank $S = 0$ and so $S = 0$. It immediately follows rank $A' < 2$.

1.2.15 Remark

It is worth noticing that totally analytic variables play an essential role in some papers. F. SOMMEN used them for a foundation of a theory of monogenic operators, which is contained in [Som1].

To establish CAUCHY's approach of quaternionic analysis, H. MALONEK in [Mal] recently also used totally analytic variables.

1.2.16 It is natural to look for possibilities to construct systems of H-regular functions. One of them consists in the action of the generalized CAUCHY-RIEMANN operator on real-valued harmonic polynomials. As it is well-known, the polynomials

$$P_k = \sum_{n\geq 0} \frac{(-1)^n \; x_3^{2n}}{(2n)!} \Delta^n(x_1^k x_2^{m-k}) , \qquad k = 0,1,\ldots,m,$$

$$Q_l = \sum_{n\geq 0} \frac{(-1)^n \; x_3^{2n+1}}{(2n+1)!} \Delta^n(x_1^l x_2^{m-l-1}) , \qquad l = 0,1,\ldots,m-1,$$

with $\Delta = \frac{\partial^2}{\partial x_1^2} + \frac{\partial^2}{\partial x_2^2}$, form the system of all linearly independent homogeneous harmonic polynomials of the degree m. Then DP_k and DQ_l belong to $A_H(R^3)$ and have the order $m - 1$. Following the considerations in [Sud] we shall describe another construction of H-regular functions, starting from real-harmonic functions by adding a purely imaginary part. A straightforward calculation leads to the

1.2.17 Proposition [Sud]

If G is a domain star-shaped with respect to the origin and $u_\emptyset : G \longrightarrow R$ is harmonic in G, then

$$u(x) = u_\emptyset(x) - \operatorname{Im} \int_0^1 t^2 \, (Du_\emptyset)(tx) \, x \, dt$$

is H-regular in G, and it holds $\operatorname{Re} u = u_\emptyset$.

R. FUETER found in his important paper [Fue4] of 1936 a method for constructing H-regular functions from complex analytic functions in the classical plane function theory. Here this assertion will be formulated without the proof, which is given in detail in [Dea] and also in [Sud].

1.2.18 Proposition [Fue4]

For $p \in H$, let $\eta_p(x+iy) = x + \frac{\operatorname{Im} p}{|\operatorname{Im} p|} y$ be an embedding of C in H and $\xi(p) = \operatorname{Re} p + i|\operatorname{Im} p|$ denote a mapping from H into the upper half-plane of C. Then, if $u:C \to C$ is an analytic function in the open set $U \subset C$ and $\tilde{u}(p) = \eta_p(u(\xi(p)))$, that $\Delta\tilde{u}$ is H-regular in the open set $\xi^{-1}(U) \subseteq H$.

The previous propositions have given us a certain feeling about the size of the set of H-regular functions.

1.2.19

Preparing a numerical approach for the solution of certain partial differential equations we shall introduce the following system of H-regular functions $\{\varphi_i\}$ in a domain $G \subset R^3$ with

$$\varphi_i(x) = \sum_{k=1}^{3} \frac{x_k - x_k^{(i)}}{|x - x^{(i)}|^3} e_k ,$$

where the points $x^{(i)}$ are lying outside of G. In Chapter 5 its linear independence and completeness in the space $W^k_{2,H} \cap A_H(G)$ is proved.

1.3 A Generalized LEIBNIZ Rule

1.3.1 The validity of a LEIBNIZ product rule is a necessary algebraic supposition in order to get BOREL-POMPEIU's formula. It is a well-known fact in the function theory over the field of complex numbers. Let u, v be complex-valued functions defined on C. The CAUCHY-RIEMANN operator

$$\frac{\partial}{\partial \bar z} = \frac{1}{2}\left(\frac{\partial}{\partial x} - i \frac{\partial}{\partial y} \right)$$

satisfies the natural multiplication law

$$\frac{\partial}{\partial \bar z}(uv) = \left(\frac{\partial}{\partial \bar z} u\right)v + u\left(\frac{\partial}{\partial \bar z} v\right) . \qquad (1.25)$$

In particular, it follows that the product of two analytic functions is again an analytic function. As mentioned above, this assertion cannot be true in the case of H-valued functions. Hence, we deduce a generalized LEIBNIZ rule for H-valued functions whose left side is similar to this one of the rule (1.25).

1.3.2 Theorem (Generalized LEIBNIZ Rule)

Let $u, v \in C^1_H(G)$.

Then $D(uv) = (Du)v + \bar u(Dv) + 2[Re(uD)]v$ (1.26)

holds.

Proof

Using the quaternionic product (1.2) it follows

$$D(uv) = D \sum_{i=0}^{3} \sum_{j=0}^{3} u_i v_j e_i e_j = \sum_{k=1}^{3} D_k e_k \left(\sum_{i=0}^{3} \sum_{j=0}^{3} u_i v_j e_i e_j \right) =$$

$$= \sum_{\substack{i,j=0 \\ k=1}}^{3} D_k(u_i v_j) e_k e_i e_j = \sum_{\substack{i,j=0 \\ k=1}}^{3} [(D_k u_i)v_j + u_i(D_k v_j)] e_k e_i e_j .$$

Setting

$$S_1 = \sum_{\substack{i,j=0 \\ k=1}}^{3} (D_k u_i) v_j e_k e_i e_j \quad \text{and} \quad S_2 = \sum_{\substack{i,j=0 \\ k=1}}^{3} u_i (D_k v_j) e_k e_i e_j$$

we obtain for $S_1 = \sum_{i=0}^{3} \sum_{k=1}^{3} (D_k u_i) e_k e_i \sum_{j=0}^{3} v_j e_j = (Du)v$.

The treatment of the term S_2 requires some algebraic calculations which will be carried out here:

$$S_2 = \sum_{\substack{j=0 \\ k,i=1}}^{3} u_i (D_k v_j) e_k e_i e_j + \sum_{\substack{j=0 \\ k=1}}^{3} u_0 (D_k v_j) e_k e_j =$$

$$= -[\sum_{i,k=1}^{3} u_i e_i D_k e_k]v - 2 \sum_{i=1}^{3} u_i D_i v + [\sum_{k=1}^{3} u_0 D_k e_k]v =$$

$$= [\sum_{k=1}^{3} uD_k e_k]v - 2\sum_{i=1}^{3} u_i D_i v = \overline{u}Dv + 2\,\mathrm{Re}(uD)v .$$

The sum $S_1 + S_2$ yields the desired formula. #

1.3.3 Corollary

Let $u_0 \in C^1_R(G)$; $u_0 v \in A_H(G)$, $v \in A_H(G)$. Then we have either $u_0 = \mathrm{const}$ or $v = 0$.

Proof

It immediately follows from Theorem 1.3.2. #

1.3.4 Corollary (Product Rules of Vector Analysis)

Let $u, v \in C^1_H(G)$. Then the following relations are true:

$$\begin{aligned}
&\text{(i)} && \mathrm{grad}\,(u_0 v_0) = (\mathrm{grad}\,u_0)v_0 + u_0(\mathrm{grad}\,v_0),\\
&\text{(ii)} && \mathrm{div}\,(u_0\hat{v}) = (\mathrm{grad}\,u_0)\hat{v} + u_0 \mathrm{div}\,\hat{v},\\
&\text{(iii)} && \mathrm{rot}\,(u_0\hat{v}) = \mathrm{grad}\,u_0 \times \hat{v} + u_0 \mathrm{rot}\,\hat{v},\\
&\text{(iv)} && \mathrm{div}\,(\hat{u} \times \hat{v}) = \hat{v}\,\mathrm{rot}\,\hat{u} - \hat{u}\,\mathrm{rot}\,\hat{v}, \qquad (1.27)\\
&\text{(v)} && \mathrm{grad}(\hat{u}\hat{v}) = \hat{u} \times \mathrm{rot}\hat{v} + \hat{v} \times \mathrm{rot}\hat{u} + (\hat{u}\cdot\mathrm{grad})\hat{v} + (\hat{v}\cdot\mathrm{grad})\hat{u},\\
&\text{(vi)} && \mathrm{rot}(\hat{u} \times \hat{v}) = \hat{u}\,\mathrm{div}\,\hat{v} - \hat{v}\,\mathrm{div}\,\hat{u} + (\hat{v}\cdot\mathrm{grad})\hat{u} - (\hat{u}\cdot\mathrm{grad})\hat{v}.
\end{aligned}$$

Proof

Setting $u = u_0 e_0$ and $v = v_0 e_0$ we immediately obtain from (1.26) the formula (i). To verify (ii) and (iii) we choose $u = u_0 e_0$ and $v = \hat{v}$. Then it follows

$$D(u_0\hat{v}) = (Du_0 e_0)\hat{v} + u_0(D\hat{v}) + 0 .$$

By making use of the notation in 1.2.4 the definition of the quaternionic product (1.10) yields

$$-(\mathrm{div}(u_0\hat{v}))e_0 + \mathrm{rot}\,u_0\hat{v} = [-(\mathrm{grad}\,u_0)\hat{v}]e_0 + \mathrm{grad}\,u_0 \times \hat{v} - $$
$$- u_0\,\mathrm{div}\,\hat{v} + u_0\,\mathrm{rot}\,\hat{v} ,$$

which proves the statements (ii) and (iii) . Now , let $u = \hat{u}$, $v = \hat{v}$. Then , by using the notation 1.2.4 after a straightforward computation , formula 1.2.8 leads to

$$-\mathrm{div}(\hat{u} \times \hat{v})e_0 - \mathrm{grad}(\hat{u}\hat{v}) + \mathrm{rot}\,\hat{u} \times \hat{v} \quad \text{on the left side} ,$$

while on the right-hand side of (1.26) it holds

$$-(\mathrm{rot}\,\hat{u})\hat{v}e_0 + (\mathrm{rot}\,\hat{u}) \times \hat{v} - (\mathrm{div}\,\hat{u})\hat{v} + (\mathrm{div}\,\hat{v})\hat{u} +$$
$$+(\hat{u}\cdot\mathrm{rot}\,\hat{v})e_0 - \hat{u} \times \mathrm{rot}\,\hat{v} - (\hat{u}\cdot\mathrm{grad})\hat{v} .$$

Interchanging the functions $\hat{u}$ and $\hat{v}$ we get the new identity

$$\mathrm{div}(\hat{u} \times \hat{v})e_0 - \mathrm{grad}(\hat{u}\hat{v}) - \mathrm{rot}(\hat{u} \times \hat{v}) = -(\mathrm{rot}\ \hat{v})\hat{u}e_0 +$$

$$+ (\mathrm{rot}\ \hat{v}) \times \hat{u} - (\mathrm{div}\ \hat{v})\hat{u} + (\mathrm{div}\ \hat{u})\hat{v} + (\hat{v}\cdot \mathrm{rot}\ \hat{u})e_0 -$$

$$- \hat{v} \times \mathrm{rot}\ \hat{u} - (\hat{v}\cdot\mathrm{grad})\hat{u} \quad .$$

By adding both identities the relation (v) aroses, while subtracting leads to (iv) and (vi). #

1.3.5 Remark

Especially Corollary 1.3.4 emphasizes the efficiency of the quaternionic analysis developed here in connection with the representation of formulas in physics and vector analysis.

1.3.6 Remark

In earlier papers A.SUDBERY [Sud] and W.SPROESSIG [Sp 1], respectively, obtained product rules, where the right-hand side was chosen similarly to (1.25) . These papers are cited in more detail in [GS 3].

1.4 BOREL-POMPEIU's Formula

1.4.1 This section deals with the derivation of a quaternionic version of a generalized BOREL-POMPEIU formula. The domain G and the boundary $\Gamma = \partial G$ fulfil the conditions described in 1.2.1 . Let

$$\alpha(y) = \sum_{i=1}^{3} \alpha_i(y)e_i \quad ,$$

where α_i is the i-th component of the outer normal on Γ at the point y . Assume $|\alpha| = 1$.

Setting

$$e(x) = \frac{-x}{4\pi|x|^3} \quad , \text{ where } \quad x = \sum_{i=1}^{3} x_i e_i \quad .$$

In [BDS, 8.8] it is proved that $e(x)$ is , in a distributional sense , a fundamental solution of the differential operator D. Introduce the operators

$$(F_\Gamma u)(x) = \int_\Gamma e(x-y)\,\alpha(y)u(y)d\Gamma_y \quad \text{and}$$

$$(T_G u)(x) = - \int_G e(x-y)u(y)dG_y \quad .$$

If there are no confusions, we shall write T, F instead of T_G , F_Γ , respectively.

1.4.2 Proposition (GAUSS' Formula)

Let $u \in C^1_H(G) \cap C_H(\bar{G})$. Then we have

$$\int_\Gamma \alpha u \, d\Gamma = \int_G Du \, dG \qquad (1.28)$$

Proof

Using GAUSS' formula for real-valued functions we obtain

$$\int_G D u \, dG = \int_G \sum_{\substack{i=1\\j=0}}^{3} D_i u_j e_i e_j \, dG = \sum_{\substack{i=1\\j=0}}^{3} \int_G D_i u_j \, dG \, e_i e_j =$$

$$= \sum_{\substack{i=1\\j=0}}^{3} \int_\Gamma \alpha_i u_j \, d\Gamma e_i e_j = \int_\Gamma \alpha u \, d\Gamma .$$

#

1.4.3 Proposition

Let $u \in C^{0,\beta}_H(G)$, $0 < \beta \leq 1$. T_G is a right-inverse operator with respect to D. More exactly it holds

$$(DT_G u)(x) = \begin{cases} u(x) , & \text{in } G \\ 0 , & \text{in } \operatorname{co} \bar{G} . \end{cases} \qquad (1.29)$$

Proof

Let $x \in G$. Using the differentiation rule for weak singular integrals ([MP IX,7(3)]) we obtain with $G_\varepsilon = G \setminus B_\varepsilon(x)$, $(DT_{G_\varepsilon} u)(x) = 0$ and

$$D(T_{B_\varepsilon(x)} u)(x) = \frac{1}{4\pi} \sum_{i,j=1}^{3} e_i e_j \, D_i \int_{B_\varepsilon(x)} \frac{x_j - y_j}{|x-y|^3} u(y) \, dB_\varepsilon =$$

$$= \frac{1}{4\pi} \sum_{i,j=1}^{3} e_i e_j \int_{B_\varepsilon} D_i \frac{(x_j - y_j)}{|x-y|^3} u(y) \, dB_\varepsilon -$$

$$- \frac{1}{4\pi} \sum_{i,j=1}^{3} e_i e_j \int_{S_1} \frac{x_j - y_j}{|x-y|} \frac{x_i - y_i}{|x-y|} dS_1 \, u(x)$$

$$= \frac{1}{4\pi} \sum_{i,j=1}^{3} e_i e_j \int_{S_1} \frac{\delta_{ij} |x-y|^2 - 3(x_j - y_j)(x_i - y_i)}{|x-y|^5} u(y) dS_1 +$$

$$+ \frac{1}{4\pi} \sum_{i=1}^{3} e_0 \int_{S_1} \frac{(x_i - y_i)^2}{|x-y|^2} dS_1 u(x) = \frac{1}{4\pi} \int_{S_1} dS_1 u(x) = u(x)$$

because $e_i e_j + e_j e_i = 0$ for $i \neq j$, $i,j \neq 0$.

In case of $x \notin G$ the integrals which occurred above have differentiable kernels, and the integral over S_1 does not appear. Therefore we have $(DT_G u)(x) = 0$. #

1.4.4 Theorem (BOREL-POMPEIU's Formula)

Let $u \in C^1_H(G) \cap C_H(\bar{G})$ .Then we have

$$(F_\Gamma u)(x) + (T_G Du)(x) = \begin{cases} u(x) & \text{in } G \\ 0 & \text{in } \operatorname{co} \bar{G} = R^3 \setminus \bar{G}. \end{cases} \tag{1.30}$$

Proof

Set in (1.26) $u = \frac{1}{4\pi|x-y|} e_0$ and $v = u$. LEIBNIZ formula now means

$$D\Big(\frac{1}{4\pi|x-y|} u\Big) = \frac{1}{4\pi}\Big(D_y \frac{1}{|x-y|}\Big) + \frac{1}{4\pi|x-y|} Du .$$

Integrating both sides over the domain $G_\varepsilon = G \setminus B_\varepsilon(x)$ we obtain the identity

$$\frac{1}{4\pi}\int_{G_\varepsilon} D_y\Big[\frac{1}{|x-y|} u(y)\Big]\, dG_{\varepsilon,y} = -\frac{1}{4\pi}\int_{G_\varepsilon}\Big[D_x \frac{1}{|x-y|}\Big]u(y)\, dG_{\varepsilon,y} +$$

$$+ \frac{1}{4\pi}\int_{G_\varepsilon} \frac{1}{|x-y|} (Du)(y)\, dG_{\varepsilon,y} . \tag{1.31}$$

Applying GAUSS' formula it follows that

$$\frac{1}{4\pi}\int_\Gamma \frac{1}{|x-y|}\alpha(y)u(y)\, d\Gamma_y - \frac{1}{4\pi\varepsilon}\int_{S_\varepsilon}\frac{y-x}{|y-x|} u(y)\, dS_\varepsilon =$$

$$= \frac{1}{4\pi}\int_{G_\varepsilon}\frac{y-x}{|x-y|^3} u(y)\, dG_{\varepsilon,y} + \frac{1}{4\pi}\int_{G_\varepsilon}\frac{1}{|x-y|}(Du)(y)\, dG_y ,$$

where $y-x = \sum_{k=1}^{3} (y_k - x_k)e_k$. Hence

$$\frac{1}{4\pi}\int_\Gamma \frac{1}{|x-y|}\alpha(y)u(y)\, d\Gamma_y - \frac{\varepsilon}{4\pi}\int_{S_1}\frac{y-x}{|y-x|} u(y)\, dS_1 =$$

$$= \frac{1}{4\pi}\int_{G_\varepsilon}\frac{1}{|x-y|}(Du)(y)\, dG_y - (T_G u)(x). \tag{1.32}$$

If ε tends to zero and we take into account (1.29) , the proof is finished application of D from the left. #

1.4.5 Corollary (BOREL-POMPEIU's Formula for Multiply Connected Domains)

Let G_i , $i=0,1,2,\dots,n$, be simply connected bounded domains in R^3 with sufficiently smooth boundaries Γ_i. The following conditions are satisfied :

(i) $G_\emptyset \supset \bigcup_{i=1}^{n} \bar{G}_i = K$,

(ii) $\bar{G}_i \cap \bar{G}_j = \emptyset$ for $i \neq j$.

If $u \in C_H^1(G_\emptyset \backslash K) \cap C_H(\bar{G}_\emptyset \backslash \text{int } K)$, then it holds

$$(F_{\Gamma_\emptyset} u)(x) - \sum_{i=1}^{n} (F_{\Gamma_i} u)(x) + (T_{G_\emptyset \backslash K} Du)(x) = \begin{cases} \emptyset , & \text{in co } \bar{G}_\emptyset \cup \text{int } K \\ u(x) , & \text{in } G_\emptyset \backslash K. \end{cases} \tag{1.33}$$

1.5 Basic Statements of H-regular Functions

This section is devoted to the development of a function theory of H-regular functions. Most of the statements can be proved in more general algebras. By reason of maintainance of the uniformity we will restrict our considerations to the case of H-valued functions.

1.5.1 Proposition (CAUCHY's Integral Theorem)

Let $\Gamma' \subset G$ be an arbitrary, sufficiently smooth boundary of a domain $\bar{G}' \subset G$. If $u \in A_H(G)$, then it holds $\int_{\Gamma'} \alpha u \, d\Gamma' = \emptyset$.

Proof

Integrating formula (1.26) over the domain G' and applying GAUSS' theorem, then we have

$$\int_{\Gamma'} \alpha (uv) \, d\Gamma' = \int_{G'} (Du)v \, dG' + \int_{G'} \bar{u}(Dv) \, dG' + 2\text{Re} \int_{G'} (uD)v \, dG'. \tag{1.34}$$

Replacing the H-valued functions u and v by $e_\emptyset$ and u, respectively, relation (1.34) is transformed itself into the desired formula.

1.5.2 Theorem (CAUCHY's Integral Formula)

Let $u \in A_H(G) \cap C_H(\bar{G})$. Then it holds

$$(F_\Gamma u)(x) = \begin{cases} u(x) , & \text{in } G \\ \emptyset , & \text{in co } \bar{G}. \end{cases}$$

Proof

It immediately follows from (1.30). #

1.5.3 Proposition (Mean Value Theorem)

Let $B_r(a) = \{ x: |x-a| < r , a \in R^3\}$. If $u \in A_H(G)$ and $a \in G$, then

$$u(a) = \frac{1}{|B_r|} \int_{B_r(a)} u(x)\, dx$$

holds for each real number $r > 0$ such that $\bar{B}_r(a) \subset G$, where $|B_r|$ denotes the volume of the ball $B_r(a)$.

Proof

Choose r such that $\bar{B}_r(a) \subset G$. From CAUCHY's integral formula we obtain

$$u(a) = \frac{1}{4\pi r^3} \int_{S_r} (a-x)\,\alpha(x) u(x)\, dS_r \qquad (1.35)$$

where $S_r = \partial B_r(a)$. Replacing u by $a-x$ in formula (1.34) and v by $u \in A_H(G)$, (1.34) may be transformed into

$$u(a) = \frac{1}{4\pi r^3} \int_{B_r(a)} D(a-x)\, u(x)\, dx = \frac{1}{|B_r|} \int_{B_r(a)} u(x)\, dx .$$

#

1.5.4 Proposition (Mean Value Formula)

If $u \in A_H(B_r(a)) \cap C_H(B_r(a) \cup S_r)$, then

$$u(a) = \frac{1}{4\pi r^2} \int_{S_r} u(x)\, dS_r .$$

Proof

Applying CAUCHY's integral formula on the ball $B_r(a)$ it follows

$$u(a) = \frac{1}{4\pi} \int_{S_r} \frac{a-x}{|a-x|^3} \alpha(x) u(x) dS_r = \frac{1}{4\pi} \int_{S_r} \frac{(a-x)(x-a)}{|x-a|^4} u(x) dS_r$$

and the proof is complete. #

1.5.5 Theorem (Maximum Modulus Theorem)

Let G be a connected domain, $u \in A_H(G)$. If there exists a point $a \in G$ such that $| u(x) | \leq | u(a) |$ for all $x \in G$, then u must be a constant H-valued function in G.

Proof

It is clear that we can find a point $a \in G$ such that there exists a sufficiently small ,closed ε- ball $\bar{B}_\varepsilon(a) \subset G$ in which for each $x \in \bar{B}_\varepsilon(a)$ the inequality $|u(a)| \geq |u(x)|$ holds. Applying the mean value formula we obtain

$$u(a) = \frac{1}{4\pi\varepsilon^2} \int_S u(x)\, dS_\varepsilon . \qquad (1.36)$$

Decompose S_ε into the components S_ε' and S_ε'' . These subsets are characterized as follows:

$S_\varepsilon' = \{x \in S_\varepsilon : |u(x)| = |u(a)|\}$,

$S_\varepsilon'' = \{x \in S_\varepsilon : |u(x)| < |u(a)|\}$.

Formula (1.36) becomes

$$|u(a)| \leq \frac{1}{4\pi\varepsilon^2} [\int_{S_\varepsilon'} |u(x)|dS_\varepsilon + \int_{S_\varepsilon''} |u(x)|dS_\varepsilon] <$$

$$< \frac{1}{4\pi\varepsilon^2} [|S_\varepsilon'|\,|u(a)| + |S_\varepsilon''|\,|u(a)|] = |u(a)|.$$

Therefore it is easy to check by consideration of the continuity of $u(x)$ that S_ε'' must be empty. That means that $|u(x)| = |u(a)|$ for all $x \in S_\varepsilon$. By choosing smaller ε- balls the same result holds.

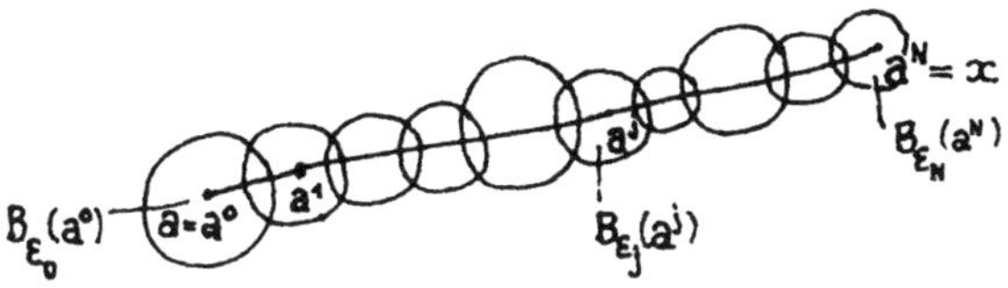

Figure 1

Hence $|u(x)| = |u(a)|$ for all $x \in \bar{B}_\varepsilon(a)$. As G is supposed to be connected, we can join the point a with an arbitrary point $x \in G$ by a continuous curve. Next, we cover this curve by small ε_j-balls $B_{\varepsilon_j}(a^j)$, $j=0,1,\ldots,N$, such that

(i) $B_\varepsilon(a) = B_{\varepsilon_0}(a^0)$,

(ii) $B_{\varepsilon_j}(a^j) \cap B_{\varepsilon_{j+1}}(a^{j+1}) \neq \emptyset$,

(iii) $x \in B_{\varepsilon_N}(a^N)$.

As for $\bar{B}_\varepsilon(a)$, the same reflection will be done for each of the balls $B_{\varepsilon_j}(a^j)$. We obtain $|u(x)| = |u(a)| = C > 0$. Obviously ,

$$\sum_{j=0}^{3} u_j^2(x) = C^2 \quad \text{for all} \quad x \in G \,.$$

Further we have

$$\sum_{j=0}^{3} u_j(x) D_i u_j = 0 \quad , \ i=1,2,3 \,.$$

A second differentiation yields

$$\sum_{j=0}^{3} u_j D_i^2 u_j + \sum_{j=0}^{3} (D_i u_j)^2 = 0 \quad , \quad i=1,2,3.$$

Since all components of an H-regular function are harmonic, we conclude

$$\sum_{i=1}^{3} (D_i u_j)^2 = 0 \quad \text{for all} \quad x \in G \text{ and } j = 0,1,2,3 \,.$$

Finally, we have $D_i u_j = 0$, $i=1,2,3$, $j=0,1,2,3$, which means that u is constant in G. #

We have the immediate

1.5.6 Corollary

If $u \in A_H(G) \cap C_H(\bar{G})$, then it follows

$$\sup_{x \in \bar{G}} |u(x)| = \sup_{x \in \Gamma} |u(x)| \,.$$

1.5.7 Proposition (LIOUVILLE's Theorem)

Let $u \in A_H(R^3)$. If $|\, u(x) \,| \leq M$ for all $x \in R^3$, then $u(x)$ is a constant H-valued function.

Proof

From CAUCHY's integral formula we obtain the representation

$$u(x) = \frac{1}{4\pi} \int_{S_R(x)} \frac{x-y}{|x-y|^3} \alpha(y) u(y) \, dS_R \quad ,$$

where R is an arbitrarily chosen positive real number. Then differentiation by D_i leads to

$$D_i u = \frac{1}{4\pi} \int_{S_R(x)} D_{i,x} \left[\sum_{j=1}^{3} \frac{x_j - y_j}{|x-y|^3} e_j\right] \alpha(y) u(y)\, dS_R \, .$$

Using the relations

$$D_{i,x} \frac{x_j - y_j}{|x-y|^3} = \begin{cases} - \dfrac{3(x_j-y_j)(x_i-y_i)}{|x-y|^5}, & i \neq j \\ - \dfrac{3(x_i-y_i)^2 + |x-y|^2}{|x-y|^5}, & i = j \end{cases}$$

we get the estimate

$$\left| D_{i,x} \sum_{j=1}^{3} \frac{x_j - y_j}{|x-y|^3} e_j \right| \leq \sum_{j=1}^{3} \left| D_{i,x} \frac{x_j - y_j}{|x-y|^3} \right| \leq \frac{9}{|x-y|^3},$$

whence

$$| D_i u | \leq \frac{1}{4\pi} \int_{S_R} \left| D_{i,x} \sum_{j=0}^{3} \frac{x_j - y_j}{|x-y|^3} e_j \right| |\alpha(y)| |u(y)| \, dS_R \leq$$

$$\leq \frac{9 M \, 4\pi R^2}{4\pi \, R^3} = \frac{9M}{R}, \qquad i=1,2,3 \, .$$

Since R is arbitrarily chosen, it follows for $R \longrightarrow \infty$ $D_i u = 0$ for $i=1,2,3$, and the assertion is checked. #

1.5.8 Remark

If in Proposition 1.5.7 $u(x) \longrightarrow 0$ for $|x| \longrightarrow \infty$, then $u(x) = 0$ for all $x \in R^3$.

1.5.9 Proposition (TAYLOR Series Expansion)

Let $u \in A_H(G)$ and x be an arbitrary point of G. Then the H-valued function u permits, in a sufficiently small neighbourhood of a, a TAYLOR series expansion of the form

$$u(x) = \sum_{k=0}^{\infty} a_k(a, \theta_0) \, |x-a|^k \quad ,$$

where $\theta_0 = \sum_{i=1}^{3} \frac{x_i - a_i}{|x-a|} e_i$.

Proof

Take $\varepsilon > 0$ such that $B_\varepsilon(a) \subset G$. CAUCHY's integral formula allows us to write

$$u(x) = \frac{1}{4\pi} \int_{S_\varepsilon(a)} \frac{x-y}{|x-y|^3} \alpha(y) u(y) \, dS_\varepsilon . \tag{1.37}$$

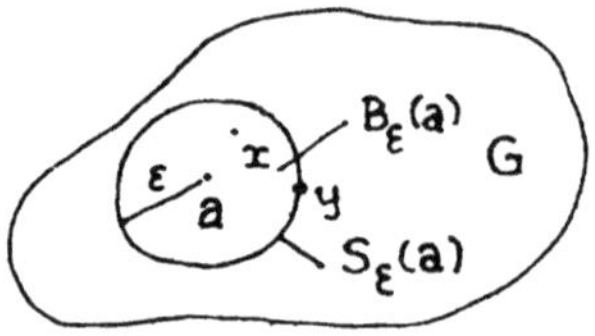

Figure 2:

In [Mue] it is proved that with $t = \theta_0 \cdot \theta_0'$, where

$\theta_0 = \sum_{i=1}^{3} \theta_{0i} e_i$, $\theta_0' = \sum_{i=1}^{3} \theta_{0i}' e_i$, $\theta_{0i} = \frac{x_i - a_i}{|x_i - a_i|}$,

$\theta_{0i}' = \frac{a_i - y_i}{|a - y|}$

the series

$$\frac{1}{|x-y|} = \sum_{n=0}^{\infty} P_n(3,t) \left|\frac{x-a}{y-a}\right|^n \cdot \frac{1}{|y-a|} \tag{1.38}$$

converges together with all series of derivatives in $|x|<r<|y|$ for every fixed $r>0$. Here $P_n(3,t)$ denotes the LEGENDRE' polynomial of the degree n in R^3 , which can be defined by RODRIGUES formula

$$P_n(3,t) = \left(\frac{-1}{2}\right)^n \left(\frac{d}{dt}\right)^n \frac{(1-t^2)^{n/2}}{n!}$$

for $n=0,1,2,$ $-1 \leq t \leq 1$. Substituting expansion (1.38) in formula (1.37) this leads to

$$u(x) = \sum_{n=0}^{\infty} \frac{1}{4\pi} \int_{S_\varepsilon} D_y \left[\frac{P_n(3,t)}{|y-a|^{n+1}} \right] \alpha(y) u(y) \, dS_\varepsilon \, |x-a|^n .$$

The footnote y of the symbol D_y means that D is acting with respect to y.

Setting

$$a_n(a, \theta_0) = \frac{1}{4\pi} \int_{S_\varepsilon} D_y \left[\frac{P_n(\mathfrak{z}, t(y))}{|y-a|^{n+1}} \right] \alpha(y) u(y) \, dS_\varepsilon \tag{1.39}$$

we have

$$u(x) = \sum_{n=0}^{\infty} a_n(a, \theta_0) \, |x-a|^n. \tag{1.40}$$

Finally, we have to show that the coefficients a_n are independent of the parameter ε. This will be proved by consideration of the uniqueness of the a_n. If x tends to a along the radius θ_0, we obtain $a_0(a, \theta_0) = u(a)$. Since the termwise derivation is permitted, so the corresponding derivation in the direction θ_0' reads

$$[\theta_0 D] \, u(x) = \sum_{n=0}^{\infty} [\theta_0 D_x] \, [a_n(a, \theta_0) \, |x-a|^n].$$

It is easy to see that $[\theta_0 D] \, a_n(a, \theta_0) = 0$ for $n = 1, 2, 3, \ldots$

and $\qquad [\theta_0 D] \, |x-a|^n = n \, |x-a|^{n-1}$,

whence it follows $\qquad [\theta_0 D] \, u(x) = \sum_{n=1}^{\infty} n a_n |x-a|^{n-1}$.

Letting $x \longrightarrow a$ along the radius we get $\lim_{x \to a} [\theta_0 D] u(x) = a_1$.

In a similar way we get

$$a_n = \frac{[\theta_0 \cdot D]^n}{n!} u(a) \quad , \tag{1.41}$$

which means that the coefficients a_n in series (1.40) are uniquely defined. Therefore the independence of the neighbourhood $B_\varepsilon(a)$ is also verified.

1.5.10 Corollary (CAUCHY's Inequality)

Let $|u(x)| \leq M$. The coefficients of the power series (1.40) fulfil the estimate

$$|a_n(a, \theta_0)| \leq \frac{C}{d^n(a, \Gamma)},$$

where $d(a,\Gamma) = \inf_{y\in\Gamma} |a-y|$ and $C > 0$.

Proof

In order to prove the assertion, we have to estimate

$$\frac{1}{4\pi}\int_{S_\varepsilon} D_y\left[\frac{P_n(3,t(y))}{|y-a|^{n+1}}\right]\alpha(y)u(y)dS_\varepsilon$$

Notice that the derivatives of the LEGENDRE polynomials are given by

$P_n^{(k)}(3,t) = c_{n,k}P_{n-k}(2k+3,t)$, where

$c_{n,k} = 2^k\Gamma(n+k+1)\Gamma(k+3/2)/\Gamma(n-k+1)\Gamma(2k+2)\Gamma(3/2)$.

Here $\Gamma(*)$ denotes the Γ-function.

Then it follows that

$$D_yP_n(3,t) = \sum_{i=1}^{3} D_i[P_n(3,t)]e_i = \sum_{i=1}^{3} c_{n,1}P_{n-1}(5,t)D_ite_i =$$

$$=\sum_{i=1}^{3} c_{n,1}\frac{P_{n-1}(5,t)}{|y-a|} \sum_{j=1}^{3}\theta_{0j}d_{ij}e_i$$

where

$$d_{ij} = \begin{cases} \theta'_{0j}\,\theta'_{0i} , & i \neq j \\ \theta^2_{0i}-1 , & i = j. \end{cases}$$

It is easy to see that now

$$D_yP_n(3,t) = c_{n,1}\frac{P_{n-1}(5,t)}{|a-y|}\sum_{i\neq j}\theta_{0j}\,\theta'_{0i}\,\theta'_{0j} ,$$

which implies

$$D_y\left[\frac{P_n(3,t)}{|y-a|^{n+1}}\right] = \sum_{i=1}^{3} D_i\left[\frac{P_n(3,t)}{|y-a|^{n+1}}\right]e_i =$$

$$= \Big[c_{n,1}P_{n-1}(5,t)\sum_{\substack{i\neq j\\ i,j=1,2,3}}\theta_{0j}\theta'_{0i}\theta_{0j}e_i - (n+1)P_n(3,t)\sum_{i=1}^{3}\theta'_{0i}e_i\Big]\frac{1}{|y-a|^{n+2}}.$$

Note that LEGENDRE polynomials satisfy the inequality $|\ P_n(k,t)\ | \leq 1$ for $-1 \leq t \leq 1$ [Mue]. Simple calculations yield by consideration of (1.39)

$$|a_n(a,\theta_0)| \leq \left[6n(n+1) + 3(n+1) \Big/ 4\pi\varepsilon^{n+2} \right] 4\pi\, \varepsilon^2 M = \frac{(2n+1)(n+1)\,3M}{\varepsilon^n} . \qquad (1.42)$$

Estimate (1.42) holds for any $\varepsilon > 0$ such that series (1.40) converges. We know from 1.5.8 that the only restriction with respect to the choice of $B_\varepsilon(a)$ is to fulfil the inclusion $B_\varepsilon(a) \subset G$. Now, set $\varepsilon = d(a,\Gamma) = \inf_{y\in\Gamma} |a-y|$, then the statement is verified.

\#

1.5.11 Proposition (Theorem of Uniqueness)

Let $u, v \in A_H(G)$. For every $a \in G$ and $\theta_0 \in S_1'$ the sets $M_{\theta_0}(a) = \{ x \in R^3 : x = a + t\,\theta_0 , t > 0 \}$ are defined, where S_1' is given by $S_1' = \{ \theta_0 \in S_1 : \theta_0$ is a rational point $\}$. Furthermore let E be such a set that a is an accumulation point for each of the sets $E \cap M_{\theta_0}(a)$. If for all $x \in E$ $u(x) = v(x)$, then $u(x)$ coincides with $v(x)$ on G.

Proof

In each direction $\theta_0 \in S_1'$ we consider a sequence which converges to a. It is denoted by $\{ a^{(n)}(\theta_0) \}$. We supposed that

$$u(a^{(n)}(\theta_0)) = v(a^{(n)}(\theta_0)), \quad n = 1,2,\dots . \qquad (1.43)$$

Developing the H-regular functions u and v in a neighbourhood $B_\varepsilon(a)$ of the point a in its corresponding TAYLOR series, we obtain

$$u(x) = \sum_{n=0}^{\infty} a_n(a,\theta_0)\, |x-a|^n ,$$

$$v(x) = \sum_{n=0}^{\infty} b_n(a,\theta_0)\, |x-a|^n . \qquad (1.44)$$

For each θ_0 there is a number $N(\theta_0)$ such that $a^{(n)}(\theta_0) \in B_\varepsilon(a)$, $n \geq N$. Substituting (1.44) in the identity (1.43) we get $a_0(a,\theta_0) = b_0(a,\theta_0)$ if n tends to infinity. It follows for all $a^{(n)}(\theta_0) \in B_\varepsilon(a)$

$$\sum_{n=1}^{\infty} a_n(a,\theta_0)|a^{(n)}(\theta_0)-a|^{n-1} = \sum_{n=1}^{\infty} b_n(a,\theta_0)|a^{(n)}(\theta_0)-a|^{n-1} .$$

This implies $a_1(a,\theta_0) = b_1(a,\theta_0)$ if n tends to infinity. Continuing this procedure we conclude $a_n(a,\theta_0) = b_n(a,\theta_0)$ for all $n \in N$ in $B_\varepsilon(a)$. For the continuity of the coefficients $a_n(a,\theta_0)$ with respect to θ_0 and the density of S_1' in S_1 $a_n(a, \theta_\emptyset) = b_n(a, \theta_\emptyset)$ follows for each direction $\theta_\emptyset \in S_1$. Making use of the same consideration as for the proof of the maximum modulus theorem we get $u(x) = v(x)$ for all $x \in G$.

#

1.5.12 Remark

The zeros of an H-regular function are not necessarily isolated [Fue5]. See, for instance,

$u(x) = (x_\emptyset e_1 - x_1 e_\emptyset) + (x_\emptyset e_2 - x_2 e_\emptyset)$.

1.5.13 Theorem (WEIERSTRASS' Theorem)

Let $(u^{(i)})_{i \in \mathbb{N}}$ be a sequence of H-valued functions in G, further let $u^{(i)} \in A_H(G)$ for $i = 1,2,3\ldots$. Assume that for each compact set $K \subset G$ and each $\varepsilon > \emptyset$ there exists a natural number $N = N(\varepsilon, K)$ with

$$\max_{x \in K} |u^{(i)}(x) - u^{(j)}(x)| < \varepsilon, \quad i,j > N,$$

then we can find an H-valued function u such that

(i) $u \in A_H(G)$,

(ii) the sequence $\partial^l u$ converges uniformly on the compact subsets of G to $\partial^l u$ for any $l = (l_1, l_2, l_3)$, where

$$\partial^l = \frac{\partial^{|l|}}{\partial_1^{l_1} \partial_2^{l_2} \partial_3^{l_3}}, \quad |l| = l_1 + l_2 + l_3 .$$

Proof

Let K be an arbitrary compact set contained in G. By the help of CAUCHY's integral formula we get the following estimate:

$$| \partial^l u^{(i)}(x) - \partial^l u^{(j)}(x) | \leq$$

$$\leq C \max_{y \in \partial K} \frac{1}{|x-y|} \max_{y \in \partial K} |u^{(i)}(y) - u^{(j)}(y)| \leq \frac{C\varepsilon}{d(K, \partial G)^{|l|+2}} .$$

As the components $u_k^{(i)}$, $k = \emptyset,1,2,3$, are harmonic functions, we conclude by using HARNACK's theorem that there exist harmonic functions u_k, $k = \emptyset,1,2,3$, such that

$$\partial^l u^{(i)} \rightrightarrows \partial^l u \quad \text{with } u = \sum_{k=\emptyset}^{3} u_k e_k \quad \text{on compact subsets of G,}$$

in particular, $Du^{(i)} \longrightarrow Du$.

Because $Du^{(i)} = \emptyset$ for $i = 1,2,\ldots$ in G, we finally obtain $u \in A_H(G)$. The last step is possible as the domain G can be exhausted by compact sets.

#

1.5.14 Corollary (2. WEIERSTRASS' Theorem)

Let $(u^{(i)})_{i \in N}$ be a sequence of H-valued functions in G , $u^{(i)}$, i = 1,2,3,..., belong to $A_H(G) \cap C_H(\bar{G})$ for all i = 1,2,3,... . If the series $\sum_{i=1}^{\infty} u_i(x)$ converges uniformly on Γ, then it converges uniformly on G .

Proof

For any $\varepsilon > 0$ there exists a natural number $N(\varepsilon)$ such that

$$\left| \sum_{k=1}^{l} u_{N+k}(x) \right| < \varepsilon$$

holds for $x \in \Gamma$ and any natural number l .

As the sum $\sum_{k=1}^{l} u_{N+k}(x) \in A_H(G) \cap C_H(\bar{G})$ it immediately follows from the maximum modulus theorem $\left| \sum_{k=1}^{l} u_{N+k}(x) \right| < \varepsilon$ for $x \in \bar{G}$, which yields the desired result.

#

1.5.15 Proposition (Extension Theorem)

Let G_1 and G_2 be bounded domains in R^3 , $G_1 \cap G_2 = \emptyset$, $\partial G_i = \Gamma_i$, i = 1,2 , $\Gamma_1 \cap \Gamma_2 = \Gamma_0$,let Γ_k ,k = 0,1,2 , be smooth LIAPUNOV surfaces . If the H-valued functions $u^{(k)}(x)$ belong to $A_H(G_k) \cap C_H(\bar{G}_k)$ for k = 1,2 and coincide on Γ_0 , then the H-valued function

$$u(x) = \begin{cases} u^{(1)}(x) & , x \in G_1 \\ u^{(2)}(x) & , x \in G_2 \\ u^{(1)}(x) = u^{(2)}(x) & , x \in \Gamma_0 \end{cases}$$

is H-regular in $G_1 \cup G_2 \cup \Gamma_0$. This extension is unique .

Proof

Let $a \in \Gamma_0$ be a fixed point , $\varepsilon > 0$ an arbitrary real number, such that $\Gamma_0 \supset B_\varepsilon(a) \cap (\Gamma_1 \cup \Gamma_2)$, where as usual $B_\varepsilon(a)$ denotes an open ball with the radius ε and the midpoint a. For $\partial(B_\varepsilon(a) \cap \Gamma_0)$ write Γ_ε .

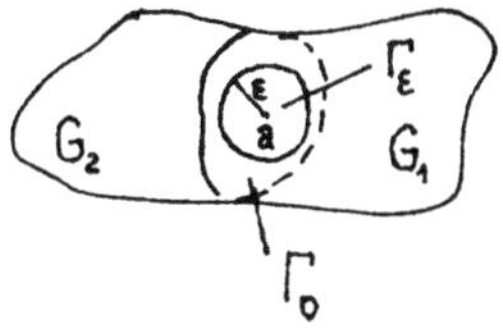

Figure 3

From CAUCHY's integral theorem it follows that

$$(F_{\Gamma_\varepsilon} u)(x) = (F_{(\Gamma_\varepsilon \cap G_1) \cup (\Gamma_0 \cap \bar{B}_\varepsilon)} u^{(1)})(x) + $$

$$+ (F_{(\Gamma_\varepsilon \cap G_2) \cup (\Gamma_0 \cap \bar{B}_\varepsilon)} u^{(2)})(x) =$$

$$= \begin{cases} u^{(1)}(x) & , \ x \in G_1 \cap B_\varepsilon(a) \\ u^{(2)}(x) & , \ x \in G_2 \cap B_\varepsilon(a) \\ u^{(1)}(x) = u^{(2)}(x) & , \ x \in \Gamma_0 \cap B_\varepsilon(a) \end{cases} = u(x) \quad , \quad x \in B_\varepsilon(a),$$

BOREL-POMPEIU's formula yields $T_{B_\varepsilon(a)}\, Du = 0$ in $B_\varepsilon(a)$. Differentiation by D from the left-hand side leads to $Du=0$ in $B_\varepsilon(a)$.

1.5.16 Corollary (LUSIN's Theorem)

Let γ be a submanifold contained in $\Gamma = \partial G$. If the H-valued function $u(x) \in A_H(G) \cap C_H(\bar{G} \cup \gamma)$ vanishes on γ , then u(x) vanishes for all $x \in G$.

Proof

Extend the H-regular function u(x) throughout the submanifold γ into a certain domain G' by writing

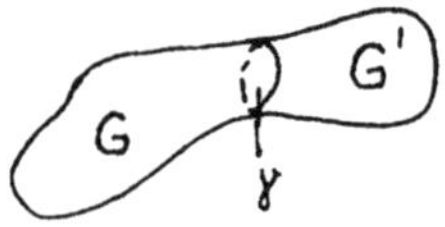

Figure 4

$$u'(x) = \begin{cases} u(x) & , \ x \in G \\ 0 & , \ x \in G' \setminus \bar{G} \\ 0 & , \ x \in \gamma \end{cases} .$$

Applying Proposition 1.5.15 and the theorem of uniqueness we can find that $u'(x) = 0$ for all $x \in G \cup G' \cup \gamma$, hence $u(x) = 0$ in G .

#

1.5.17 Theorem (MORERA's Theorem)

Let $u \in C^1_H(G)$, $Du \in L_{r,H}(G)$, $r > 3$. If for any smooth LIAPUNOV manifold-without-boundary $\gamma \subset G$ $\int_\gamma \alpha\, u\, d\gamma = 0$, then $Du=0$ in G.

Proof

Let $\{ G_k \}_{k \in N}$ be a regular sequence of domains , which is contracting to the point $x \in G$. Denote $\partial G_k = \Gamma_k$. Obviously , we have $x \in \bigcap_{k \geq 0} G_k$.

Using LEBESGUE's theorem then we obtain for each H-valued function $w \in L_{r,H}(G)$, $r > 3$, that there exists

$$\lim_{n\to\infty} \frac{1}{|G_k|} \int_{G_k} w(y)\, dG_{k,y} =: \tilde{w}(x)$$

and $w=\tilde{w}$ in the space $L_{r,H}(G)$. Putting $w=Du$ it follows by GAUSS' formula $\int_{G_k} Du\, dG_k = \int_{\Gamma_k} \alpha u\, d\Gamma_k$.

From the supposition we get $\frac{1}{|G_k|}\int_{G_k} Du\, dG_k = 0$ for $k=0,1,2,\ldots$

and therefore $w = Du = 0$ almost everywhere in G . #

1.5.18 Remark

Under the assumption $u \in C_H(G)$, MORERA's theorem can also be verified. The proof is considerably more extensive than the above mentioned . The reader can find it in [Bi1] and [BDS] .

To continue, we recall some facts about GEGENBAUER polynomials. For this reason we formulate

1.5.19 Lemma [Len]

Suppose $\alpha \in [0,1)$, $t \in [-1,1]$, $\nu \in R$, then we expand the function $(1-2\alpha t+\alpha^2)^{-\nu}$ in a power series , where the coefficients are the GEGENBAUER polynomials $C_k(t)$, namely

$$(1-2\alpha t+\alpha^2)^{-\nu} = \sum_{k=0}^{\infty} C_k^{\nu}(t)\,\alpha^k \quad .$$

The GEGENBAUER polynomials may be defined by the following formula :

$$C_{k+1}^{\nu}(t) = \frac{2(k+\nu)}{k+1}\, tC_k^{\nu}(t) - \left(\frac{k+1+2\nu-2}{k+1}\right) C_{k-1}^{\nu}(t) \quad \text{and}$$

$$C_0^{\nu}(t) = 1 \quad , \quad C_1^{\nu}(t) = 2\nu t \quad .$$

1.5.20 Proposition (CAUCHY's Integral Formula Outside of a Bounded Domain)

Let $u \in A_H(R^3\setminus\bar{G}) \cap C_H(R^3\setminus G)$ and suppose $\lim_{|x|\to\infty} u(x) = u(\infty)$. Then we have

$$(Fu)(x) = \int_{\Gamma} e(x-y)\,\alpha(y)u(y)\,d\Gamma_y =$$

$$= \begin{cases} u(\infty) - u(x) & , \; x \in R^3 \setminus \bar{G} \\ u(\infty) & , \; x \in G \, . \end{cases}$$

Proof

Let $B_R(\emptyset)$ be such a ball that $B_R(\emptyset) \supset \bar{G}$. Setting

$t = \sum_{i=1}^{3} \frac{x_i y_i}{|x||y|}$, then $|x-y|^{-2} = \frac{1}{|y|^2}\left(1-2t\left|\frac{x}{y}\right| + \left|\frac{x}{y}\right|^2\right)^{-1}$.

Applying Lemma 1.5.19 it follows for $|x| < |y|$

$$|x-y|^{-2} = \frac{1}{|y|^2} \sum_{k=\emptyset}^{\infty} C_k^1(t)\left|\frac{x}{y}\right|^k . \tag{1.45}$$

Using CAUCHY's integral formula we obtain

$$\frac{1}{4\pi} \int_{S_R} \frac{x-y}{|x-y|^3}\,\alpha(y)u(y)\,dS_R - \frac{1}{4\pi}\int_{\Gamma} \frac{x-y}{|x-y|^3}\,\alpha(y)u(y)\,d\Gamma =$$

$$= \begin{cases} \emptyset & , \; x \in co\,(\overline{B_R(a)\setminus\bar{G}}) \\ u(x) & , \; x \in B_R(a)\setminus\bar{G} \; . \end{cases}$$

Now, let us consider the first integral. On the sphere S_R the unit vector of the outer normal α becomes $\frac{y-x}{|y-x|}$.

By consideration of formula (1.45) the transformation of coordinates takes the integral

$$\frac{1}{4\pi} \int_{S_R} \frac{u(y)}{|x-y|^2}\,dS_R$$

to the sum

$$\sum_{k=\emptyset}^{\infty} \frac{1}{4\pi} \int_{S_1} \frac{C_k^1(t)}{R^k}\,u(\theta R)\,d\theta\,|x|^k \; .$$

The limit for $R \longrightarrow \infty$ leads to

$$\frac{C_0^1(t)}{4\pi} \int_{S_1} u(\infty)\,d\theta = u(\infty) \; .$$

which gives the result required. #

1.5.21 Theorem (LAURENT Series Expansion)

Assume that u is H-regular in the open annulus $B_{rR}(a) = \{ x \in R^3 : r < |x-a| < R , \; a \in R^3 , \; r \geq \emptyset \}$

and continuous on its closure $B_{rR}(a)$. Then $u(x)$ allows the following series expansion:

$$u(x) = \sum_{n=-\infty}^{\infty} a_n(a, \theta_\emptyset)\ |a-x|^n \quad ,$$

where

$$a_n(a,\theta) = \frac{1}{4\pi} \int_{S_{R'}} \frac{\left[\theta_0 C_{n-1}^{3/2}(t_0) - \theta_0' C_n^{3/2}(t_0)\right]}{|y-a|^{n+2}}\, \theta_0' u\, dS_{R'} . \tag{1.46}$$

R' is arbitrarily chosen in (r,R) and $t_\emptyset = \sum_{i=1}^{3} \frac{(a_i - x_i)(y_i - a_i)}{|a-x||y-a|}$.

Furthermore we put $C_{-1}^{3/2}(t_\emptyset)=0$ and $C_{-k-2}^{3/2}(t_\emptyset)=C_{k-1}^{3/2}(t_\emptyset)$.

Proof

Let $B_{r_1 R_1}(a) = \{x \in B_{rR}(a):\ r < r_1 < |a-x| < R_1 < R\}$ be a further annulus inside $B_{rR}(a)$. Making use of CAUCHY's integral formula we get for all points $x \in B_{r_1 R_1}(a)$

$$u(x) = \int_{S_R} e(x-y)\alpha(y)u(y)dS_{R_1,y} - \int_{S_r} e(x-y)\alpha(y)u(y)dS_{r_1,y}. \tag{1.47}$$

As we can easily see, the identity $\theta_\emptyset|a-x| + \theta|x-y| + \theta_\emptyset'|y-a| = 0$, where $\theta_0 = (a-x)/|a-x|$, $\theta = (x-y)/|x-y|$, $\theta' = (y-a)/|y-a|$ holds . Inserting the corresponding expression for θ into the formula (1.47) it follows

$$u(x) = \frac{1}{4\pi}\left(\int_{S_{r_1}} dS_{r_1} - \int_{S_{R_1}} dS_{R_1}\right)\frac{\theta_0'\theta_0'}{|x-y|^3}|y-a|\,u(y) - \frac{1}{4\pi}\left(\int_{S_{r_1}} dS_{r_1} - \int_{S_{R_1}} dS_{R_1}\right)\frac{\theta_0\theta_0'}{|x-y|^3}|x-a|\,u(y) . \tag{1.48}$$

Clearly Lemma 1.5.19 yields

$$|x-y|^{-3} = \begin{cases} |y-a|^{-3} \sum_{k=0}^{\infty} C_k^{3/2}(t_0)\left|\frac{x-a}{y-a}\right|^k , & |x-a| < |y-a| \\ |x-a|^{-3} \sum_{k=0}^{\infty} C_k^{3/2}(t_0)\left|\frac{y-a}{x-a}\right|^k , & |x-a| > |y-a| . \end{cases}$$

Inserting this expansion in (1.48) we get

$$u(x) = \frac{1}{4\pi}\sum_{k=0}^{\infty}\Bigg\{\int_{S_{R_1}} \frac{\theta_0\theta_0'}{|a-y|^{k+3}} C_k^{3/2}(t_0)\, u\, dS_{R_1} |x-a|^{k+1} - \int_{S_{R_1}} \frac{\theta_0'\theta_0'}{|a-y|^{k+2}} C_k^{3/2}(t_0)\, u\, dS_{R_1} |x-a|^{k} +$$

$$+ \int_{S_{r_1}} \frac{\theta_0'\theta_0'}{|a-y|^{-k-1}} C_k^{3/2}(t_0)\, u\, dS_{r_1} |x-a|^{-k-3} - \int_{S_{r_1}} \frac{\theta_0\theta_0'}{|y-a|^{-k}} C_k^{3/2}(t_0)\, u\, dS_{r_1} |x-a|^{-k-2}\Bigg\}.$$

Define $C_{(-1)}(t_\emptyset) = 0$. The coefficients $a_k(a, \theta_\emptyset)$, which are in front of the positive powers , may be computed by

$$a_k(a,\theta_\emptyset) = \frac{1}{4\pi} \int_{S_{R_1}} \frac{[\theta_\emptyset C^{3/2}_{k-1}(t_\emptyset) - \theta'_\emptyset C^{3/2}_k(t_\emptyset)]\, \theta'_\emptyset\, u}{|a-y|^{k+2}} \, dS_{R_1} \quad , \tag{1.49}$$

while the coefficients $a_k(a,\theta_\emptyset)$,which stand in front of the negative powers , are given by the formula

$$a_{-k-2}(a,\theta_\emptyset) = \frac{1}{4\pi} \int_{S_{r_1}} \frac{[-\theta'_\emptyset C^{3/2}_{k-1}(t_\emptyset) + \theta_\emptyset C^{3/2}_k(t_\emptyset)]\, \theta'_\emptyset\, u}{|a-y|^{-k}} \, dS_{r_1} . \tag{1.50}$$

It is clear that $a_{-1} = \emptyset$. We define a new index by

$$n = \begin{cases} k & \text{in } (1.49) \\ -k-2 & \text{in } (1.50) \end{cases} \qquad \text{for } k=\emptyset,1,2,\ldots .$$

Together with the determination $C^{3/2}_{-1}(t_\emptyset)=\emptyset$ and $C^{3/2}_{-k-2}(t_\emptyset) = C^{3/2}_{k-1}(t_\emptyset)$ we obtain a more unified representation of the coefficients as follows :

$$a_n(a,\theta_\emptyset) = \begin{cases} \dfrac{1}{4\pi} \displaystyle\int_{S_{R_1}} \dfrac{[\theta_\emptyset C^{3/2}_{n-1}(t_\emptyset) - \theta'_\emptyset C^{3/2}_n(t_\emptyset)]\, \theta'_\emptyset\, u}{|a-y|^{n+2}} \, dS_{R_1} \quad , & n \geq \emptyset \\ \dfrac{1}{4\pi} \displaystyle\int_{S_{r_1}} \dfrac{[\theta_\emptyset C^{3/2}_{n-1}(t_\emptyset) - \theta'_\emptyset C^{3/2}_n(t_\emptyset)]\, \theta'_\emptyset\, u}{|a-y|^{n+2}} \, dS_{r_1} \quad , & n < \emptyset . \end{cases}$$

In order to complete this proof, we have to show that the coefficients do not depend on the spheres S_{R_1} and S_{r_1} . For it we have to verify that from

$$\sum_{n=-\infty}^{\infty} a_n(a,\theta_\emptyset)|x-a|^n = \emptyset \tag{1.51}$$

$a_n = \emptyset$, $n=\emptyset, \pm 1, \pm 2, \ldots$, follows. Multiplying series (1.51) successively by arbitrarily chosen spherical functions $y_{j-1,2}(\theta_\emptyset)\theta_\emptyset$, $j = 1,2,3,\ldots$, of the degree j and integrating over the unit sphere S_1 with respect to $\theta_\emptyset$ we obtain the equations

$$\int_{S_1} y_{j-1,2}(\theta_\emptyset)\theta_\emptyset a_n(a,\theta_\emptyset) dS_1 |x-a|^{n-1} +$$

$$+ \int_{S_1} y_{j-1,2}(\theta_\emptyset)\theta_\emptyset a_{-n-2}(a,\theta_\emptyset) dS_1 |x-a|^{-n-2} = \emptyset$$

for all $j=1,2,\ldots$. To obtain the last relation, we made use of the formulas

(i) $y_{n,2}(\theta_0) = \frac{C(n)}{4\pi} \int_{S_1} y_{n,2}(\theta_0')C_n^{3/2}(t_0)dS_1$, where $C(n)>0$,

(ii) $\int_{S_1} y_{n,2}(\theta_0')C_j^{3/2}(t_0)dS_1 = 0$ for $j \neq n$,

which the reader can find in [Mi1, §14]. Using the completeness of the system $\{y_{j-1,2}(\theta_0)\}$ in the space $L_{2,R}(S_1)$ we have the equality

$a_n(a, \theta_0) = -a_{-n-2}(a, \theta_0)|x-a|^{-2n-2}$, [MP] .

If x runs on the right-hand side along the radius which is defined by a fixed θ_0 also the left-hand side should be changed. This is in contrast to the definition of the coefficients $a_n(a, \theta_0)$, and so the uniqueness is proved. Therefore it is always possible to choose for their computation an arbitrary sphere inside the annulus $B_{rR}(a)$. #

1.5.22 Remark

As it is not our main aim to explain a relatively complete theory of H-regular functions , we shall finish Section 1.5 with some definitions ,which may be used for a certain 3 -dimensional residue theory and a theory of meromorphic H-valued functions.

1.5.23 Definition (LAURENT Series Expansion)

The series

$$u(x) = \sum_{n=-\infty}^{\infty} a_n(a, \theta_0)|x-a|^n \tag{1.52}$$

with coefficients (1.46) is called LAURENT series expansion of the H-valued function $u(x)$, which is H-regular in $B_{0R}(a)$. This series may be decomposed into two items, namely the regular part

$$u_1(x) = \sum_{n=0}^{\infty} a_n(a, \theta_0)|x-a|^n$$

and the so-called principal part of $u(x)$ at

$$u_2(x) = \sum_{n=-\infty}^{-2} a_n(a, \theta_0)|x-a|^n.$$

1.5.24 Definition (Isolated Singular Points)

A point $a \in R^3$ is called an isolated singular point of the H-valued function $u(x)$ iff for any $r>0$ $u \notin A_H(B_r(a))$, and there exists a number $r>0$ with $u \in A_H(B_r(a)\setminus\{a\})$. If the principal part of the LAURENT series expansion either vanishes , only consists of a finite number of items or does not break off, then the isolated singular point a is called removable singularity , pole or essential singularity , respectively . In case of a pole the largest index $-m$ with $a_m(a, \theta_0) \neq 0$ for all θ_0 is called the order of the pole $u(x)$ at a.

1.5.25 Remark

Each pole has the order two at least, which immediately follows from formula (1.46).

1.5.26 Remark

The previously given definitions deviate somewhat from those formulated in [BDS].

1.5.27 Definition (Residue)

Let a be an isolated singularity.

Then the coefficient $a_{-2}(a, \theta_0)$ of the LAURENT series expansion is called residue of $u(x)$ at the point a in the direction θ_0. It may be denoted by $res[u,a, \theta_0]$.

1.5.28 Proposition

Let $u \in A_H(B_{0R}(a))$. For any ε with $0 < \varepsilon < R$ it holds

$$res[u,a, \theta_0] = \frac{\theta_0}{4\pi} \int_{S_\varepsilon} \theta_0' u(\varepsilon\theta_0')\, d\theta_0' \quad .$$

Proof

Immediately from 1.5.21 we get for sufficiently small $\varepsilon > 0$

$$a_{-2}(a, \theta_0) = \frac{1}{4\pi} \int_{S_\varepsilon} \theta_0 \theta_0' C_0^{3/2}(t)\, u(\varepsilon\theta_0')\, d\theta_0' \, .$$

As $C_0^{3/2}(t)=1$ the proof is complete. #

1.5.29 Proposition

Let a be a pole of the order $m \geq 2$ of the H-valued function $u(x) \in A_H(B_{0R}(a))$, then there exists a sufficiently small ball B (a) such that

$$|u(x)| \geq \frac{\varepsilon}{|x-a|^m} \, .$$

Proof

The theorem of the LAURENT series expansion yields

$$u(x) = \sum_{n=0}^{\infty} a_n(a, \Theta_0)|x-a|^n + \sum_{n=2}^{m} a_{-n}(a, \Theta_0)|x-a|^{-n} .$$

From Definition 1.5.24 it follows

$$\lim_{\substack{x \to a \\ x \neq a}} |x-a|^m u(x) = a_{-m}(a, \Theta_0) \neq 0 .$$

For any ε with $\varepsilon < \min_{S_1} |a_{-m}(a, \Theta_0)|$ we find a real number $\delta(\varepsilon) > 0$ such that

$$|x-a|^m |u(x)| \geq \varepsilon$$

for all points x with $0 < |x-a| < \delta$. #

2. OPERATORS

2.1 Above all in this section we intend to discuss some functional analytic properties of the operators D, T_G and F_Γ introduced in the foregoing Section 1. We restrict our considerations to such topics which are needed later.

To ensure that the class of all H-regular functions $A_H(G)$ is a subspace in certain functional spaces we prove the

2.2 Proposition

The classes $A_H(G) \cap C_H^{m,\beta}(\bar{G})$, $0<\beta\leq 1$, $m\geq 0$, $m \in N$, and $A_H(G) \cap L_{p,H}(G)$, $p\geq 1$, are closed subspaces in $C_H^{m,\beta}(\bar{G})$ and $L_{p,H}(G)$, respectively.

Proof

Because it is easy to see the assertion concerning the space $C_H^{m,\beta}(\bar{G})$, we can restrict our proof to the space $L_{p,H}(G)$. Let $\{\phi_l\} \subset A_H(G)$ be a CAUCHY sequence. Obviously, $\{\phi_l\}$ converges to a certain H-valued function $\phi \in L_{p,H}(G)$. It remains to show that $\phi \in A_H(G)$. From the mean value theorem (Proposition 1.5.3) we find

$$\phi_l(x) - \phi_k(x) = \frac{1}{|B_r|} \int_{|x-y|\leq r} [\phi_m(y) - \phi_k(y)] dB_{r,y} ,$$

where r is chosen sufficiently small. It follows that

$$\sup_{x \in G_r} |\phi_m(x) - \phi_k(x)| \leq C \|\phi_m - \phi_k\|_{L_{p,H}}$$

with $G_r = \{x \in G : \mathrm{dist}(x, \partial G) > r\}$. Hence the sequence converges uniformly for any compact subset of G. Using WEIERSTRASS' theorem (Theorem 1.5.13) we get to an H-regular limit function ϕ_0. Consequently, we have $\phi = \phi_0$, and the proof is complete. #

2.3 Properties of the T-Operator

2.3.1 Theorem

Let $u \in L_{1,H}(G)$. Then

(i) the integral $(T_G u)(x)$ exists for all $x \in R^3 \setminus \bar{G}$ and tends to zero for $|x| \longrightarrow \infty$; besides $(T_G u)(x)$ belongs to $A_H(R^3 \setminus \bar{G})$,

(ii) the integral $(T_G u)(x)$ exists almost everywhere on $\mathbf{R}^3$ and belongs to $L_{q,H}(G')$, $1<q<3/2$, where $G'\subset \mathbf{R}^3$ denotes any bounded domain.

Proof

(i) For the proof it is only necessary to note that always $x\neq y$ and $De(x-y)=0$. Further the estimate

$$|T_G u(x)| \leq \frac{1}{4\pi}\int_G \frac{1}{|x-y|^2}|u(y)|\,dG_y \leq \frac{1}{4\pi}\max_{y\in G}\frac{1}{|x-y|^2}\int_G |u(y)|\,dG_y ,$$

show that $|(T_G u)(x)| \longrightarrow 0$ as $|x| \longrightarrow \infty$.

(ii) Let $v\in L_{p,H}(G)$ with $p>3$. From [Sm ,p.287] we know that

$$v^*(x) = \int_G \frac{|v(y)|}{|x-y|^2}\,dG_y$$

is continuous on G.

Therefore $|u|v^* \in L_{1,\mathbf{R}}(G)$. Write

$$u^*(x) = \int_G \frac{|u(y)|}{|x-y|^2}\,dG_y .$$

FUBINI's theorem leads to the identity for any $v\in L_{p,H}(G)$

$$\int_G |u(y)|\,v^*(y)\,dG_y = \int_G |v(y)|\,u^*(y)\,dG_y .$$

Thence $u^* \in L_{q,\mathbf{R}}(G)$, with $1/p+1/q=1$, which means that also $T_G u \in L_{q,H}(G)$. Together with (i) we find that $T_G u \in L_{q,H}(G')$ for any bounded domain G'. It is immediately clear that $1<q<3/2$.

#

2.3.2 Theorem

Let $u \in L_{p,H}(G)$, $p>3$, then $(T_G u)(x)$ satisfies the inequalities

(i) $$|(T_G u)(x)| \leq C_1(G,p)\,\|u\|_{L_{p,H}} , \qquad (2.1)$$

(ii) $$|(T_G u)(x)-(T_G u)(x')| \leq C_2(G,p)\,\|u\|_{L_{p,H}}\,|x-x'|^{\frac{p-3}{3}},\quad x\neq x'. \qquad (2.2)$$

Proof

(i) HOELDER's inequality clearly yields

$$|(T_G u)(x)| \leq \frac{1}{4\pi}\left(\int_G \frac{1}{|x-y|^{2q}}\,dG_y\right)^{1/q} \|u\|_{L_{p,H}} .$$

As $2q<3$ the above integral converges, and it is easy to find such a constant $C_1(G,p)$.

(ii) Using HOELDER s inequality again we get the estimate

$$|(Tu)(x)-(Tu)(x')|^q \le \frac{1}{4\pi}\int_G \frac{|(x-y)|x'-y|^3-(x'-y)|x-y|^3|^q}{|x-y|^{3q}\,|x'-y|^{3q}}\,dG_y\,\|u\|^q_{L_{p,H}}\,.$$

Under the conditions (i) $0<\alpha,\beta<3$, (ii) $\alpha+\beta>3$ the well-known estimate

$$\int_G \frac{dG_y}{|x-y|^\alpha\,|x'-y|^\beta} \le \frac{C_0}{|x-x'|^{\alpha+\beta-3}} \qquad \text{for } x\neq x'$$

is valid, which the reader can find in [Sm ,p.312]. Now we obtain

$$|(Tu)(x)-(Tu)(x')|^q \le \frac{C_0}{(4\pi)^q}\left(\int_G \frac{|(x-y)|x'-y|^3-(x'-y)|x-y|^3|^q}{|x-y|^{\frac{3}{2}q}\,|x'-y|^{\frac{3}{2}q}}\,dG_y\right)\|y\|^q_{L_{p,H}}\,|x-x'|^{3-3p}. \tag{2.3}$$

It remains to show that

$$|(x-y)|x'-y|^3-(x'-y)|x-y|^3|^q \le f(x,x',y)\,|x-x'|^q \tag{2.4}$$

with a certain continuous positive function f. To prove this we carry out the following short computation. First, set $a=x-y$ and $b=x'-y$. Then

$$|a|b|^3-b|a|^3| \le |a|\,||b|^3-|a|^3| + |a-b|\,|a|^3 \le$$
$$\le |a-b|\,|a|\,[|b|^2+|a|\,|b|+2|a|^2].$$

Because $x-x'=a-b$, the wanted function is given by

$$f(x,x',y) = |x-y|^q[|x'-y|^2+|x'-y|\,|x-y|+2|x-y|^2]^q\ .$$

Using (2.4) we get

$$|(Tu)(x)-(Tu)(x')| \le \frac{C_0^{1/q}}{4\pi}\left(\int_G \frac{f(x,x',y)}{|x-y|^{\frac{3}{2}q}\,|x'-y|^{\frac{3}{2}q}}\,dG_y\right)^{1/q}\|u\|_{L_{p,H}}\,|x'-x|^{\frac{p-3}{3}},$$

where $1/p+1/q=1$. It is easy to see that $M = \sup_{x,x'\in G}\int \frac{f(x,x',y)}{|x-y|^{\frac{3}{2}q}|x'-y|^{\frac{3}{2}q}}\,dG_y < \infty$.

Now we can set

$$C_2(G,p) = \frac{(C_0 M)^{1/q}}{4\pi}$$

and the proof is finished. ∎

The following assertion immediately yields.

<u>2.3.3 Corollary</u>

Let $p>3$.

The operator $T_G : L_{p,H}(G) \longrightarrow C_H^{0,(p-3)/p}(\bar{G})$

is continuous, and it holds

$$\|T_G u\|_{C_H^{0,\frac{p-3}{3}}} \le C\,\|u\|_{L_{p,H}}\ ,\quad C>0\ .$$

In an analogous manner we may deduce

2.3.4 Corollary

Let $u \in C_H(\bar{G})$, then

(i) $|(T_G u)(x)| \leq C_0 |u|_{C_H}$,

(ii) $|(T_G u)(x)-(T_G u)(x')| \leq \{C_1 |\ln|x-x'|+C_2\}|x-x'||u|_C$,

where C_i, $i=0,1,2$, denote positive constants.

2.3.5 Theorem

Let $u \in L_{p,H}(G)$, $1<p<3$. Then it holds for $p'<\frac{3p}{3-p}$:

(i) $\|T_G u\|_{L_{p',H}} \leq C\|u\|_{L_{p,H}}$, $C>0$.

(ii) The operator $T_G : L_{p,H}(G) \longrightarrow L_{p',H}(G')$ is compact.

(iii) For each $\varepsilon>0$ there is a real number $\delta>0$ such that

$$\|(T_G u)(x)-(T_G u)(x')\|_{L_{p',H}} \leq \varepsilon \|u\|_{L_{p,H}}$$

if $|x-x'|<\delta$.

Proof

For the proof we refer to [Sm ,p.309].

2.3.6 Theorem

(i) Let $1<p<\infty, k=0,1,\ldots$. Then we have

$$T_G : W^k_{p,H}(G) \longrightarrow W^{k+1}_{p,H}(G) . \tag{2.5}$$

(ii) Let $0<\beta<1$. Then it holds

$$T_G : C^{0,\beta}_H(\bar{G}) \longrightarrow C^{1,\beta}_H(\bar{G}) . \tag{2.6}$$

Proof

(i) It is clear that $e(x)$ is homogeneous of the degree -2. Therefore it follows, by the aid of Theorem 9.3, Chapt.IX in [MP] that

$\partial_i T_G u \in W^k_{p,H}(G)$ if $u \in W^k_{p,H}(G)$.

(ii) For the proof of (2.6) we may refer to [Sp2],[Gir]. *

2.3.7 Corollary

Let $1<p<\infty$, $k=0,1,\ldots$. Then

$$T_G : W^{k,loc}_{p,H}(G) \longrightarrow W^{k+1,loc}_{p,H}(G) .$$

Proof

Let K be a compact subset of G. $T_G u$ allows the

representation

$$(T_G u)(x) = \int_K e(x-y)u(y)dy + \int_{G\setminus K} e(x-y)u(y)dy =$$
$$= (T_K u)(x) + (T_{G\setminus K} u)(x) .$$

Theorem 2.3.1 yields $T_{G\setminus K} \in A_H(K)$. Applying Theorem 2.3.6 to $(T_K u)(x)$ we obtain the desired result. #

2.3.8 Proposition

Let $u \in L_{p,H}(G)$, $p>3$, then we have the estimate

$$\|T_G u\|_{C_H} \leq \frac{p-1}{p-3} \left(\frac{3}{4\pi}\right)^{\frac{p-3}{3(p-1)}} |G|^{\frac{p-3}{3(p-1)}} \|u\|_{L_{p,H}} .$$

Proof

By virtue of HOELDER's inequality we obtain with $\theta = \sum_{i=1}^{3} \frac{x_i - y_i}{|x-y|} e_i$

$$|(Tu)(x)| \leq \frac{1}{4\pi} \int_G \frac{|\theta|}{|x-y|^2} |u(y)| dG_y \leq \|u\|_{L_{p,H}} \cdot \frac{1}{4\pi} \left(\int_G \frac{1}{|x-y|^{2q}} dG_y\right)^{1/q}$$

where $1/p+1/q=1$. As in the proof of SCHMIDT's inequality of classical complex analysis we replace the domain G by a certain ball with the radius $R=(|G|/|S_1|)^{1/3}$ and the centre x. That means that G and $B_R(x)$ have the same volume. It holds

$$\frac{1}{4\pi} \int_G \frac{1}{|x-y|^{2q}} dG_y \leq \frac{1}{4\pi} \int_{B_R(x)} \frac{1}{|x-y|^{2q}} dG_y \leq \int_0^R r^{2-2q} dr \leq \frac{R^{3-2q}}{3-2q} = \frac{|G|^{\frac{p-3}{3(p-1)}} \frac{3}{4\pi}^{\frac{p-3}{3(p-1)}}}{\frac{p-3}{p-1}},$$

whence the assertion follows. #

This proposition yields the following

2.3.9 Corollary

Let $u \in C_H(\bar{G})$, then it holds

$$\|T_G u\|_{C_H} \leq \sqrt[3]{\frac{3}{4\pi}} \sqrt[3]{|G|} \|u\|_{C_H} .$$

2.3.10 Proposition

The operator $T_G : L_{p,H}(G) \longrightarrow L_{q,H}(G)$ is continuous for any $q<\frac{3p}{3-p}$, and it holds the estimate

$$\|T_G\|_{[L_{p,H}, L_{q,H}]} \leq (\operatorname{diam} G)^{1+3/q-3/p}\, 2^{4+2/q-3/p}\, \pi^{1-1/p+1/q} (1+3/p-3/q)^{1/p-1-1/q} \left(\frac{p}{p-1}\right)^{1-1/p} q^{1/q} .$$

Proof

The proof is made in [Sm]. #

Special case (p=2,q=2)

$$\|T_G\|_{L(L_{2,H})} \leq \operatorname{diam} G \cdot 32\pi .$$

2.4 VEKUA's Theorems

In this subsection we shall indicate the main topics of a three-dimensional VEKUA theory. The first papers in this direction were established by V.IFTIMIE [I1] and W.I.SCHEWTSCHENKO [Sh1]. The authors have also added some results, [Sp6].

2.4.1 Definition

Denote by $\overset{\circ}{M}{}^1_H(G)$ the space of all H-valued functions which belong to $C^1_H(G) \cap C_H(\bar{G})$ and vanish on the boundary $\Gamma = \partial G$.
Let $u, v \in L_{1,H}(G)$, $\varphi \in \overset{\circ}{M}{}^1_H(G)$. If it is satisfied the relation

$$\int_G \bar{v}\, D\varphi\, dG = -\int_G \bar{u}\, \varphi\, dG \tag{2.7}$$

then the H-valued function u is called generalized derivative of the H-valued function v with respect to the operator D.
We fix the notation $u = D_g v$.
The class of all H-valued functions which possess a generalized derivative $D_g u \in L_{p,H}(G)$ is assigned by $M^1_{p,H}(G)$ for $k=0,1,\dots$.

2.4.2 Proposition

Let $u \in L_{p,H}(G)$, then $T_G u \in M^1_{p,H}(G)$. Furthermore the identity

$$D_g T_G u = u$$

holds.

Proof

Let $\varphi \in \overset{\circ}{M}{}^1_H(G)$. Using Definition (2.7) and FUBINI's theorem we carry out the transformations

$$\int_G \overline{T_G u}\, D\varphi\, dG_x = \frac{1}{4\pi}\int_G \Big(\int_G \overline{\frac{x-y}{|x-y|^3}\, u\, dG_y}\, D\varphi \Big) dG_x =$$

$$= \frac{1}{4\pi}\int_G \bar{u}(y) \int_G \overline{\frac{x-y}{|x-y|^3}}\, D\varphi\, dG_x\, dG_y = -\int_G \bar{u}\, T_G D\varphi\, dG_y .$$

BOREL-POMPEIU's formula finally yields

$$\int_G \overline{T_G u}\, D\varphi\, dG_x = -\int_G \bar{u}\, \varphi\, dG_y .$$

The assertion follows from Definition 2.4.1. #

It is natural to raise the question if for all H-valued functions $u \in C^1_H(G)$ the derivatives Du and $D_g u$ may be identified.

2.4.3 Proposition

Let $u \in W^1_{p,H}(G)$. We have

$$D_g u = Du.$$

Proof

It is sufficient to assume $u \in C^1_H(G)$, $\varphi \in \overset{\circ}{M}{}^1_{p,H}(G)$. Consider LEIBNIZ' formula (1.26)

$$D(u\varphi) = \bar{u}\, D\varphi + (Du)\varphi - 2(\hat{u}\cdot \mathrm{grad})\varphi. \tag{2.8}$$

A straightforward evaluation yields

$$-\int_G (\hat{u}\cdot \mathrm{grad})\varphi\, dG_y = \int_G (\mathrm{div}\ \hat{u})e_0 \varphi\, dG_y.$$

Integrating (2.8) over G and using GAUSS' formula we get

$$0 = \int_G \bar{u}\, D\varphi\, dG_y + \int_G [Du - 2(\mathrm{div}\ \hat{u})e_0]\,\varphi\, dG_y,$$

which implies

$$\int_G \bar{u}\, D\varphi\, dG_y = \int_G \overline{Du}\,\varphi\, dG_y$$

and so $Du = D_g u$. #

2.4.4 Remark

As we work in $W^1_{p,H}(G)$, it is not necessary to distinguish between the componentwise approach of the definition of the derivative and the generalized definition. Usually we shall use the same notation.

2.4.5 Proposition

(i) Suppose $u \in M^1_{p,H}(G)$ and $v = Du$, then u can be represented by $u = w + T_G v$, where $w \in A_H(G)$.

(ii) Conversely, given $w \in A_H(G)$ and $v \in L_{p,H}(G)$, then we have $u = w + T_G v \in M^1_{p,H}(G)$, and u satisfies the equation $Du = v$.

Proof

(i) We have $u = [u - T_G Du] + T_G Du$. As $v = Du \in L_{p,H}(G)$ from Theorem 2.3.6 it follows $T_G v \in W^1_{p,H}(G)$. It remains to show that $u - T_G Du \in A_H(G)$. Using Proposition 2.4.2 we obtain

$$Du - DT_G Du = Du - Du = 0.$$

(ii) Because w and $T_G v$ each have a generalized derivative, then their sum has one, too. We consider Proposition 2.4.2 and get $Du = v$. #

2.4.6 Corollary

The class $M^1_{p,H}(G)$ allows the decomposition as a direct sum

(i) $M^1_{p,H}(G) = A_H(G) \oplus T_G L_{p,H}(G)$.

Besides it holds

(ii) $W^1_{p,H}(G) \subset M^1_{p,H}(G) \subset W^{loc}_{p,H}(G)$.

Proof

(i) Proposition 2.4.5 guarantees that each element $u \in M^1_{p,H}(G)$ allows the representation $u=w+T_G v$ with $w \in A_H(G)$ and $v \in L_{p,H}(G)$. Suppose now that there exists a non-vanishing H-valued function $z \in A_H(G) \cap T_G L_{p,H}(G)$, then we have $Dz=0$. On the other hand, it holds $z=T_G f$ with $f \in L_{p,H}(G)$. Hence $0=Dz=DT_G f=f$, therefore $z=0$.

(ii) To prove the inclusion (ii) we remark that $A_H(G) \subset C^\infty_H(G')$ with $\bar{G}' \subset G$. #

2.4.7 Corollary

Let $G' \subset G$ be a subdomain. Assume $v \in M^1_{p,H}(G)$, then $v \in M^1_{p,H}(G')$.

Proof

As $v \in M^1_{p,H}(G)$ we get $v=w+T_G u=w+T_{G\setminus G'}u+T_{G'}u$, where $u \in L_{p,H}(G)$. From Theorem 2.3.1 we obtain $T_{G\setminus G'}u \in A_H(G')$. Setting $w'=w+T_{G\setminus G'}u$ and the assertion follows from Proposition 2.4.5.

2.4.8 Corollary

For any $x \in G$ there exists a real number $\varepsilon>0$ such that $\overline{B_\varepsilon(x)} \subset G$. If $u \in M^1_{p,H}(B_\varepsilon(x))$ for any $x \in G$, then we have $Du \in L^{loc}_{p,H}(G)$.

Proof

Let $\overline{G'}$ be an arbitrary compact subdomain of G. Then there exists a finite system of open balls $B_i=B_{\varepsilon_i}(x_i)$, i=1,2,...,n, which overlaps $\overline{G'}$ and is contained in G. In B_i u allows the representation $u=w_i+T_B v_i$, i=1,...,n, with $w_i \in A_H(B_i)$ and $v_i \in L_{p,H}(B_i)$. It is easy to show that for points in $B_i \cap B_j$ $v_i=v_j$ holds. Setting $w=v_i$ in B_i. If $x \in B_i$ then we conclude

$$D(u-T_{G'}w)=D(u-T_{B_i}v_i-T_{G'\setminus B_i}w)=0$$

because $u-T_{B_i}v_i$ and $T_{G'\setminus B_i}w$ are H-regular functions. #

2.5 Some Integral Operators on the Manifold Γ

For a better understanding we recall the meaning of some notations which relate to functional spaces on manifolds.

folds.

2.5.1 Definition

Let Γ be a closed compact smooth LIAPUNOV manifold (i.e. a manifold of the class $C^{k,\alpha}$ ($0<\alpha\leq 1$)) of the dimension 2 which bounds a certain domain $G\subset R^3$. Assume $\{V_i\}_{i=1}^n$ a finite covering of Γ and $\{\varphi_i\}_{i=1}^n$ the corresponding partition of unity. Denote by s_i the diffeomorphism $V_i \longrightarrow \Omega_i \subset R^2$. An H-valued function $u(x)$ defined on Γ belongs to the space $C_H^{0,\beta}(\Gamma)$ $(0<\beta\leq 1)$ if and only if

$\varphi_i(s_i^{-1}(x'))u(s_i^{-1}(x'))\in C_H^{0,\beta}$ (Ω_i) for each i. The norm in

$C_H^{0,\beta}(\Gamma)$ is determined by

$$\|u\|_{C_H^{0,\beta}(\Gamma)} = \sum_{i=1}^{n} \|(\varphi_i u)(s_i^{-1}(x'))\|_{C_H^{0,\beta}(\Omega_i)} .$$

On the manifold SOBOLEV spaces $W_{p,H}^k(\Gamma)$, $k=0,1,\ldots$, (in particular $L_{p,H}(\Gamma)$) may be defined in an analogous manner. Therefore the space $W_{p,H}^k(\Gamma)$ consists of all functions u with the finite norm

$$\|u\|_{W_{p,H}^k(\Gamma)}^p = \sum_{|\lambda|=0}^{k} \sum_{i=1}^{n} \|u_i^{(\lambda)}\|_{L_{p,H}(\Omega_i)}^p .$$

2.5.2 Proposition

Denote by V_α the potential of the single layer

$$(V_\alpha u)(x) = \frac{1}{4\pi}\int_\Gamma \frac{\alpha(y)}{|x-y|}u(y)\,d\Gamma_y , \tag{2.9}$$

where $\alpha(y)$ denotes the outer normal at the point y and denote by K the volume potential

$$(Ku)(x) = \frac{1}{4\pi}\int_G \frac{u(y)}{|x-y|}\,dy . \tag{2.10}$$

Let (i) $u\in C_H^1(G)\cap C_H(\bar{G})$

or (ii) $u\in W_{p,H}^1(G)$, $p>1$.

Then the identity

$$V_\alpha u = T_G u + KDu \tag{2.11}$$

is valid.

Proof

(i) Applying the generalized LEIBNIZ formula we obtain

$$D_y\left[\frac{1}{|x-y|}u(y)\right] = \frac{x-y}{|x-y|^3}u(y) + \frac{1}{|x-y|}[Du](y) .$$

Integrating over the domain $G_\varepsilon = G\setminus\overline{B_\varepsilon(x)}$ with respect to y we obtain from GAUSS' formula

$$\frac{1}{4\pi}\int_\Gamma \frac{\alpha(y)}{|x-y|}u(y)d\Gamma_y = \frac{1}{4\pi}\int_{G_\varepsilon}\frac{x-y}{|x-y|^3}u(y)\,dG_y + \frac{1}{4\pi}\int_{G_\varepsilon}\frac{1}{|x-y|}(Du)(y)\,dG_y -$$

$$-\frac{1}{4\pi}\int_{S_\varepsilon(x)}\frac{x-y}{|x-y|^2}u(y)\,dS_\varepsilon$$

, where $S_\varepsilon(x)=\partial B_\varepsilon(x)$.

Since

$$\frac{1}{4\pi}\int_{S_\varepsilon}\frac{x-y}{|x-y|^2}u(y)dS_\varepsilon = \frac{\varepsilon}{4\pi}\int_{S_1}\theta\, u(x+\theta\varepsilon)\,dS_1$$

and because all integrals in identity (2.11) exist. If ε tends to zero, the wanted formula is deduced.

Using Proposition 2.4.2 we obtain the case (ii).

#

2.5.3 Proposition

The operator F introduced in 1.4.1 allows the factorization

$$F_\Gamma = DV_\alpha . \qquad (2.12)$$

Proof

Formula (2.12) may be immediately achieved by changing the order of differentiation and integration.

#

2.5.4 Corollary

Let $u \in W^1_{p,H}(G)$, p>1 Then the formula

$$F_\Gamma u + T_G Du = u \quad \text{holds in } G. \qquad (2.13)$$

Proof

This identity may be proved using Propositions 2.4.2 and 2.5.2.

#

2.5.5 Theorem

Let $1<p<\infty$, $k>0$, $k \in \mathbb{N}$. Then we have

$$F_\Gamma : W^{k-1/p}_{p,H}(\Gamma) \longrightarrow W^k_{p,H}(G) \cap \ker D(G). \qquad (2.14)$$

Proof

Let $u \in W^{k-1/p}_p(\Gamma)$. Then there exists an H-valued function

$v \in W^k_{p,H}(G)$ with tr v=u, and (2.13) yields $F_\Gamma u = v - T_G Dv$. From Theorem 2.3.6 it follows $v - T_G Dv \in W^k_{p,H}(G)$. #

In order to obtain a similar result in spaces $C_H^{\emptyset,\beta}$, some preparatory considerations are necessary.

<u>2.5.6 Proposition</u>

$$\frac{1}{4\pi}\int_\Gamma \frac{\theta\,\alpha(y)}{|x-y|}\,d\Gamma_y = \begin{cases} e_\emptyset, & x\in G, \\ \tfrac{1}{2}e_0, & x\in\Gamma, \\ \emptyset, & x\in\mathbb{R}^3\setminus\overline{G}, \end{cases}$$

where $\theta = \frac{x-y}{|x-y|}$ and $\alpha(y)$ denotes the outer normal on Γ at the point y. In case of $x\in\Gamma$, the above integral exists in the sense of CAUCHY's principal value.

<u>Proof</u>

Taking the algebraic properties in H into account we get

$$\theta\alpha = (\theta\cdot\alpha)e_\emptyset + \sum_{i>j}(\theta_i\alpha_j - \theta_j\alpha_i)e_ie_j .$$

This leads to

$$\frac{1}{4\pi}\int_\Gamma \frac{\theta\alpha}{|x-y|^2}\,d\Gamma_y = -\frac{1}{4\pi}\int_\Gamma \frac{\theta\cdot\alpha}{|x-y|^2}\,d\Gamma_y\, e_\emptyset + \frac{1}{4\pi}\int_\Gamma\sum_{i>j}\frac{\theta_i\alpha_j - \theta_j\alpha_i}{|x-y|^2}\,d\Gamma_y\, e_ie_j .$$

The first integral may be considered as the potential of the double layer with the density 1. It is well-known that the following formula is valid

$$\frac{1}{4\pi}\int_\Gamma \frac{\theta\cdot\alpha(y)}{|x-y|^2}\,d\Gamma_y = \begin{cases} -e_\emptyset, & x\in G \\ -\tfrac{1}{2}e_\emptyset, & x\in\Gamma \\ \emptyset, & x\in\mathbb{R}^3\setminus\overline{G} \end{cases} .$$

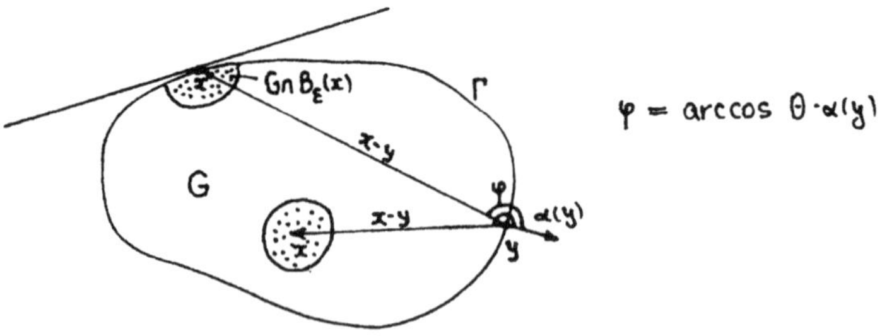

<u>Figure 5:</u>

Proof

By applying GAUSS' formula we deduce

$$\int_{\Gamma} \frac{\theta_i \alpha_j - \theta_j \alpha_i}{|x-y|^2} d\Gamma_y = \lim_{\varepsilon \to 0} \int_{\Gamma_\varepsilon} \frac{\theta_i \alpha_j - \theta_j \alpha_i}{|x-y|^2} d\Gamma_y = \lim_{\varepsilon \to 0} \Big[\frac{\partial}{\partial x_j} \int_{\Gamma_\varepsilon} \frac{\alpha_i}{|x-y|} d\Gamma_{\varepsilon,y} -$$

$$- \frac{\partial}{\partial x_i} \int_{\Gamma_\varepsilon} \frac{\alpha_j}{|x-y|} d\Gamma_{\varepsilon,y} = \lim_{\varepsilon \to 0} \Big[\frac{\partial^2}{\partial x_i \partial x_j} \int_{G_\varepsilon} \frac{1}{|x-y|} dG_{\varepsilon,y} - \frac{\partial^2}{\partial x_j \partial x_i} \int_{G_\varepsilon} \frac{1}{|x-y|} dG_{\varepsilon,y} \Big] = 0,$$

where $\Gamma_\varepsilon = \partial G_\varepsilon$ and $G_\varepsilon = \bar{G} \setminus \overline{B_\varepsilon(x)}$. #

2.5.7 Corollary

Let $u \in C_H^{0,\beta}(\Gamma)$, $0 < \beta \leq 1$. Then there exists the integral

$$(S_\Gamma u)(x) = \frac{1}{2\pi} \int_{\Gamma} \frac{\theta\, \alpha(y)}{|x-y|^2} u(y)\, d\Gamma_y$$

for all points $x \in \Gamma$ in the sense of CAUCHY's principal value.

The singular integral operator S_Γ may be continuously extended onto the whole space $L_{2,H}(\Gamma)$, see [Mi2].

2.5.8 Theorem

The singular integral operator S_Γ acts within the spaces $C_H^{0,\beta}(\Gamma)$ and $L_{2,H}(\Gamma)$ and is continuous there.

Proof

G. GIRAUD shows in [Gir] for more general integral operators that the image of an H-valued function $u \in C_H^{0,\beta}(\Gamma)$ of the operator S_Γ is included in $C_H^{0,\beta}(\Gamma)$ again. The continuity of S_Γ in the space $C_H^{0,\beta}(\Gamma)$ is proved in [Ge]. To verify the validity of our assertion in $L_{2,H}(\Gamma)$, we refer to [Sp3]. #

2.5.9 Remark

In [MP,p.421] the continuity of S_Γ in the scale of spaces $W^s_{p,H}(\Gamma)$, $1<p<\infty$, $s>0$ is shown.

2.5.10 Theorem (PLEMELJ-SOKHOTZKIJ's Formula)

Let $u \in C_H^{0,\beta}(\Gamma)$, $0 < \beta \leq 1$. Then we have

$$\lim_{\substack{x \to x_0 \\ x \in G^\pm,\, x_0 \in \Gamma}} (F_\Gamma u)(x) = \tfrac{1}{2}\{\pm u(x_0) + (S_\Gamma u)(x_0)\},$$

where $G^+=G$ and $G^-=\mathbb{R}^3\setminus\bar{G}$.

Proof

Let $x_\emptyset$ be a fixed point at Γ. Proposition 2.5.6 yields

$$(Fu)(x)=\frac{1}{4\pi}\int_\Gamma \frac{\Theta\alpha(y)}{|x-y|^2}[u(y)-u(x_\emptyset)]\,d\Gamma_y+u(x_\emptyset)\quad,\quad x\in G,$$

and

$$(Fu)(x)=\frac{1}{4\pi}\int_\Gamma \frac{\Theta\alpha(y)}{|x-y|^2}[u(y)-u(x_\emptyset)]\,d\Gamma_y\quad,\quad x\in \mathbb{R}^3\setminus\bar{G}.$$

We shall write for brevity

$$a(x)=\frac{1}{4\pi}\int_\Gamma \frac{\Theta\alpha(y)}{|x-y|^2}[u(y)-u(x_\emptyset)]\,d\Gamma_y,$$

assume $x\neq x_\emptyset$ and estimate as follows:

$$|a(x)-a(x_\emptyset)|\leq \frac{1}{4\pi}\int_\Gamma \frac{|(x_\emptyset-y)|x-y|^3-(x-y)|x_\emptyset-y|^3|}{|x_\emptyset-y|^3|x-y|^3}|u(y)-u(x_\emptyset)|\,d\Gamma_y.$$

Using (2.4) one obtains

$$|a(x)-a(x_\emptyset)|\leq\frac{1}{4\pi}\int_{\Gamma_\varepsilon}\frac{|x_\emptyset-y|^{-2+\beta}[|x-y|^2+|x_\emptyset-y||x-y|+2|x_\emptyset-y|^2]}{|x-y|^3}\ \frac{|u(y)-u(x_\emptyset)|}{|y-x_\emptyset|^\beta}\,d\Gamma_{\varepsilon,y}\,|x-x_\emptyset|+$$

$$+\frac{1}{4\pi}\int_{\Gamma'}\frac{|x_\emptyset-y|^{-2+\beta}[|x-y|^2+|x_\emptyset-y||x-y|+2|x_\emptyset-y|^2]}{|x-y|^3}\ \frac{|u(y)-u(x_\emptyset)|}{|y-x_\emptyset|^\beta}\,d\Gamma'_y\,|x-x_\emptyset|,$$

where ε is a sufficiently small positive number, $\Gamma_\varepsilon=\{y\in\Gamma:|x_\emptyset-y|\leq\varepsilon\}$ and $\Gamma'=\{y\in\Gamma:|x_\emptyset-y|>\varepsilon\}$. For smooth LIAPUNOV manifolds we have inside of the ball $B_\varepsilon(x_\emptyset)$ the estimate

$$2|x_\emptyset-y|\geq|x-y|\geq\frac{1}{2}|x_\emptyset-y|. \tag{2.15}$$

Hence $|x-x_\emptyset|\leq|x_\emptyset-y|+|y-x|\leq3|x_\emptyset-y|$. Thus we get

$$|a(x)-a(x_\emptyset)|\leq\left\{\frac{11}{2\pi}3^\delta\int_{\Gamma_\varepsilon}|x_\emptyset-y|^{-3+\beta+\delta}\,d\Gamma_{\varepsilon,y}|x-x_\emptyset|^{1-\delta}+\frac{2}{\pi\varepsilon^3}\int_{\Gamma'}|x_\emptyset-y|^{-2+\beta}[\ldots]\,d\Gamma'|x-x_\emptyset|\right\}\|u\|_{C_H^{\emptyset,\beta}},$$

where δ is chosen such that $0<\delta<1$ and $\beta+\delta>1$. The integrals occurring exist in the usual sense, whence the estimate

$$|a(x)-a(x_\emptyset)|\leq \text{const.}\,|x-x_\emptyset|^{1-\delta}\,\|u\|_{C_H^{\emptyset,\beta}}$$

follows.

It is clear that $a(x)$ converges to $a(x_\emptyset)$ if x tends to $x_\emptyset$. The definition of the operator S_Γ and Proposition 2.5.6 yield the desired result. #

2.5.11 Remark

To formulate a corresponding theorem in the space $L_{2,H}(\Gamma)$, we refer to the interesting paper of CL. MUELLER [Mue2,p.35]. Theorem 2.5.10 is also fulfilled in $L_{p,H}(\Gamma)$ if the point x which runs to x_0 is not lying on a tangent of Γ at the point $x_\emptyset$. The proof may be carried out similarly to the 2-dimensional case.

2.5.12 Proposition

Let $u \in C_H^{0,\beta}(\Gamma)$, $0<\beta\le 1$.

(i) The equality $S_\Gamma u=u$ is sufficient and necessary that $u(x)$ are boundary values of an H-valued function which is H-regular in G.

(ii) The condition $S_\Gamma u=-u$ is sufficient and necessary that $u(x)$ are boundary values of a function which is H-regular in $R^3\backslash\bar{G}$ and vanishes at infinity.

Proof

(i) First let U be the H-regular extension of the H-valued function $u(x)$, $x\in\Gamma$, into the domain G. From CAUCHY's integral theorem it follows that $F_\Gamma u=u$. PLEMELJ-SOKHOTZKIJ's formulas lead to

$$u(x_\emptyset) = \lim_{x\to x_\emptyset, x\in G, x_\emptyset\in\Gamma} (F_\Gamma u)(x) = \frac{1}{2}(S_\Gamma u)(x_\emptyset) + \frac{1}{2}u(x_\emptyset) ,$$

whence $S_\Gamma u=u$. Conversely, assume $(S_\Gamma u)(x)=u(x)$ for any point on Γ, then it is clear that $F_\Gamma u$ is the wanted H-regular extension.

(ii) This assertion may be proved by using Proposition 1.5.20.

\#

2.5.13 Corollary

Let $u \in C_H^{0,\beta}(\Gamma)$, $0<\beta\le 1$, then the identity $S_\Gamma^2 u=u$ holds.

Proof

Setting $\lim_{x\to x_\emptyset\in\Gamma, x\in G}(F_\Gamma u)(x)=F^+(x_\emptyset)$ and $\lim_{x\to x_\emptyset\in\Gamma, x\in R^3\backslash\bar{G}}(F_\Gamma u)(x)=F^-(x_\emptyset)$.

Using the generalized formulas of PLEMELJ-SOKHOTZKIJ, then it follows from Proposition 2.5.12 $(S_\Gamma u)(x_\emptyset)=F^+(x_\emptyset)+F^-(x_\emptyset)$, whence $(S_\Gamma^2 u)(x_\emptyset)=(S_\Gamma F^+)(x_\emptyset)+(S_\Gamma F^-)(x_\emptyset)=F^+(x_\emptyset)-F^-(x_\emptyset)=u(x_\emptyset)$. #

2.5.14 Definition

Denote with P_Γ the projection onto the space of all H-valued functions which may be H-regular extended into the domain G. Q_Γ denotes the projection onto the space of all

H-valued functions which may be H-regular extended into the domain $\mathbb{R}^3\backslash\bar{G}$ and vanish at infinity.

2.5.15 Corollary

The projections P_Γ and Q_Γ may be represented by

$$P_\Gamma = \frac{1}{2}(I + S_\Gamma) \quad \text{and} \quad Q_\Gamma = \frac{1}{2}(I - S_\Gamma).$$

Obviously, the algebraic properties

$$P_\Gamma^2 = P_\Gamma, \; Q_\Gamma^2 = Q_\Gamma, \; P_\Gamma Q_\Gamma = Q_\Gamma P_\Gamma = 0$$

are fulfilled.

Proof

It can be proved immediately by making use of Proposition 2.5.12. #

2.5.16 Theorem

Let $0<\beta<1$. Then $F_\Gamma : C_H^{0,\beta}(\Gamma) \longrightarrow C_H^{0,\beta}(\bar{G})$ holds.

Proof

With $u \in C_H^{0,\beta}(\Gamma)$ we get $P_\Gamma u \in C_H^{0,\beta}(\Gamma)$, whence follows that $F_\Gamma u = F_\Gamma P_\Gamma u$ is the harmonic extension of $P_\Gamma u$ into the domain G. Now, the assertion may be proved by using a result of SMOLITZKI [Smo] on the smoothness of harmonic extensions. #

2.5.17 Remark

Proposition 2.5.12 and its corollaries are also valid for continuous extensions of the operators S_Γ, P_Γ and Q_Γ onto the space $L_{2,H}(\Gamma)$.

2.5.18 Theorem

Let $v, w \in A_H(G) \cap W_H^1(G)$, $\varrho \in C_H^{0,\beta}(G)$, $0<\beta\leq 1$. Besides, let the H-valued function u allow the quite different representations

(i) $u = v + T_G w$,

(ii) $u = \int_\Gamma \frac{\varrho(y)}{|x-y|} d\Gamma_y$.

Then there exists a unique connection between the H-regular functions v and w on the one hand and the vector density ϱ on the other hand. These relations can be described by the integral equations

(iii) $\operatorname{tr} v = \operatorname{tr} V_\alpha Q_\Gamma(\bar{\alpha}\varrho)$,

(iv) $\operatorname{tr} w = P_\Gamma(\bar{\alpha}\varrho)$.

Proof

Proposition 2.4.2 and formula (2.11) yield $Du=w=DV_{\alpha}(\bar{\alpha}\varrho)$. Using Proposition 2.5.4 and Theorem 2.5.10 we obtain the integral equation (iv). Obviously, we have

$$v=u-T_G w=V_{\alpha}(\bar{\alpha}\varrho)-V_{\alpha}P_{\Gamma}(\bar{\alpha}\varrho)=V_{\alpha}Q_{\Gamma}(\bar{\alpha}\varrho).$$

Out of this follows (iii).

#

2.5.19 Remark

The function u has to be smooth in such a way that $\frac{\partial u}{\partial \alpha}$ exists, for instance $u \in W^2_{2,H}$.

2.5.20 Remark

Using the preceding assertions we can deduce a generalization of the well-known POINCARE´-BERTRAND´s formula.

Let $u \in C_H^{0,\beta}(\Gamma \times \Gamma)$, then we have

$$\int_\Gamma \frac{x-x_0}{|x-x_0|^3}\alpha(x)\left(\int_\Gamma \frac{x-y}{|x-y|^3}\alpha(y)u(x,y)\,d\Gamma_y\right)d\Gamma_x=(4\pi)^2 u(x_0,y_0)+\int_\Gamma d\Gamma_y\int_\Gamma \frac{(x-x_0)\alpha(x)(x-y)\alpha(y)u(x,y)}{|x-y|^3\,|x-x_0|^3}\,d\Gamma_x .$$

3. ORTHOGONAL DECOMPOSITION OF THE SPACE $L_{2,H}(G)$

Let R be an operator which is defined by

$$Ru = \sum_{i=\emptyset}^{3} r_i u_i e_i ,$$

where the scalar functions $r_i \in C_R^\infty(\bar{G})$, $i=\emptyset,1,2,3$, are positive in $L_{2,H}(G)$. It is clear that $R^{-1}u = \sum_{i=\emptyset}^{3} r_i^{-1} u_i e_i$.

The following special cases are important for further considerations:

(i) $r_\emptyset = r_1 = r_2 = r_3 = 1$,

(ii) $r_\emptyset = r_1 = r_2 = r_3 = r(x)$,

(iii) $r_1 = r_2 = r_3 = 1$ and $r_\emptyset$ a fixed constant.

We may introduce in $L_{2,H}(G)$, considered as a real vector space, the inner product

$$[u,v]_R = \int_G \overline{R^{-1}u}\, R^{-1}v \, dG \in H. \qquad (3.1)$$

Note that the values of $[u,v]_R$ are not necessarily real numbers, but $[u,u]_R \geq \emptyset$.

3.1 Theorem

The HILBERT space $L_{2,H}(G)$ allows the orthogonal decomposition

$$L_{2,H}(G) = R\, A_H(G) \cap L_{2,H}(G) \oplus_R D(\mathring{W}^1_{2,H}(G)),$$

where $\oplus_R$ denotes an orthogonal sum according to (3.1).

Proof

The right linear sets $X_1 = L_{2,H}(G) \cap R\, A_H(G)$ and $X_2 = L_{2,H}(G) \ominus X_1$ are subspaces of $L_{2,H}(G)$. Let $u \in X_2$. As for any $u \in L_{2,H}(G)$, $T_G R^{-1} \in W^1_{2,H}(G)$, it follows that $u = RDv$ with $v \in W^1_{2,H}(G)$. Since $u \in X_2$ we have for any $g \in X_1$

$$\int_G \overline{Dv}\, R^{-1} g \, dG = \emptyset$$

and, in particular, for any $l \in \mathbb{N}$

$$\int_G \overline{Dv}\, R^{-1} g_l \, dG = \emptyset$$

with $g_l=R\{(x-y_l)/|x-y_l|^3\}\in RA_H(G)$, $y_l\in\mathbb{R}^3\setminus\bar{G}$. We assume that the set $\{y_l,\ l\in\mathbb{N}\}$ is dense in $\mathbb{R}^3\setminus\bar{G}$. By componentwise considerations we obtain

$$0=\sum_{\substack{i=1\\j=0}}^{3}\int_G \overline{e_i\frac{\partial}{\partial x_i}v_je_j}\,R^{-1}g_l\,dG=\sum_{\substack{i=1\\j=0}}^{3}\int_G \overline{e_j}\,\overline{e_i}\left(\frac{\partial}{\partial x_i}v_j\right)R^{-1}g_l\,dG=$$

$$=-\sum_{i=1}^{3}\int_G \overline{e_j}\,\overline{e_i}\,v_j\frac{\partial}{\partial x_i}R^{-1}g_l\,dG+\sum_{\substack{i=1\\j=0}}^{3}\int_\Gamma \overline{e_j}\,\overline{e_i}\,v_j\alpha_i\,R^{-1}g_l\,d\Gamma=$$

$$=-\int_G \bar{v}\,DR^{-1}g_l\,dG+\int_\Gamma \bar{v}\,\bar{\alpha}\,\overline{\overline{R^{-1}g_l}}\,d\Gamma=$$

$$=\overline{\int_\Gamma \overline{R^{-1}g_l}\,\alpha\,v\,d\Gamma}=$$

$$=\overline{\int_\Gamma \frac{y_l-x}{|x-y_l|^3}\,\alpha(x)\,v(x)\,d\Gamma}=$$

$$=-4\pi\,\overline{\int_\Gamma e(y_l-x)\,\alpha(x)\,v(x)\,d\Gamma}=-4\pi\,\overline{(F_\Gamma(\mathrm{tr}\,v))(y_l)}$$

for any y_l. Hence $F_\Gamma(\mathrm{tr}\,v)=0$ in $\mathrm{co}\,\bar{G}$, and it follows that $\mathrm{tr}\,v\in\mathrm{im}\,P_\Gamma\cap W^{1/2}_{2,H}(\Gamma)$. Consequently, there exists an H-valued function $h\in W^1_{2,H}(G)\cap A_H(G)$ with the property that $\mathrm{tr}\,h=\mathrm{tr}\,v$. Taking the function $w=v-h\in\mathring{W}^1_{2,H}(G)$ we get that $u=Dv=Dw\in D(\mathring{W}^1_{2,H}(G))$. #

3.2 Corollary

There exist two orthoprojections P_R and Q_R with

$$P_R:L_{2,H}(G)\longrightarrow R(A_H(G))\cap L_{2,H}(G),$$

$$Q_R=I-P_R:L_{2,H}(G)\longrightarrow D(\mathring{W}^1_{2,H}(G))\cap L_{2,H}(G).$$

In case of $R=I$ we set $P_R=P$ and $Q_R=Q$.

Proof

This is a direct consequence of Theorem 3.1. #

3.3 Corollary

The differentiation rule

$$DR^{-1}Q_R u = DR^{-1}u \ , \quad u \in W^1_{2,H}(G).$$

holds.

Proof

We have $DR^{-1}Q_R u = DR^{-1}u - DR^{-1}P_R u$ because

$R^{-1} \text{im } P_R \subset \ker D \subset C^\infty_H(G)$. #

4. SOME BOUNDARY VALUE PROBLEMS OF DIRICHLET'S TYPE

The aim of Chapter 4 consists in the study of problems of mathematical physics of various kinds by the methods of quaternionic analysis. Many mathematicians have dealt with this subject, but up to now it has not been possible to consider it in such a self-contained form. An application of the quaternionic analysis enables us to formulate a unified approach for solving all questions arising in the consideration of boundary value problems. Therefore treatment is carried out in a well-defined order (existence of a solution, uniqueness, representation of the solution, regularity). A corresponding quaternionic numerical mathematics is given in Chapter 5.

4.1 LAPLACE Equation

4.1.1 Theorem

Let $f \in W^k_{2,H}(G)$. DIRICHLET's problem

$$-\Delta u = f \quad \text{in} \quad G \tag{4.1}$$

$$u = 0 \quad \text{on} \quad \Gamma \tag{4.2}$$

has a solution $u \in W^{k+2,loc}_{2,H}(G)$, which may be represented by the formula

$$u = T_G Q T_G f \ .$$

Proof

For Theorem 3.1 there exists an H-valued function $u \in \overset{\circ}{W}{}^1_{2,H}(G)$ with $QT_G f = Du$. Theorem 2.3.6 and Corollary 2.3.7 yield

$T_G f \in W^{k+1}_{2,H}(G)$ and $QT_G f \in W^{k+1,loc}_{2,H}(G)$. By the help of BOREL-POMPEIU's formula

$$T_G Q T_G f = T_G Du = u$$

follows, and therefore $u \in W^{k+2,loc}_{2,H}(G)$. Finally, Corollary 3.3 implies $-\Delta u = DDu = DQT_G f = f$. #

4.1.2 Corollary

Let $g \in W^{k+1/2}_{2,H}(\Gamma)$, $k \geq 0$. DIRICHLET's problem

$$\Delta u = 0 \quad \text{in} \quad G, \tag{4.3}$$

$$u = g \quad \text{on} \quad \Gamma \tag{4.4}$$

has a solution $u \in W_{2,H}^{k+1,loc}(G)$ of the form

$$u = F_{\Gamma} g + T_G Q D h, \tag{4.5}$$

where h is a $W_{2,H}^{k+1}(G)$-extension of g.

Proof

As $g \in W_{2,H}^{k+1/2}(\Gamma)$ there exists a $W_{2,H}^{k+1}$-extension h with $tr\ h=g$. Put $u=v+h$. The problem (4.3)-(4.4) will be transformed into

$$\Delta v = \Delta h \quad \text{in } G,$$
$$v = 0 \quad \text{on } \Gamma.$$

For the validity of Theorem 4.1.1 there exists the solution $v \in W_{2,H}^{k+1,loc}(G)$ with

$$v = T_G Q T_G \Delta h.$$

Using BOREL-POMPEIU's formula, $P=I-Q$ and $DD=-\Delta$ we find

$$v = -T_G Q D h + T_G Q F_{\Gamma} D h = -T_G D h + T_G P D h = -h + T_G P D h + F_{\Gamma} h.$$

With $u=v+h$ we get (4.5). #

4.1.3 Theorem (Existence)

Let $f \in W_{2,H}^{k}(G)$, $g \in W_{2,H}^{k+3/2}(\Gamma)$, $k \geq 0$. DIRICHLET's problem

$$-\Delta u = f \quad \text{in } G, \tag{4.6}$$
$$u = g \quad \text{on } \Gamma \tag{4.7}$$

has the solution

$$u = F_{\Gamma} g + T_G P D h + T_G Q T_G f \in W_{2,H}^{k+2,loc}(G). \tag{4.8}$$

Here h denotes a $W_{2,H}^{k+2}$-extension of g.

Proof

Let u_1, u_2 be solutions of the problems (4.1)-(4.2), (4.3)-(4.4), respectively, then $u=u_1+u_2$ solves the boundary value problem (4.6)-(4.7) in the space $W_{2,H}^{k+2,loc}(G)$. #

4.1.4 Theorem

Let $f \in W_{2,H}^{k}(G)$, $g \in W_{2,H}^{k+3/2}(\Gamma)$, $k \geq 0$. The H-valued function u of (4.8) is the only solution of the boundary value problem (4.6)-(4.7).

Proof

Consider the boundary value problem $\{\Delta u=0,\ tr\ u=0\}$. BOREL-POMPEIU's formula and the fact that $Du \in im\ Q$ lead, to a solution u of this problem

$$u = T_G Du = T_G QDu \ .$$

On the other hand, because of $Du \in A_H(G)$,

$$u = T_G Du = T_G PDu$$

follows. Therefore

$$PDu = QDu,$$

whence

$$Du = \emptyset$$

and finally

$$u = \emptyset.$$

#

4.1.5 Proposition

Let $k \geq \emptyset$, $k \in \mathbb{N}$. Then the operator

$$\mathrm{tr}\ T_G F_\Gamma : W_{2,H}^{k+1/2}(\Gamma) \cap \mathrm{im}\ P_\Gamma \longrightarrow W_{2,H}^{k+3/2}(\Gamma) \cap \mathrm{im}\ Q_\Gamma$$

is an isomorphism.

Proof

By the aid of Theorem 2.3.6, Theorem 2.5.5 and trace theorems we obtain

$$(\mathrm{tr}\ T_G F_\Gamma)(W_{2,H}^{k+1/2}(\Gamma)) \subset W_{2,H}^{k+3/2}(\Gamma).$$

Let $v \in W_{2,H}^{k+1/2}(\Gamma) \cap \mathrm{im}\ P_\Gamma$ and $\mathrm{tr}\ T_G F_\Gamma v = \emptyset$. For the validity of Theorem 4.1.3, Theorem 4.1.4 we have $T_G F_\Gamma v = \emptyset$ and $T_G F_\Gamma v \in (\ker \Delta)(G)$, whence $F_\Gamma v = \emptyset$ follows and therefore $v = \emptyset$, because for any $v \in \mathrm{im}\ P_\Gamma$ $F_\Gamma v = v$.
Let $w \in \mathrm{im}\ Q_\Gamma$. Using Corollary 2.5.14 we have $F_\Gamma w = \emptyset$. Then there exists (Theorem 4.1.3) $u \in (\ker \Delta)(G)$ with $w = \mathrm{tr}\ u$. BOREL-POMPEIU's formula (Theorem 1.4.4) now yields $u = T_G Du$. Obviously, $v = Du \in \mathrm{Ker}\ D$. Applying Corollary 2.5.4 we find $v = F_\Gamma \mathrm{tr}\ v$ for any $v \in (\ker D)(G)$, whence $u = T_G F_\Gamma \mathrm{tr}\ v$ and so $w = \mathrm{tr}\ T_G F_\Gamma(\mathrm{tr}\ v) \in \mathrm{im}\ \mathrm{tr}\ T_G F$.

#

4.1.6 Proposition

The orthoprojections P and Q have the algebraic representations

$$P = F_\Gamma(\mathrm{tr}\ T_G F_\Gamma)^{-1} \mathrm{tr}\ T_G,$$

$$Q = I - F_\Gamma(\mathrm{tr}\ T_G F_\Gamma)^{-1} \mathrm{tr}\ T_G.$$

For $u \in W^k_{2,H}(G)$ we have $Pu,\ Qu \in W^k_{2,H}(G)$.

Proof

Put $P' = F_\Gamma(\mathrm{tr}\ T_G F_\Gamma)^{-1}\mathrm{tr}\ T_G$. Let $u \in W^k_{2,H}(G)$, $k \geq 1$. Following Theorem 2.3.6 and using a trace theorem we have $\mathrm{tr}\ T_G u \in W^{k-1/2}_{2,H}(\Gamma)$. By Proposition 2.5.12 $\mathrm{tr}\ T_G u \in \mathrm{im}\ Q$. Proposition 4.1.5 yields

$$(\mathrm{tr}\ T_G F_\Gamma)^{-1}\mathrm{tr}\ T_G u \in W^{k-1/2}_{2,H}(\Gamma) \cap \mathrm{im}\ P_\Gamma .$$

Consequently, by using Theorem 2.5.5 we obtain
$P'u \in W^k_{2,H}(G) \cap \ker D$.

It is easy to verify

$$P'^2 u = (F_\Gamma(\mathrm{tr}\ T_G F_\Gamma)^{-1}\mathrm{tr}\ T_G)(F_\Gamma(\mathrm{tr}\ T_G F_\Gamma)^{-1}\mathrm{tr}\ T_G u =$$

$$= F_\Gamma(\mathrm{tr}\ T_G F_\Gamma)^{-1}\mathrm{tr}\ T_G u = P'.$$

It remains to show that $I-P'$ is a projection onto the subspace $D(\overset{\circ}{W}{}^1_{2,H}(G)) \cap W^2_{2,H}(G)$. Obviously, $(I-P')^2 = I-P'$ holds. Furthermore we have
$(I-P')u = u - P'u = D(T_G u - T_G F_\Gamma(\mathrm{tr}\ T_G F_\Gamma)^{-1}\mathrm{tr}\ T_G u)$.
It is clear that
$\mathrm{tr}\ [T_G u - T_G F_\Gamma(\mathrm{tr}\ T_G F_\Gamma)^{-1}\mathrm{tr}\ T_G u] = 0$.
For the uniqueness of the projections onto $A_H(G) \cap L_{2,H}(G)$ and $D(\overset{\circ}{W}{}^1_{2,H}(G))$ we obtain $P = P'$ and $Q = I - P'$. #

4.1.7 Corollary

The solutions $u \in W^{k+2,loc}_{2,H}(G)$ which are expressed in Theorem 4.1.1, Theorem 4.1.2 and Theorem 4.1.3 belong to the space $W^{k+2}_{2,H}(G)$.

Proof

This follows because of the special structure of the projection P. #

4.1.8 Proposition

Let $1 < p < \infty$, $k > 0$, $0 < \beta \leq 1$. The operators P and Q are acting within the spaces $W^k_{p,H}(G)$ and $C_H^{0,\beta}(G)$.

Proof

The application of Theorem 2.3.6, Theorem 2.5.15 and Proposition 4.1.5 leads to

$$\mathrm{tr}\ T_G F_\Gamma : W^{k-1/p}_{p,H}(\Gamma) \cap \mathrm{im}\ P_\Gamma \longrightarrow W^{k+1-1/p}_{p,H}(\Gamma) \cap \mathrm{im}\ Q_\Gamma ,$$

$$\mathrm{tr}\ T_G F : C_H^{0,\beta}(\Gamma) \cap \mathrm{im}\ P_\Gamma \longrightarrow C_H^{1,\beta}(\Gamma) \cap \mathrm{im}\ Q_\Gamma ,$$

from which the assertion immediately follows. #

4.1.9 Corollary (Regularity)

Let $f \in W^k_{p,H}(G)$ $(f \in C_H^{0,\beta}(\bar{G}))$, $g \in W^{k+2-1/p}_{p,H}(\Gamma)$ $(g \in C_H^{2,\beta}(\Gamma))$. DIRICHLET's problem

$$-\Delta u = f \quad \text{in } G,$$
$$u = g \quad \text{on } \Gamma$$

has the unique solution

$$u = F_\Gamma g + T_G PDh + T_G Q T_G f ,$$

which belongs to $W^{k+2}_{p,H}(G)$ $(C_H^{2,\beta}(\bar{G}))$. h denotes a $W^{k+2}_{p,H}$-extension ($C_H^{2,\beta}$-extension) of g into G.

Proof

By making use of Theorem 2.3.6, Theorem 2.5.5, Theorem 2.5.15 and Proposition 4.1.8 we may obtain the proof. #

4.1.10 Remark

Sometimes it is useful to express the solution of the problem (4.6)-(4.7) by the formula

$$u = F_\Gamma g + T_G F_\Gamma (tr\ T_G F_\Gamma)^{-1} Q_\Gamma g + T_G^2 f - T_G F_\Gamma (tr\ T_G F_\Gamma)^{-1} tr\ T_G^2 f,$$

which apparently follows by PLEMELJ-SOKHOTZKIJ's formula.

4.1.11 Theorem (VEKUA's Representation)

An arbitrary H-valued function $u \in W^k_{p,H}(G)$ whose components are real harmonic functions allows a representation

$$u = f_1 + T_G f_2,$$

where $f_1 \in \ker D \cap W^k_{p,H}(G)$ and $f_2 \in \ker D \cap W^{k-1}_{p,H}(G)$.

Proof

The proof of Corollary 4.1.2 yields the representation

$$u = F_\Gamma(tr\ h) + T_G PDh .$$

Set $f_1 = F_\Gamma(tr\ h)$, $f_2 = PDh$. The corresponding smoothness conditions follow by using Theorem 2.3.6, Theorem 2.5.5, Proposition 4.1.6 and Proposition 4.1.8. #

Now we shall deal with a new possibility to express the first eigenvalue λ_1 of DIRICHLET's problem of the LAPLACE equation. Furthermore we compute a lower bound for λ_1, which only depends on the volume $|G|$ of the domain G. It seems to be essential to note that no geometrical restrictions and only very weak smoothness requirements are assumed. Even holes in G are permitted.

4.1.12 Proposition

Let P_G be a right parallelepiped with the property $G \subset P_G$.

The length of the edges will be denoted by $c^{(i)}(G)$, i=1,2,3. We define a real number c by $c=\inf_{i=1,2,3,\, P_G\supset G} c^{(i)}(G)$. For any H-regular function $u\in \overset{\circ}{W}{}^1_{2,H}(G)$ we have

$$\|Du\|_{L_{2,H}} \geq c^{-1} \|u\|_{L_{2,H}}. \tag{4.9}$$

Proof

Let $u\in W^2_{2,H}(G)$. It follows from the properties (1.5), (2.7) and GAUSS' formula that

$$\|Du\|^2_{L_{2,H}} = \int_G \overline{Du}\,Du\,dG = \sum_{i=1}^{3}\sum_{j=0}^{3}\int_G \overline{e_i\frac{\partial}{\partial x_i}u_j e_j}\,Du\,dG_x =$$

$$= -\sum_{i=1}^{3}\sum_{j=0}^{3}\int_G \overline{e_j}\frac{\partial}{\partial x_i}u_j e_i\,Du\,dG_x = \sum_{i=1}^{3}\sum_{j=0}^{3}\int_G \overline{e_j}u_j e_i\frac{\partial}{\partial x_i}(Du)\,dG_x =$$

$$= \sum_{j=0}^{3}\int_G \overline{e_j}\,u_j(-\Delta u)\,dG_x = \sum_{j=0}^{3} e_j\overline{e_j}\int_G |\operatorname{grad} u_j|^2\,dG_x .$$

POINCARE's inequality yields

$$\sum_{j=0}^{3}\int_G |\operatorname{grad} u_j|^2 dG_x \geq c^{-2}\sum_{j=0}^{3}\int_G |u_j|^2 dG_x = c^{-2}\|u\|_{L_{2,H}}.$$

Completion of $\overset{\circ}{W}{}^2_{2,H}(G)$ in $\overset{\circ}{W}{}^1_{2,H}(G)$ implies the desired inequality. #

4.1.13 Corollary

The operator $D:\overset{\circ}{W}{}^1_{2,H}(G) \longrightarrow D(\overset{\circ}{W}{}^1_{2,H}(G))$ is invertible, and the inverse operator is given by T_G. It holds the inequality

$$\|T_G v\|_{L_{2,H}} \leq c\|v\|_{L_{2,H}} \quad ,\ v\in \operatorname{im} Q .$$

Proof

For H-valued functions $u\in \overset{\circ}{W}{}^1_{2,H}(G)$ the equation $F_\Gamma u=0$ holds. Using Corollary 2.5.4 and Proposition 2.4.2 it follows $u=DT_G u=T_G Du$. As $v\in \operatorname{im} Q$ $v=Du$ with $u\in \overset{\circ}{W}{}^1_{2,H}(G)$, whence follows $u=T_G v$. Proposition 4.1.12 implies

$$\|v\|_{L_{2,H}} = \|Du\|_{L_{2,H}} \geq c^{-1}\|u\|_{L_{2,H}} = c^{-1}\|T_G v\|_{L_{2,H}} \quad .$$ #

4.1.14 Theorem

The first eigenvalue λ_1 of DIRICHLET's problem may be represented by the formulae

(i) $$\lambda_1(G) = \frac{1}{\|T_G\|^2_{[\mathrm{im}\,Q,\,L_{2,H}]}} ,$$

(ii) $$\lambda_1(G) = \frac{1}{\|T_G\|^2_{[\mathrm{im}\,Q,\,\mathring{W}^1_{2,H}]} - 1} .$$

Proof

It is readily seen that $\lambda_1(G)=\inf_{u\in\mathring{W}^1_{2,H}} \|Du\|^2_{L_{2,H}} / \|u\|^2_{L_{2,H}}$. Hence it follows from Corollary 4.1.13

$$\|T_G\|_{[\mathrm{im}\,Q,L_{2,H}]} \le (\lambda_1)^{-1/2} .$$

Now let u be an eigenfunction of $-\Delta$ corresponding to the eigenvalue λ_1 and let w=Du. Then

$$\|T_G w\|^2_{L_{2,H}} = \|T_G Du\|^2_{L_{2,H}} = \|u\|^2_{L_{2,H}} = (-\Delta u,u)_{L_{2,H}} \lambda_1^{-1} =$$
$$= \lambda_1^{-1} \|Du\|^2_{L_{2,H}} = \lambda_1^{-1} \|w\|^2_{L_{2,H}} ,$$

and our assertion (i) is proved. The statement (ii) is obtained from the following estimations:

$$\|T_G\|^2_{[\mathrm{im}Q,\mathring{W}^1_{2,H}]} = \sup_{u\in \mathrm{im}Q} \frac{\|T_G u\|^2_{\mathring{W}^1_{2,H}}}{\|u\|^2_{L_{2,H}}} = \sup_{v\in\mathring{W}^1_{2,H}} \frac{\|T_G Dv\|^2_{\mathring{W}^1_{2,H}}}{\|Dv\|^2_{L_{2,H}}} =$$

$$= \sup_{v\in\mathring{W}^1_{2,H}} \frac{\|v\|^2_{\mathring{W}^1_{2,H}}}{\|Dv\|^2_{L_{2,H}}} \leqslant \sup_{v\in\mathring{W}^1_{2,H}} \frac{\|Dv\|^2_{L_{2,H}}}{\|Dv\|^2_{L_{2,H}}} + \sup_{v\in\mathring{W}^1_{2,H}} \frac{\|v\|^2_{L_{2,H}}}{\|Dv\|_{L_{2,H}}} = 1 + \frac{1}{\lambda_1} .$$

Taking w=Du and proceeding as in the proof of (i) we verified (ii).

\#

4.1.15 Corollary

A lower bound of the first eigenvalue of DIRICHLET's problem is given by

$$\lambda_1(G) \geqslant \sqrt[3]{\left(\frac{4\pi}{3}\right)^2}\, |G|^{-\frac{2}{3}} .$$

Proof

Using the result (i) of Theorem 4.1.14 we obtain

$$\|T_G\|_{[L_{2,H},L_{2,H}]} \geq \lambda_1^{-1/2}.$$

On the other hand, from a Lemma by P.LAX which has been generalized in [Te,p.36] we get $\|T_G\|_{[C(G),C(G)]} \geq \lambda_1^{-1/2}$. Corollary 2.3.9 yields the formulated result. #

4.1.16 Remark

In his book "The theory of sound" [Ray] published in 1896 I.W.S. RAYLEIGH formulated the statement that in the set of all fixed membranes the disc has the smallest fundamental frequence. Independently of each other G.FABER (1923) [Fab] and E.KRAHN (1924) [Kra] found a proof for RAYLEIGH's statement in the case of domains with analytical boundary. Starting from these papers in her book [Ban] C.BANDLE proved the estimate

$$\lambda_1(G) \geq \sqrt[n]{\frac{|B_1|^2}{|G|^2}}\, j_{\frac{n-2}{2}},$$

where j_k denotes the first positive zero of the BESSEL function J_k. Essential results have been obtained by using several geometrical considerations. In case of a convex domain $G \subset R^2$ J.HERSCH [Her] in 1960 proved

$$\lambda_1(G) \geq \frac{\pi^2}{4\varrho^2},$$

where ϱ means the radius of the largest disc contained in G. This estimate has been generalized in multidimensional domains. In 1981 M.H.PROTTER [Pro] gave the following bound:

$$\lambda_1(G) \geq \frac{\pi^2}{4}\left(\frac{1}{\varrho^2} + \frac{n-1}{d^2}\right)$$

with $d = \sup_{x,y\in G} \operatorname{dist}(x,y)$. Many other mathematicians have dealt with the computing of lower bounds for the first eigenvalue. Not all of them could be taken in our consideration. For further references we refer to [GS4], [Ka] and [KS].

4.1.17 Proposition

Let $\partial G \in C^\infty$, then an arbitrary eigenfunction of DIRICHLET's problem belongs to $C_H^\infty(G)$.

Proof

Let $u \in L_{2,H}(G)$ be an eigenfunction of the eigenvalue problem $\{-\Delta u = \lambda u,\ \ \operatorname{tr} u = 0\}$. Using Theorem 4.1.1 we obtain its representation $u=\lambda T_G Q T_G u$. Corollary 4.1.7 states that then $u \in W^2_{2,H}(G)$. Repeating this conclusion we get $u \in W^{2k}_{2,H}(G)$ for any $k \in \mathbb{N}$, $k \geq 0$, which implies our statement.

#

4.1.18 Theorem (Decomposition)

Let $g \in W^{k+3/2}_{2,H}(\Gamma)$, $k \geq 0$, $k \in \mathbb{N}$. The solution $u \in W^{k+2}_{2,H}(G)$ of DIRICHLET's problem may be represented by $u=v+T_G w$, where $v \in W^{k+2}_{2,H}(G)$ and $w \in W^{k+1}_{2,H}(G)$ are uniquely defined solutions of two boundary value problems of the first order:

$$\text{(i)} \quad \begin{aligned} Dv &= 0 && \text{in } G \\ v &= P_\Gamma g && \text{on } \Gamma \end{aligned}, \qquad \text{(ii)} \quad \begin{aligned} Dw &= 0 && \text{in } G \\ \operatorname{tr} T_G w &= Q_\Gamma g && \text{on } \Gamma \end{aligned}.$$

Proof

First we show that problem (i) has a solution $v \in W^{k+2}_{2,H}(G)$ iff $g \in W^{k+3/2}_{2,H}(\Gamma) \cap \operatorname{im} P_\Gamma$. Indeed, let $v \in W^{k+2}_{2,H}(G)$ be a solution of (i), then it follows from BOREL-POMPEIU's formula $v=F_\Gamma \operatorname{tr} v$ and so $\operatorname{tr} v = P_\Gamma \operatorname{tr} v$.
If $g \in W^{k+3/2}_{2,H}(\Gamma) \cap \operatorname{im} P_\Gamma$ then $u=F_\Gamma g \in \ker D \cap W^{k+2}_{2,H}(G)$ (Theorem 2.5.5). BOREL-POMPEIU's formula implies the uniqueness of the solution. From the unique solvability of the problem $\{\Delta z=0,\ \operatorname{tr} z=Q g\}$ we deduce

$$z = F_\Gamma \operatorname{tr} z + T_G Dz = F_\Gamma Q_\Gamma g + T_G Dz = T_G w,$$

with $w=Dz$. Because $z \in (\ker \Delta)(G)$, we get $w \in (\ker D)(G)$, and w is a solution of (ii).

#

4.2 HELMHOLTZ EQUATION

The first boundary value problem of HELMHOLTZ' equation may be written by

$$\Delta u_0 + \lambda^2 u_0 = f_0 \quad \text{in } G, \tag{4.10}$$

$$u_0 = g_0 \quad \text{on } \Gamma, \tag{4.11}$$

where $\lambda \in \mathbb{R}$, $\lambda^2 \geq 0$. To express it in the quaternionic calculus, we define the action of the operator $(\Delta + \lambda^2)I$ on $\mathbb{H}$-valued functions in this way that it acts on each component. Formal factorization of HELMHOLTZ' operator is given by

$$(\Delta + \lambda^2)I = -D_{-\lambda} D_\lambda ,$$

where $D_\lambda = \lambda e_0 + D$.

Writing $u = \sum_{i=0}^{3} u_i e_i$, $f = \sum_{i=0}^{3} f_i e_i$ and $g = \sum_{i=0}^{3} g_i e_i$, we may formulate the problem

$$-D_\lambda D_{-\lambda} u = f \quad \text{in } G,$$

$$u = g \quad \text{on } \Gamma.$$

The knowledge of fundamental solutions of the differential operators of first order D_λ and $D_{-\lambda}$ is necessary for defining suitable integral operators similarly to $\lambda = 0$. For the sake of simplicity we restrict our considerations to the operator D_λ, $\lambda > 0$.

4.2.1 Theorem

Let $x, y \in \mathbb{R}^3$, $x \neq y$, $\lambda \in \mathbb{R}$ and

$$e_\lambda(x) = e(x)[\cos \lambda |x| + |x| \lambda \sin \lambda |x|] + \frac{\lambda \cos \lambda |x|}{4\pi |x|} e_0 .$$

Then

$$D_{\lambda, x} e_\lambda(x-y) = -\delta(x-y). \tag{4.12}$$

Proof

Using the relation $D_x e(x-y) = -\delta(x-y)$ (see [Gol4]) we get (4.12) by a straightforward computation. #

4.2.2 Definition

In analogy to the operators T_G, F_Γ and S_Γ defined above we here determine the operators T_λ, F_λ and S_λ in the following way:

(i) $$(T_\lambda u)(x) = -\int_G e_\lambda(x-y) u(y) dG_y \quad , \; x \in \mathbb{R}^3,$$

(ii) $$(F_\lambda u)(x) = \int_\Gamma e_\lambda(x-y) \alpha(y) u(y) d\Gamma_y \quad , \; x \in \mathbb{R}^3 \setminus \Gamma,$$

(iii) $(S_\lambda u)(x) = 2\int_\Gamma e_\lambda(x-y)\,\alpha(y)u(y)d\Gamma_y \quad , \quad x\in\Gamma,$

(iv) $(P_\lambda u)(x) = \frac{1}{2}[(I + S_\lambda)u](x) \; ,$

$(Q_\lambda u)(x) = \frac{1}{2}[(I - S_\lambda)u](x) \; ,$

where in (iii) and (iv) the integrals are to be understood in the sense of CAUCHY's principal value.

4.2.3 Proposition (BOREL-POMPEIU's Formula)

Let $u\in C_H^1(G)\cap C_H(\bar{G})$. Then it holds in G

$$u = F_\lambda u + T_\lambda D_\lambda u \; .$$

Proof

We have $(T_\lambda D_\lambda u)(x) = -\int_G e_\lambda(x-y)D_{\lambda,y}u(Y)dG_y =$

$$= \lim_{\varepsilon\to 0} -\int_{G_\varepsilon} e_\lambda(x-y)D_{\lambda,y}u(y)dG_y$$

with $G_\varepsilon = G\setminus\{y: |x-y|\le\varepsilon\} = G\setminus B_\varepsilon(x)$. Let $S_\varepsilon = \partial B_\varepsilon$, $\Gamma = \partial G$. Then

$$-\int_{G_\varepsilon} e_\lambda(x-y)D_{\lambda,y}u(y)dG_y = -\int_{G_\varepsilon}[e_\lambda(x-y)D_{-\lambda,y,r}]u(y)dG_y -$$

$$-\int_\Gamma e_\lambda(x-y)\,\alpha(y)u(y)d\Gamma_y + \int_{\Gamma_\varepsilon} e(x-y)\,\alpha(y)u(y)d\Gamma_y \; ,$$

where $D_{-\lambda,r}$ means that the operator is acting from the right-hand side. It is readily seen that

$$D_{\lambda,x}e_\lambda(x-y) = e_\lambda(x-y)D_{-\lambda,x,r} \; .$$

Therefore we get from (4.12) our statement. #

4.2.4 Corollary (CAUCHY's Integral Formula)

Let $u\in C_H^1(G)\cap C_H(\bar{G})\cap \ker D_\lambda$. Then we have in G.

$$u = F_\lambda \operatorname{tr} u.$$

Proof

It apparently follows from Proposition 4.2.3. #

4.2.5 Corollary

Let $u\in C_H^1(G)\cap C_H(\bar{G})$. The operator T_λ is the algebraic right-inverse to D_λ. It holds in G

$$D_\lambda T_\lambda u = u.$$

Proof

By using (4.12) the proof is analogous to Proposition 1.4.2. #

4.2.6 Proposition (PLEMELJ-SOKHOTZKIJ's Formula)

Let $u \in C_H^{0,\beta}(\Gamma)$. The following formulae are valid:

(i) $\lim\limits_{\substack{x \to x_0 \in \Gamma \\ x \in G}} (F_\lambda u)(x) = (P_\lambda u)(x_0)$ (4.13)

(ii) $\lim\limits_{\substack{x \to x_0 \in \Gamma \\ x \notin G}} (F_\lambda u)(x) = (-Q_\lambda u)(x_0)$ (4.14)

for any $x_0 \in \Gamma$.

Proof

The function $\cos \lambda |x-y|$ allows the following TAYLOR series expansion

$$\cos \lambda |x-y| = \sum_{k=0}^{\infty} \frac{(-1)^k}{(2k)!} |x_0-y|^{2k} \lambda^{2k}, \quad x_0 \in \Gamma.$$

Making use of some properties of weakly singular integral operators we find

$$\lim_{x \to x_0} (F_\lambda u)(x) = \frac{1}{4\pi} \int_\Gamma \Big[\frac{\lambda \cos \lambda |x_0-y|}{|x_0-y|} + \frac{x_0-y}{|x_0-y|^2} \lambda \sin \lambda |x_0-y| +$$

$$+ \frac{x_0-y}{|x_0-y|^3} \sum_{k=1}^{\infty} (-1)^k \frac{1}{(2k)!} |x_0-y|^{2k} \lambda^{2k} \Big] \alpha(y)\, u(y)\, d\Gamma_y +$$

$$+ \lim_{\substack{x \to x_0 \\ x \in G}} \frac{1}{4\pi} \int_\Gamma \frac{x-y}{|x-y|^3} \alpha(y)\, u(y)\, d\Gamma_y . \qquad (4.15)$$

Combining (4.15) with Theorem 2.5.10 gives (4.13). Analogous arguments yield (4.14). #

4.2.7 Corollary

Let $u \in C_H^{0,\beta}(\Gamma)$. Then

(i) $S_\lambda^2 u = u$, (ii) $F_\lambda P_\lambda u = F_\lambda u$, (iii) $P_\lambda^2 u = P_\lambda u$.

Proof

It is clear that $S_\lambda^2 u = u$ iff $P_\lambda^2 u = P_\lambda u$. Formula (iii) can be verified by applying BOREL-POMPEIU's formula to $F_\lambda u \in \ker D_\lambda$. Using (iii) and applying Proposition 4.2.3 to $F_\lambda u - F_\lambda P_\lambda u$ we arrive at (ii). #

4.2.8 Remark

The above obtained results remain also true under weaker smoothness conditions for the H-valued function u.

4.2.9 Proposition

Let $1<p<\infty$, $k\in\mathbb{N}$, $k>0$, $\lambda\in\mathbb{R}$. We have

(i) $T_\lambda : W^k_{p,H}(G) \longrightarrow W^{k+1}_{p,H}(G)$,

(ii) $F_\lambda : W^{k-1/p}_{p,H}(\Gamma) \longrightarrow W^k_{p,H}(G) \cap \ker D_\lambda$.

Proof

(i) The operator T_λ allows the representation

$$(T_\lambda u)(x) = \frac{-1}{4\pi}\int_G D_{-\lambda}\frac{\cos\lambda|x-y|}{|x-y|}\,u(y)dG_y \ .$$

Using the TAYLOR series expansion of $\cos\lambda|x-y|$ with respect to λ we get

$$\begin{aligned}(T_\lambda u)(x) = \frac{-1}{4\pi}\int_G [D_\lambda \frac{1}{|x-y|}]u(y)dG_y \ - \\ -\frac{1}{4\pi}\sum_{k=1}^{\infty}\frac{(-1)^k 2k}{(2k)!}\lambda^{2k}\int_G [D_{-\lambda}|x-y|^{2k-1}]u(y)dG_y .\end{aligned} \tag{4.16}$$

Therefore we may reduce investigations concerning the smoothness of the first integral, because all the other integrals occurring in (4.16) have bounded kernels at most. The result follows from Theorem 2.3.6.

(ii) It is clear that $T_\lambda D_\lambda u \in W^k_{p,H}(G)$ also holds. BOREL-POMPEIU's formula leads to $F_\lambda \operatorname{tr} u \in W^k_{p,H}(G)$ for $u\in W^k_{p,H}(G)$. A trace theorem implies our statement. #

4.2.10 Proposition

The singular integral operators S_λ, P_λ and Q_λ are continuous within the spaces $W^k_{p,H}(\Gamma)$ for real $k\geq 0$.

Proof

For the proof we refer to Proposition 4.2.9, Theorem 2.5.8 and Remark 2.5.8. #

4.2.11 Theorem

The HILBERT space $L_{2,H}(G)$ allows the orthogonal decomposition

$$L_{2,H}(G) = \ker D_\lambda \cap L_{2,H}(G) \oplus D_\lambda(\overset{\circ}{W}{}^1_{2,H}(G)) \ ,$$

where $\oplus$ denotes an orthogonal sum with respect to 3.1 in case of $r_o=r_1=r_2=r_3=1$.

Proof

The proof runs in the same way as it is done in Theorem

3.1. We consider the right-linear subspaces $X_1=L_{2,H}(G)\cap \ker D_\lambda$ and $X_2=L_{2,H}(G)\ominus X_1$. Let $u\in X_2$. For any $u\in L_{2,H}(G)$ we find $u=D_\lambda v$ with $v\in W^1_{2,H}(G)$. Since $u\in X_2$, we have for any $\varphi\in X_1$

$$\int_G \overline{D_\lambda v}\,\varphi\, dG = 0$$

and, in particular, for any $l\in \mathbb{N}$

$$\int_G \overline{D_\lambda v}\,\varphi_l\, dG = 0$$

with $\varphi_l(x)=e_\lambda(x-y_l)$, $y_l\in \mathbb{R}^3\setminus\bar G$. Obviously, $\varphi_l\in\ker D_\lambda$. We assume that the set $\{y_l\}$ is dense in $\mathbb{R}^3\setminus\bar G$. By componentwise considerations we obtain

$$0=\sum_{\substack{i=1\\ j=0}}^{3}\int_G \overline{\frac{\partial}{\partial x_i} e_i v_j e_j}\,\varphi_l\, dG + \int_G \overline{\lambda v}\,\varphi_l\, dG =$$

$$= \int_G \overline{\lambda v}\,\varphi_l dG - \sum_{\substack{i=1\\ j=0}}^{3}\int_G \bar e_j \bar e_i v_j \frac{\partial}{\partial x_i}\varphi_l\, dG + \sum_{\substack{i=1\\ j=0}}^{3}\int_\Gamma \bar e_j \bar e_i v_j \alpha_i \varphi_l\, d\Gamma =$$

$$= \int_G \bar v \lambda \varphi_l dG + \int_G \bar v D\varphi_l dG + \int_\Gamma \bar v \bar\alpha\, \varphi_l\, d\Gamma =$$

$$= \int_G \bar v D_\lambda \varphi_l\, dG + \overline{\int_\Gamma \bar\varphi_l\, \alpha v\, d\Gamma} = 4\pi\, \overline{F_\lambda(\mathrm{tr}\, v)(y_l)}$$

for any y_l. Hence $F_\lambda(\mathrm{tr}\ v)=0$ in $\mathrm{co}\,\bar G$, and so it follows that $\mathrm{tr}\ v\in \mathrm{im}\ P_\lambda\cap W^{1/2}_{2,H}(\Gamma)$. Consequently, there exists an H-valued function $h\in W^1_{2,H}(G)\cap\ker D_\lambda$ with the property that $\mathrm{tr}\ h=\mathrm{tr}\ v$. Taking the function $w=v-h\in \mathring W^1_{2,H}(G)$ we get that $u=D_\lambda v=D_\lambda w\in D_\lambda(\mathring W^1_{2,H}(G))$. #

<u>4.2.12 Corollary</u>

There exist two orthoprojections P^λ and Q^λ with

$$P^\lambda : L_{2,H}(G) \longrightarrow \ker D_\lambda \cap L_{2,H}(G),$$

$$Q^\lambda = I - P^\lambda : L_{2,H}(G) \longrightarrow D_\lambda(\mathring W^1_{2,H}(G))\cap L_{2,H}(G).$$

<u>Proof</u>

This immediately follows from Theorem 4.2.11. #

4.2.13 Corollary

The differentiation rule

$D_\lambda Q^\lambda u = D_\lambda u$, $u \in W^1_{2,H}(G)$, holds.

Proof

We have $D_\lambda Q^\lambda u = D_\lambda u - D_\lambda P^\lambda u = D_\lambda u$, as $\operatorname{im} P^\lambda \subset \ker D_\lambda \subset C^\infty_H(G)$. #

4.2.14 Theorem

Let $f \in W^k_{2,H}(G)$. DIRICHLET's problem

$$(\Delta + \lambda^2)u = f \quad \text{in } G, \qquad (4.17)$$

$$u = 0 \quad \text{on } \Gamma \qquad (4.18)$$

has a solution $u \in W^{k+2,loc}_{2,H}(G)$ which may be represented by the formula

$$u = -T_{-\lambda} Q^\lambda T_\lambda f \ .$$

Proof

The validity of Theorem 4.2.11 shows that there exists an H-valued function $u \in \overset{\circ}{W}{}^1_{2,H}(G)$ with $Q^\lambda T_\lambda f = D_\lambda u$. Proposition 4.2.9 and Corollary 4.2.13 yield $T_\lambda f \in W^{k+1}_{2,H}(G)$ and $Q^\lambda T_\lambda f \in W^{k+2,loc}_{2,H}(G)$. By the aid of BOREL-POMPEIU's formula it follows

$$-T_{-\lambda} Q^\lambda T_\lambda f = -T_{-\lambda} D_\lambda u = u,$$

and so $u \in W^{k+2,loc}_{2,H}(G)$. Finally, using Corollary 4.2.13 we obtain

$$\Delta u + \lambda^2 u = -D_\lambda D_{-\lambda} u = D_\lambda Q^\lambda T_\lambda f = f.$$ #

4.2.15 Corollary

Let $g \in W^{k+3/2}_{2,H}(\Gamma)$, $k \geq 0$. The first boundary value problem

$$\Delta u + \lambda^2 u = 0 \quad \text{in } G , \qquad (4.19)$$

$$u = g \quad \text{on } \Gamma \qquad (4.20)$$

has a solution $u \in W^{k+2,loc}_{2,H}(G)$ of the form

$$u = F_\lambda g + T_{-\lambda} P^\lambda D_\lambda h \ ,$$

where h is a $W^{k+2}_{2,H}(G)$-extension of g.

Proof

As $g \in W^{k+3/2}_{2,H}(\Gamma)$, there exists a $W^{k+2}_{2,H}(G)$-extension h with $\operatorname{tr} h = g$. With $u = v + h$ the boundary value problem (4.19)-(4.20) will be transformed into

$$\Delta v + \lambda^2 v = -\Delta h - \lambda^2 h \quad \text{in } G,$$

$$v = 0 \quad \text{on } \Gamma.$$

Therefore it is clear that

$$v = T_{-\lambda} Q^{\lambda} T_{\lambda} (\Delta h + \lambda^2 h).$$

Furthermore we get

$$v = -T_{-\lambda} Q^{\lambda} D_{\lambda} h + T_{-\lambda} Q^{\lambda} F_{\lambda} D_{\lambda} h = -T_{-\lambda} D_{\lambda} h + T_{-\lambda} P^{\lambda} D_{\lambda} h =$$

$$= -h + T_{-\lambda} P_{\lambda} D_{\lambda} h + F_{\lambda} h.$$

With $u=v+h$ we gain the assertion. #

4.2.16 Theorem (Existence)

Let $f \in W^k_{2,H}(G)$, $g \in W^{k+3/2}_{2,H}(\Gamma)$. The first boundary value problem

$$\Delta u + \lambda^2 u = f \quad \text{in } G, \tag{4.21}$$

$$u = g \quad \text{on } \Gamma \tag{4.22}$$

has the solution

$$u = F_{\lambda} g + T_{-\lambda} P^{\lambda} D_{\lambda} h - T_{-\lambda} Q^{\lambda} T_{\lambda} f \in W^{k+2,loc}_{2,H}(G), \tag{4.23}$$

where h denotes a $W^{k+2}_{2,H}(G)$-extension of g.

Proof

The sum of the solutions of the boundary value problems (4.17)-(4.18) and (4.19)-(4.20) solves the problem (4.21)-(4.22). #

4.2.17 Theorem (Uniqueness)

Let $f \in W^k_{2,H}(G)$, $g \in W^{k+3/2}_{2,H}(\Gamma)$, $k \geq 0$, and suppose λ^2 is not an eigenvalue of $\{-\Delta, tr\}$. Solution (4.23) is the only solution of the boundary value problem (4.21)-(4.22).

Proof

The proof may be carried out in a completely analogous way to Theorem 4.1.4, replacing T_G by T_{λ}, D by D_{λ}, Q by Q^{λ} and P by P^{λ}. #

4.2.18 Proposition

Let $k \geq 0$, $k \in \mathbb{N}$. Then the operator

$$tr\, T_{\lambda} F_{\lambda} : W^{k+1/2}_{2,H}(\Gamma) \cap \operatorname{im} P_{\lambda} \longrightarrow W^{k+3/2}_{2,H}(\Gamma) \cap \operatorname{im} Q_{\lambda}$$

is an isomorphism if λ^2 is not an eigenvalue of $\{-\Delta, tr\}$.

Proof

By the aid of Proposition 4.2.9 and a trace theorem we obtain

$$(tr\, T_{\lambda} F_{\lambda})(W^{k+1/2}_{2,H}(\Gamma)) \subset W^{k+3/2}_{2,H}(\Gamma) .$$

Let $v \in W^{k+1/2}_{2,H}(\Gamma) \cap \operatorname{im} P_{\lambda}$ and $tr\, T_{\lambda} F_{\lambda} v = 0$. Theorem 4.2.16 and Theorem 4.2.17 yield $T_{\lambda} F_{\lambda} v = 0$ and $T_{\lambda} F_{\lambda} v \in \ker(\Delta + \lambda^2)$, whence follows $v=0$, as for any $v \in \operatorname{im} P_\lambda$, $F_\lambda v=v$.

Now let $w \in \operatorname{im} Q_\lambda$. Making use of Corollary 4.2.7 (ii) we have $F_\lambda w=0$. Then there exists an H-valued function $u \in \ker(\triangle + \lambda^2)$ with $w=\operatorname{tr} u$. BOREL-POMPEIU's formula (Proposition 4.2.3) gives us $u=T_\lambda D_\lambda u$. It is clear that $v=D_\lambda u \in \ker D_\lambda$. Applying Proposition 4.2.3 and Remark 4.2.8 we find $v=F_\lambda \operatorname{tr} v$, whence $u=T_\lambda F_\lambda \operatorname{tr} v$ and so

$$w = \operatorname{tr} T_\lambda F_\lambda(\operatorname{tr} v) \in \operatorname{im} \operatorname{tr} T_\lambda F_\lambda .$$

#

4.2.19 Corollary

Let $k \geq 1$. Then for $u \in W^k_{2,H}(G)$ we have $P^\lambda u, Q^\lambda u \in W^k_{2,H}(G)$ and the orthoprojections P^λ and Q^λ allow the representations

$$P^\lambda = F_\lambda(\operatorname{tr} T_\lambda F_\lambda)^{-1} \operatorname{tr} T_\lambda ,$$

$$Q^\lambda = I - F_\lambda(\operatorname{tr} T_\lambda F_\lambda)^{-1} \operatorname{tr} T_\lambda .$$

Proof

Let $u \in W^k_{2,H}(G)$, $k \geq 1$. For the sake of brevity we put $P'=F_\lambda(\operatorname{tr} T_\lambda F_\lambda)^{-1} \operatorname{tr} T_\lambda$. Using Proposition 4.2.8 and a trace theorem we get $\operatorname{tr} T_\lambda u \in W^{k+1/2}_{2,H}(\Gamma)$. Proposition 4.2.18 yields $\operatorname{tr} T_\lambda u \in \operatorname{im} Q_\lambda$, whence $(\operatorname{tr} T_\lambda F_\lambda)^{-1} \operatorname{tr} T_\lambda u \in W^{k-1/2}_{2,H}(\Gamma) \in \operatorname{im} P_\lambda$. Consequently, now using Proposition 4.2.9, we obtain $P'u \in W^k_{2,H}(G) \cap \ker D_\lambda$. It is easy to see that $P'^2=P'$. Obviously, $(I-P')^2=I-P'$. Furthermore we have

$$(I-P')u=u-P'u=D_\lambda(T_\lambda u-T_\lambda F_\lambda(\operatorname{tr} T_\lambda F_\lambda)^{-1} \operatorname{tr} T_\lambda u)=D_\lambda w$$

with $\operatorname{tr} w=0$. Owing to the uniqueness of the projections P^λ and Q^λ, we obtain $P^\lambda=P'$ and $Q'=I-P'=Q^\lambda$. #

4.2.20 Corollary

The solutions $u \in W^{k+2,loc}_{2,H}(G)$ which are expressed in Theorem 4.2.14, Theorem 4.2.15 and Theorem 4.2.16 belong to the space $W^{k+2}_{2,H}(G)$.

Proof

The special structure of P^λ yields the proof. #

4.2.21 Theorem (Regularity)

Let $1<p<\infty$, $k \geq 0$, and suppose λ^2 is not an eigenvalue of $\{-\triangle, \operatorname{tr}\}$. Suppose $f \in W^k_{p,H}(G)$, $g \in W^{k+2-1/p}_{p,H}(\Gamma)$, then the boundary value problem (4.21)-(4.22) has the unique solution

$$u = F_\lambda g + T_\lambda P^\lambda D_\lambda h + T_\lambda Q^\lambda T_\lambda f,$$

which belongs to $W^{k+2}_{p,H}(G)$. h denotes a $W^{k+2}_{p,H}(G)$-extension of g in G.

Proof

From Proposition 4.2.9 it follows that

$$\operatorname{tr} T_\lambda F_\lambda : W^{k+1-1/p}_{p,H}(\Gamma) \cap \operatorname{im} P_\lambda \longrightarrow W^{k+2-1/p}_{p,H}(\Gamma) \cap \operatorname{im} Q_\lambda ,$$

and therefore P^λ and Q^λ are acting within the spaces $W^{k}_{p,H}(G)$. Using this proposition once more we obtain the desired result. #

4.2.22 Remark

An analogous theorem can be formulated for $f \in C^{0,\beta}_H(\bar{G})$, $g \in C^{2,\beta}_H(\Gamma)$. Then the solution u belongs to $C^{2,\beta}_H(\bar{G})$.

4.2.23 Proposition (VEKUA's Representation)

Every $u \in W^k_{p,H}(G)$ whose components are real solutions of HELMHOLTZ' equation has the representation

$$u = f_1 + T_{-\lambda} f_2 ,$$

where $f_1 \in \ker D_{-\lambda} \cap W^k_{p,H}(G)$, $f_2 \in \ker D_\lambda \cap W^{k-1}_{p,H}(G)$.

Proof

In Corollary 4.2.15 we achieved the representation

$$u = F_\lambda(\operatorname{tr} h) + T_{-\lambda} P^\lambda D_\lambda h .$$

Set $f_1 = F_\lambda (\operatorname{tr} h)$, $f_2 = P^\lambda D_\lambda h$. The corresponding smoothness conditions follow from Proposition 4.2.9 and Theorem 4.2.21. #

4.2.24 Theorem (Decomposition)

Let $g \in W^{k+3/2}_{2,H}(\Gamma)$, $k \geq 0$, $k \in N$, and suppose λ^2 is not an eigenvalue of $\{-\Delta, \operatorname{tr}\}$. Then the boundary value problem (4.19)-(4.20) allows factorization into two boundary value problems of first order

$$\begin{aligned} D_\lambda v &= 0 \quad \text{in } G \\ v &= P_\lambda g \quad \text{on } \Gamma \end{aligned} \qquad (4.24), \qquad \begin{aligned} D_{-\lambda} w &= 0 \quad \text{in } G \\ \operatorname{tr} T_\lambda w &= Q_\lambda g \quad \text{on } \Gamma \end{aligned} \qquad (4.25)$$

and the solution u of (4.19)-(4.20) is given by $u = v + T_{-\lambda} w$.

Proof

Problem (4.24) has a solution $v \in W^{k+2}_{2,H}(G)$ iff $g \in W^{k+3/2}_{2,H}(\Gamma) \cap \operatorname{im} P_\lambda$. This follows similarly to the proof of Theorem 4.1.18. It is clear that the solution exists uniquely. The unique solvability of the problem $(\Delta + \lambda^2) z = 0$ and $\operatorname{tr} z = Q_\lambda g$ leads to

$$z = F_\lambda \operatorname{tr} z + T_\lambda D_\lambda z = F_\lambda Q_\lambda g + T_\lambda D_\lambda z = T_\lambda w,$$

with $w=D_\lambda z$. Because $z \in \ker(\Delta + \lambda^2)$ we get $w \in \ker D_{-\lambda}$, and w is a solution of (4.25).

#

4.3 Equations of Linear Elasticity

The equations of linear elasticity may be written by

$$\Delta\hat{u} + \frac{m}{m-2} \operatorname{grad} \operatorname{div} \hat{u} = -\hat{f} \quad \text{in } G, \tag{4.26}$$

$$\hat{u} = \hat{g} \quad \text{on } \Gamma \tag{4.27}$$

where $\hat{u}=\sum_{i=1}^{3} u_i e_i$, $f=\sum_{i=1}^{3} \hat{f}_i e_i$, $\hat{g}=\sum_{i=1}^{3} g_i e_i$, $m \neq 2$, $m \in R$. For $m>2$ the constant m denotes the POISSON number. After addition of the scalar DIRICHLET problem of LAPLACE's equation $\{\Delta u_o = -f_o,\ \operatorname{tr} u_o = g_o\}$ we get by a straightforward computation the quaternionic form

$$DM^{-1}Du = f \quad \text{in } G, \tag{4.28}$$

$$u = g \quad \text{on } \Gamma \tag{4.29}$$

with $u=u_o e_o+\hat{u}$, $f=f_o e_o+\hat{f}$, $g=g_o e_o+\hat{g}$, $Mu=\frac{1}{2}\,\frac{m-2}{m-1}\, u_o e_o+\hat{u}$, $m \neq 1$.

Obviously $M^{-1}u=\frac{2(m-1)}{m-2}\, u_o e_o+\hat{u}$ holds. By the aid of the operator M we introduce in $L_{2,H}(G)$, considered as a real vector space the following inner product

$$[u,v]_m = \int_G M^{-1}u\, M^{-1}v \, dG. \tag{4.30}$$

Theorem 4.3.1:

Let $f \in W_{2,H}^{k}(G)$. The boundary value problem (4.28)-(4.29) with $g=0$ has a solution $u \in W_{2,H}^{k+2,loc}(G)$ which may be represented by the formula

$$u = T_G Q_m M T_G f \ ,$$

where $Q_m := Q_G$ and $P_m := P_M$.

Proof

By Theorem 3.1 there exists an H-valued function $u \in \mathring{W}_{2,H}^{1}(G)$ with

$$Q_m M T_G f = Du \ .$$

Theorem 2.3.6 yields $T_G f \in W_{2,H}^{k+1}(G)$ and $Q_m R T_G f \in W_{2,H}^{k+1,loc}(G)$. By the aid of BOREL-POMPEIU's formula it follows that

$$T_G Q_m M T_G f = T_G Du = u \ ,$$

and therefore $u \in W_{2,H}^{k+2,loc}(G)$. Finally, Corollary 3.3 implies

$$DM^{-1}Du = DM^{-1}Q_{\mathbf{m}}MT_G f = f - DM^{-1}P_{\mathbf{m}}MT_G f = f - DM^{-1}Mw,$$

with $w \in \ker D$. Hence we get

$$DM^{-1}Du = f.$$

#

4.3.2 Corollary

Let $g \in W_{2,H}^{k+3/2}(\Gamma)$, $k \geq 0$. The boundary value problem (4.28)-(4.29) with $f=0$ has a solution $u \in W_{2,H}^{k+2,loc}(G)$ of the form

$$u = F_\Gamma g + T_G P_{\mathbf{m}} Dh ,$$

where h is a $W_{2,H}^{k+2}(G)$-extension of the H-valued function g.

Proof

As $g \in W_{2,H}^{k+3/2}(\Gamma)$ there exists a $W_{2,H}^{k+2}$-extension h with $tr\, h=g$. Put $u=v+h$. The problem (4.28)-(4.29) with $f=0$ shall be transformed into the boundary value problem

$$-DM^{-1}Dv = DM^{-1}Dh \quad \text{in } G,$$

$$v = 0 \quad \text{on } \Gamma.$$

Applying Theorem 4.3.1 we find the solution

$$v = -T_G Q_{\mathbf{m}} MT_G DM^{-1}Dh .$$

Using BOREL-POMPEIU's formula we obtain

$$v = -T_G Q_{\mathbf{m}} Dh + T_G Q_{\mathbf{m}} MF\, M^{-1}Dh =$$

$$= -h + F_\Gamma h - T_G P_{\mathbf{m}} Dh + T_G MF\, M^{-1}Dh - T_G P_{\mathbf{m}} MF\, M^{-1}Dh .$$

As $P_{\mathbf{m}} MF\, M^{-1}Dh = MF_\Gamma M^{-1}Dh$ our statement follows from $u=v+h$. #

4.3.3 Theorem (Existence)

Let $f \in W_{2,H}^{k}(G)$, $g \in W_{2,H}^{k+3/2}(\Gamma)$. The boundary value problem (4.28)-(4.29) has the solution

$$u = F_\Gamma g + T_G P_{\mathbf{m}} Dh + T_G Q_{\mathbf{m}} MT_G f \in W_{2,H}^{k+2,loc}(G), \tag{4.31}$$

where h denotes a $W_{2,H}^{k+2}(G)$-extension of g.

Proof

The sum of the solutions which are given by Theorem 4.3.1 and its Corollary yields the solution (4.31).

#

4.3.4 Theorem (Uniqueness)

Let $f \in W_{2,H}^{k}(G)$, $g \in W_{2,H}^{k+3/2}(\Gamma)$, $k \geq 0$. In (4.31) expressed H-valued function u is the only solution of the boundary value problem (4.28)-(4.29).

Proof

Consider the boundary value problem $\{DM^{-1}Du=0,\ tr\, u=0\}$.

BOREL-POMPEIU's formula and the fact that $Du \in \operatorname{im} Q_{\mathbf{m}}$ leads to a solution of this problem .

$$u = T_G Du = T_G Q_{\mathbf{m}} Du .$$

Besides we have $Du \in \operatorname{im} P_{\mathbf{m}}$

$$u = T_G Du = T_G P_{\mathbf{m}} M^{-1} Du .$$

Therefore it holds

$$Q_{\mathbf{m}} Du = P_{\mathbf{m}} M^{-1} Du,$$

whence $Du=\emptyset$ and finally (Theorem 4.1.3, Theorem 4.1.4) $u=\emptyset$.

#

<u>4.3.5 Proposition</u>

Let $k \geq \emptyset$, $k \in \mathbb{N}$. Then the operator

$$\operatorname{tr} T_G M F_\Gamma : W^{k+1/2}_{2,H}(\Gamma) \cap \operatorname{im} P_\Gamma \longrightarrow W^{k+3/2}_{2,H}(\Gamma) \cap \operatorname{im} Q_\Gamma$$

is an isomorphism.

<u>Proof</u>

By the aid of Theorem 2.3.6, Theorem 2.5.5 and trace theorems we obtain

$$(\operatorname{tr} T_G M F_\Gamma)(W^{k+1/2}_{2,H}(\Gamma)) \subset W^{k+3/2}_{2,H}(\Gamma).$$

Let $v \in W^{k+1/2}_{2,H}(\Gamma) \cap \operatorname{im} P_\Gamma$ and $\operatorname{tr} T_G M F_\Gamma v=\emptyset$. From Theorem 4.3.4, Theorem 4.3.3 we have $T_G M F_\Gamma v=\emptyset$ and $T_G M F_\Gamma v \in \ker DM^{-1}D$, whence follows $F_\Gamma v=\emptyset$ and therefore $v=\emptyset$. Now let $w \in \operatorname{im} Q$, Corollary 2.5.14 states $F_\Gamma w=\emptyset$. Then there exists (Corollary 4.3.2) $u \in \ker DM^{-1}D$ with $\operatorname{tr} u=w$. Furthermore we immediately conclude now that $u=T_G P_{\mathbf{m}} Du$. Obviously, $v=Du \in MA_H(G)$. It follows that $P_{\mathbf{m}} v=Ms$ for any $s \in \ker D$, whence $u=T_G MF \operatorname{tr} s$ and so $w=\operatorname{tr} T_G MF \operatorname{tr} s \in \operatorname{im} \operatorname{tr} T_G MF$.

#

<u>4.3.6 Proposition</u>

The orthoprojections $P_{\mathbf{m}}$ and $Q_{\mathbf{m}}$ have the representations

$$P_{\mathbf{m}} = MF_\Gamma(\operatorname{tr} T_G M F_\Gamma)^{-1} \operatorname{tr} T_G ,$$

$$Q_{\mathbf{m}} = I - MF_\Gamma(\operatorname{tr} T_G M F_\Gamma)^{-1} \operatorname{tr} T_G .$$

For $u \in W^k_{2,H}(G)$ we have $P_{\mathbf{m}} u, Q_{\mathbf{m}} u \in W^k_{2,H}(G)$.

<u>Proof</u>

Put $P_{\mathbf{m}}' = F_\Gamma(\operatorname{tr} T_G M F_\Gamma)^{-1} \operatorname{tr} T_G$. Let $u \in W^k_{2,H}(G)$, $k > \emptyset$. Using Proposition 4.3.5 and the proof of Proposition 4.1.6 we get

$(tr\ T_G MF_\Gamma)^{-1} tr\ T_G u \in W^{k-1/2}_{2,H}(\Gamma) \cap im\ P_\Gamma$ and consequently $P'_m u \in W^k_{2,H}(G) \cap \ker D$. It may be readily seen that $(P'_m)^2 = P'_m$ and $(Q'_m)^2 = Q'_m$. Furthermore we have

$$(I-P'_m)u = u - P'_m u = D(T_G u - T_G MF_\Gamma (tr\ T_G MF_\Gamma)^{-1} tr\ T_G u).$$

It is clear that $tr[T_G u - T_G MF_\Gamma (tr\ T_G MF_\Gamma)^{-1} tr\ T_G u] = 0$. Owing to the uniqueness of the projections it holds $P'_m = P_m$ and $Q'_m = Q_m$. #

4.3.7 Corollary

The solutions $u \in W^{k+2,loc}_{2,H}(G)$ which are expressed in Theorem 4.3.1, Corollary 4.3.2 and Theorem 4.3.3 belong to the space $W^{k+2}_{2,H}(G)$.

Proof

The proof immediately follows from Corollary 4.1.6. #

4.3.8 Proposition

Let $1<p<\infty$, $k>0$, $0<\beta\leq 1$. The operators P_m and Q_m are acting within the spaces $W^k_{p,H}(G)$ and $C^{0,\beta}_H(\bar{G})$.

Proof

The proof can be carried out similarly to Proposition 4.1.8. #

4.3.9 Corollary (Regularity)

Let $f \in W^k_{p,H}(G)$ $(f \in C^{0,\beta}_H(\bar{G}))$, $g \in W^{k+2-1/p}_{p,H}(\Gamma)$ $(g \in C^{2,\beta}_H(\Gamma))$. The boundary value problem (4.28)-(4.29) has the unique solution

$$u = F_\Gamma g + T_G P_m Dh + T_G Q_m M T_G f \in W^{k+2}_{p,H}(G) \quad (\in C^{2,\beta}_H(\bar{G})).$$

h denotes a $W^{k+2}_{p,H}(G)$-extension ($C^{2,\beta}_H(\bar{G})$-extension) of g in G.

Proof

Using the special structure of solution (4.31) of problem (4.28)-(4.29) and Proposition 4.3.8 we may find out the proof. #

4.3.10 Remark

Sometimes it is useful to express the solution of problem (4.28)-(4.29) by the formula

$$u = F_\Gamma g + T_G MF\ (tr\ T_G MF_\Gamma)^{-1} Q_\Gamma g + T_G M[T_G f - F_\Gamma (tr\ T_G MF_\Gamma)^{-1} tr\ T_G M T_G f].$$

4.3.11 Theorem (VEKUA's Representation)

An arbitrary H-valued function $u \in W^k_{p,H}(G)$ whose first component fulfils LAPLACE equation and whose other components satisfy the equations of linear elasticity with $f=0$ allows a representation

$$u = f_1 + T_G M f_2 ,$$

where $f_1 \in \ker D \cap W^k_{p,H}(G)$ and $f_2 \in \ker D \cap W^{k-1}_{p,H}(G)$.

Proof

The proof of Corollary 4.3.2 yields the representation

$$u = F_\Gamma(\mathrm{tr}\ h) + T_G P_m Dh .$$

Set $f_1 = F_\Gamma(\mathrm{tr}\ h)$, $f_2 = F_\Gamma(\mathrm{tr}\ T_G M F_\Gamma)^{-1} Q_\Gamma g$ ($Q_\Gamma g = \mathrm{tr}\ T_G Dh$!). The corresponding smoothness conditions are obtained by the aid of Theorem 2.3.6, Theorem 2.5.5, Proposition 4.3.6 and Proposition 4.3.8.

■

4.3.12 Theorem (Lower Eigenvalue Bounds)

Lower bounds for the first eigenvalue of the boundary value problem (4.28)-(4.28) are given by

(i) $\lambda_1 \geq \frac{1}{\|T_G\|^2_{L(L_{2,H})}}$, (ii) $\lambda_1 \geq \frac{1}{\|T_G\|^2_{L(C_H)}}$, (iii) $\lambda_1 \geq \left(\frac{4\pi}{3}|G|^{-1}\right)^{2/3}$.

Proof

The eigenvalue problem means: An H-valued function $u \in W^1_{2,H}(G)$, $u \neq 0$, and a real parameter λ are to be found such that

$$DM^{-1}Du = \lambda u .$$

Using Theorem 4.3.3 we have the integral representation

$$u - \lambda T_G Q_m M T_G u = 0 . \qquad (4.32)$$

If $|\lambda| < \|T_G Q_m M T_G\|^{-1}_{L(L_{2,H})}$, then the only solution is the function which is zero everywhere. Therefore for the first eigenvalue λ_1 the inequality

$$\lambda_1 \geq \|T_G Q_m M T_G\|^{-1}_{L(L_{2,H})} \geq \|T_G\|^{-2}_{L(L_{2,H})},$$

must be satisfied because $\|Q_m\|_{L(L^M_{2,H})} = 1$, $\|M\|_{L(L_{2,H})} \leq 1$ and (i) is proved.

To check (ii) we have to use a result of H.TESCHKE in [Te], whence

$$\|T_G\|_{L(L_{2,H})} \leq \|T_G\|_{L(C_H)} ,$$

and (ii) follows.

Using Corollary 2.3.9 we obtain (iii). #

4.3.13 Remark

Lower bounds of the first eigenvalue of the equations of linear elasticity have not been studied comprehensively. Essential papers were written by H.A.LEVINE/M.H.PROTTER [LP] and B.KAWOHL [Ka].

4.3.14 Proposition

Let $\partial G \in C^{\infty}_{R}$. Then an arbitrary eigenfunction of the first eigenvalue problem of linear elasticity belongs to $C^{\infty}_{H}(G)$.

Proof

Let $u \in L_{2,H}(G)$ be an eigenfunction of the problem $\{DM^{-1}Du = \lambda u,\ \text{tr}\ u=0\}$. Using Theorem 4.3.1 we obtain the representation

$$u = \lambda T_G Q_m M T_G u \ .$$

Corollary 4.3.7 states that then $u \in W^2_{2,H}(G)$. Repeating this conclusion we get $u \in W^{2k}_{2,H}(G)$ for any $k \in \mathbf{N}$, $k \geq 0$, which implies our statement. #

4.3.15 Theorem (Decomposition)

Let $g \in W^{k+3/2}_{2,H}(\Gamma)$, $k \geq 0$, $k \in \mathbf{N}$. The solution $u \in W^{k+2}_{2,H}(G)$ of the first boundary value problem may be represented by

$$u = v + T_G M w \ ,$$

where $v \in W^{k+2}_{2,H}(G)$ and $w \in W^{k+1}_{2,H}(G)$ are uniquely defined solutions of the boundary value problems of first order

$$\text{(i)}\quad \begin{array}{ll} Dv = 0 & \text{in } G \\ v = P_\Gamma g & \text{on } \Gamma \end{array} , \qquad \text{(ii)}\quad \begin{array}{ll} Dw = 0 & \text{in } G \\ \text{tr}\ T_G M w = Q_\Gamma g & \text{on } \Gamma \end{array} .$$

Proof

Similarly to Theorem 4.1.18 it is to be shown that (i) has a $W^{k+2}_{2,H}(G)$-solution v. For the unique solvability of the problem $\{DM^{-1}Dz=0,\ \text{tr}\ z=Q_\Gamma g\}$ we deduce

$$z = F_\Gamma \text{tr}\ z + T_G D z = F_\Gamma Q_\Gamma g + T_G M w = T_G M w,$$

with $w=M^{-1}Dz$. Since $z \in \ker DM^{-1}D$, we get $w \in \ker D$, and w solves the boundary value problem (ii). #

4.4 Time-Independent MAXWELL Equations

An electric charge is distributed over a three-dimensional domain G with the density ϱ . We shall compute the electric field $\hat{E}=\sum_{i=1}^{3} E_i e_i$ and the magnetic field $\hat{H}=\sum_{i=1}^{3} H_i e_i$, where the electric conductivity $\varkappa$, dielectric constant ε , the permeability μ and the magnetic field $\hat{g}=\operatorname{tr}\hat{H}$ are given on the boundary Γ. In the time-independent case MAXWELL equations read as follows:

$$\operatorname{div}\varepsilon\hat{E} = \varrho, \qquad \operatorname{div}\mu\hat{H} = 0 ,$$
$$\operatorname{rot}\hat{E} = 0, \qquad \operatorname{rot}\hat{H} = \varkappa\hat{E} .$$

Using Corollary 1.3.4 it holds

$$\operatorname{div}\varepsilon\hat{E} = \varrho , \qquad \operatorname{div}\mu\hat{H} = 0 , \tag{4.33}$$
$$\operatorname{rot}\varepsilon\hat{E} = \operatorname{grad}\varepsilon \times \hat{E}, \qquad \operatorname{rot}\mu\hat{H} = \mu\varkappa\hat{E} + \operatorname{grad}\mu \times \hat{H}.$$

Put $E=\varepsilon\hat{E}$, $H=\mu\hat{H}$, $g=\operatorname{tr} H$, $a=\varepsilon^{-1}\operatorname{grad}\varepsilon$, $b=\mu^{-1}\operatorname{grad}\mu$, $f=-\varrho e_0$ and $\gamma=\varepsilon^{-1}\mu\varkappa$. Equations (4.33) can be given the quaternionic structure

$$DE = f + \operatorname{Im}(aE), \tag{4.34}$$
$$DH = \gamma E + \operatorname{Im}(bH).$$

First let ε and μ be constants. Then system (4.34) is reduced to

$$DE = f ,$$
$$DH = \gamma E .$$

Finally, we have to consider the boundary value problem

$$D\gamma^{-1}DH = f \quad \text{in } G, \tag{4.35}$$
$$H = g \quad \text{on } \Gamma, \tag{4.36}$$

where the real function $\gamma=\gamma(x)$ is not necessarily constant. By reason of non-vanishing electric conductivity , we have $\gamma(x)>0$ for $x\in\bar{G}$.

4.4.1 Theorem

Let $f\in W^k_{2,H}(G)$, $\gamma\in C^\infty_R(\bar{G})$. The first boundary value problem

$$D\gamma^{-1}DH = f \quad \text{in } G, \tag{4.37}$$
$$H = 0 \quad \text{on } \Gamma \tag{4.38}$$

has a solution $H\in W^{k+2,loc}_{2,H}(G)$, which may be represented by

$$H = T_G Q_\gamma \gamma T_G f .$$

Proof

For Theorem 3.1 there exists an H-valued function $H\in \overset{\circ}{W}{}^1_{2,H}(G)$ with $Q_\gamma\gamma T_G f=DH$, where R is identified with the operator of multiplication by the scalar function γ. Theorem 2.3.6 yields $T_G f \in W^{k+1}_{2,H}(G)$ and $Q_\gamma \gamma T_G f \in W^{k+1,loc}_{2,H}(G)$, as $\gamma \in C^\infty_R(G)$. An application of BOREL-POMPEIU's formula leads to

$$T_G Q_\gamma \gamma T_G f = T_G DH = H,$$

and therefore $H \in W^{k+2,loc}_{2,H}(G)$. Finally, Corollary 3.3 implies

$$D\gamma^{-1}DH = D\gamma^{-1}Q_\gamma \gamma T_G f = D\gamma^{-1}\gamma T_G f - D\gamma^{-1}P_\gamma \gamma T_G f =$$

$$= f - D\gamma^{-1}\gamma w,$$

with $w \in \ker D$. Therefore

$$D\gamma^{-1}DH = f.$$ #

4.4.2 Corollary

Let $g \in W^{k+3/2}_{2,H}(\Gamma)$, $k \geq 0$. The first boundary value problem

$$D\gamma^{-1}DH = 0 \quad \text{in } G, \tag{4.39}$$

$$H = g \quad \text{on } \Gamma \tag{4.40}$$

has a solution $H \in W^{k+2,loc}_{2,H}(G)$ of the form

$$H = F_\Gamma g + T_G P_\gamma Dh,$$

where h is a $W^{k+2}_{2,H}(G)$-extension of g.

Proof

Using Theorem 4.4.1 the proof is analogous to Corollary 4.1.2. #

4.4.3 Theorem (Existence)

Let $f \in W^k_{2,H}(G)$, $g \in W^{k+3/2}_{2,H}(\Gamma)$. The first boundary value problem (4.35)-(4.36) has the solution

$$H = F_\Gamma g + T_G P_\gamma Dh + T_G Q_\gamma \gamma T_G f \in W^{k+2,loc}_{2,H}(G), \tag{4.41}$$

where h denotes a $W^{k+2}_{2,H}(G)$-extension of g.

Proof

The sum of the solutions of problems (4.37)-(4.38) and (4.39)-(4.40) solves (4.35)-(4.36) and allows the representation (4.41). #

4.4.4 Theorem (Uniqueness)

Let $f \in W^k_{2,H}(G)$, $g \in W^{k+3/2}_{2,H}(\Gamma)$, $k \geq 0$. The H-valued function H expressed in (4.41) is the only solution of the boundary value problem (4.35)-(4.36).

Proof

Substituting M by γ in the proof of Theorem 4.3.4 we have a proof of our assertion. #

4.4.5 Proposition

Let $k \geq 0$, $k \in \mathbb{N}$. Then the operator

$$\operatorname{tr} T_G \gamma F_\Gamma : W_{2,H}^{k+1/2}(\Gamma) \cap \operatorname{im} P_\Gamma \longrightarrow W_{2,H}^{k+3/2}(\Gamma) \cap \operatorname{im} Q_\Gamma$$

is an isomorphism.

Proof

Using Theorem 4.4.4, Theorem 4.4.3 and Corollary 4.4.2 in this order the proof is complete analogously to Proposition 4.3.5. #

4.4.6 Proposition

The orthoprojections P_γ and Q_γ have the representations

$$P_\gamma = \gamma F_\Gamma (\operatorname{tr} T_G \gamma F_\Gamma)^{-1} \operatorname{tr} T_G ,$$

$$Q_\gamma = I - \gamma F_\Gamma (\operatorname{tr} T_G \gamma F_\Gamma)^{-1} \operatorname{tr} T_G .$$

For $u \in W_{2,H}^k(G)$, $k \geq 1$, there holds $P_\gamma u, Q_\gamma u \in W_{2,H}^k(G)$.

Proof

The proof is similar to the proof of Proposition 4.3.6, if we replace M by γ. #

4.4.7 Corollary

The solutions $H \in W_{2,H}^{k+2,loc}(G)$ which are expressed in Theorem 4.4.1, Corollary 4.4.2 and Theorem 4.4.3 belong to the space $W_{2,H}^{k+2}(G)$.

Proof

Corollary 4.1.6 yields the proof. #

4.4.8 Proposition

Let $1<p<\infty$, $k>0$, $0<\beta<1$. The operators P_γ and Q_γ are acting within the spaces $W_{p,H}^k(G)$ and $C_H^{0,\beta}(\bar{G})$.

Proof

For the proof we may refer to the proof of Proposition 4.1.8. #

4.4.9 Corollary (Regularity)

Let $f \in W_{p,H}^k(G)$ $(f \in C_H^{0,\beta}(\bar{G}))$, $g \in W_{p,H}^{k+2-1/p}(\Gamma)$ $(g \in C_H^{2,\beta}(\Gamma))$. The boundary value problem (4.35)-(4.36) has the unique solution

$$H = F_\Gamma g + T_G P_\gamma Dh + T_G Q_\gamma \gamma T_G f \in W^{k+2}_{p,H}(G) \quad (\in C^{2,\beta}_H(\bar{G})).$$

h denotes a $W^{k+2}_{p,H}(G)$-extension ($C^{2,\beta}_H(\bar{G})$-extension) of g in G.

Proof

The special structure of the solution proves our statement. #

4.4.10 Corollary

In the original notation for the corresponding expressions of the electric and magnetic fields we obtain

$$\hat{H} = \frac{1}{\mu} F_\Gamma \hat{g} + \frac{1}{\mu} T_G \varkappa F_\Gamma (\operatorname{tr} T_G \varkappa F_\Gamma)^{-1} Q_\Gamma \hat{g} -$$

$$- \frac{1}{\varepsilon} T_G \varkappa F_\Gamma (\operatorname{tr} T_G \varkappa F_\Gamma)^{-1} \varkappa \operatorname{tr} T_G^2 \varrho +$$

$$+ \frac{1}{\varepsilon} T_G \varkappa T_G \varrho,$$

$$\hat{E} = \frac{1}{\mu^2} F_\Gamma (\operatorname{tr} T_G \varkappa F_\Gamma)^{-1} Q_\Gamma \hat{g} - \frac{1}{\varepsilon\mu} F_\Gamma (\operatorname{tr} T_G \varkappa F_\Gamma)^{-1} \varkappa \operatorname{tr} T_G \varrho +$$

$$+ \frac{1}{\mu} T_G \varrho .$$

In the case of a constant $\varkappa$ these formulae may be simplified. It holds

$$\hat{H} = \frac{1}{\mu} F_\Gamma \hat{g} + \frac{1}{\mu} T_G F_\Gamma (\operatorname{tr} T_G F_\Gamma)^{-1} Q_\Gamma \hat{g} -$$

$$- \frac{\varkappa}{\varepsilon} T_G F_\Gamma (\operatorname{tr} T_G F_\Gamma)^{-1} T_G \varrho + \frac{\varkappa}{\varepsilon} T_G^2 \varrho,$$

$$\hat{E} = \frac{1}{\mu^2 \varkappa} F_\Gamma (\operatorname{tr} T_G F_\Gamma)^{-1} Q_\Gamma \hat{g} - \frac{1}{\varepsilon\mu} F_\Gamma (\operatorname{tr} T_G F_\Gamma)^{-1} T_G \varrho - \frac{1}{\mu} T_G \varrho .$$

Proof

A straightforward computation leads to these formulae. #

4.4.11 Theorem (VEKUA's Representation)

An arbitrary H-valued function $H \in W^k_{p,H}(G)$ which satisfies the equation

$$D\, \gamma^{-1} DH = 0$$

allows the representation

$$H = f_1 + T_G \gamma f_2 ,$$

where $f_1 \in \ker D \cap W^k_{p,H}(G)$ and $f_2 \in \ker D \cap W^{k-1}_{p,H}(G)$.

Proof

For the proof we refer to the proof of Theorem 4.3.11. #

4.4.12 Theorem (Lower Eigenvalue Bounds)

A lower bound for the first eigenvalue of the boundary value problem (4.35)-(4.36) is given by

$$\text{(i)}\quad \lambda_1 \geqslant \|T_G\|^{-2}_{L(L_{2,H})}\gamma^{-1}, \quad \text{(ii)}\quad \lambda_2 \geqslant \|T_G\|^{-2}_{L(C_H)}\gamma^{-1}, \quad \text{(iii)}\quad \lambda_1 \geqslant \left[\left(\tfrac{4\pi}{3}\right)|G|^{-1}\right]^{2/3}\gamma^{-1}.$$

Proof

The proof may be carried out in a similar way to that of Theorem 4.3.12. It follows that

$$\lambda_1 \geqslant \|T_G Q_\gamma \gamma T_G\|^{-1}_{L(L_{2,H})} \geqslant \|T_G\|^{-2}_{L(L_{2,H})}\gamma^{-1},$$

whence

$$\|T_G\|_{L(L_{2,H})} \leq \|T_G\|_{L(C_H)},$$

and so (ii) and finally (iii). #

4.4.13 Remark

In original notation the eigenvalue problem has the formulation

$$\operatorname{div}\varepsilon\hat{E} = 0, \qquad \operatorname{div}\mu\hat{H} = 0,$$
$$\operatorname{rot}\hat{E} = \lambda\mu\hat{H}, \qquad \operatorname{rot}\hat{H} = \varkappa\hat{E},$$

where λ denotes the parameter of the eigenvalue.

4.4.14 Proposition

Any eigenfunction of the first boundary problem of MAXWELL's equations belongs to $C^\infty_H(G)$.

Proof

Replacing M by γ then the proof is the same as for Proposition 4.3.14. #

4.4.15 Theorem (Decomposition)

Let $g \in W^{k+3/2}_{2,H}(\Gamma)$, $k \geq 0$, $k \in \mathbb{N}$. The solution $H \in W^{k+2}_{2,H}(G)$ of the first boundary value problem may be represented by $u = v + T_G\gamma w$, where $v \in W^{k+2}_{2,H}(G)$ and $w \in W^{k+1}_{2,H}(G)$ are uniquely determined solutions of the boundary value problems of first order

$$\text{(i)}\quad \begin{array}{lll} Dv = 0 & \text{in} & G \\ v = P_\Gamma g & \text{on} & \Gamma \end{array}, \qquad \text{(ii)}\quad \begin{array}{lll} Dw = 0 & \text{in} & G \\ \operatorname{tr} T_G\gamma w = Q_\Gamma g & \text{on} & \Gamma. \end{array}$$

Proof

The proof is similar to Theorem 4.3.15 by replacing M by γ. #

4.4.16 Theorem (Generalization)

Let $f \in L_{2,H}(G)$, $g \in W^{3/2}_{2,H}(\Gamma)$; $\mu, \varepsilon, \varkappa \in C^\infty_R(G)$.

(i) The iteration procedure $\{E_n=T_G f+T_G \operatorname{Im}(aE_{n-1}),\ E_o=0\}$ converges in $L_{2,H}(G)$ for $n\longrightarrow\infty$ to a unique solution E if there holds the inequality $\lambda_1^{-1/2}\|a\|_{C_H}\le r<1$.

(ii) The iteration procedure

$$\{H_n=F_\Gamma g+T_G P_\gamma Dh+T_G Q_\gamma \gamma T_G(f+\operatorname{Im}(aE))+T_G Q_\gamma \operatorname{Im}(bH_{n-1}),\ H_o=0\}$$

converges in $L_{2,H}(G)$ for $n\longrightarrow\infty$ to a unique solution H if $\lambda_1^{-1/2}\|b\|_{C_H}\le r<1$. Besides we have $\operatorname{tr} H=g$.

<u>Proof</u>

(i) It holds $E_k-E_{k-1}=(T_G \operatorname{Im} a)(E_{k-1}-E_{k-2})=(T_G \operatorname{Im} a)^{k-1}T_G f$.

As $\operatorname{Im} a(E_1-E_{1-1})=\hat{a}\times(\hat{E}_1-\hat{E}_{1-1})\le|a||E_1-E_{1-1}|$ we find

$\|E_k-E_{k-1}\|_{L_{2,H}}\le\|T_G\|^{k-1}_{L(L_{2,H})}\|a\|^{k-1}_{C_R}\|T_G f\|_{L_{2,H}}$. To verify the convergence of $\sum_{k=1}^{\infty}(E_k-E_{k-1})$ it is sufficient to show that the sum $\sum_{k=1}^{\infty}\|E_k-E_{k-1}\|_{L_{2,H}}$ is convergent. We have

$$\Big\|\sum_{k=1}^{n}(E_k-E_{k-1})\Big\|_{L_{2,H}}\le\sum_{k=1}^{n}\|E_k-E_{k-1}\|_{L_{2,H}}\le$$

$$\le\sum_{k=1}^{n}\|T_G\|^{k-1}_{L(L_{2,H})}\|a\|^{k-1}_{C_R}\|T_G f\|_{L_{2,H}}\le$$

$$\le\frac{1-r^n}{1-r}\|T_G f\|_{L_{2,H}}\le\frac{1}{1-r}\|T_G f\|_{L_{2,H}}.$$ Because $E_n=\sum_{k=1}^{n}(E_k-E_{k-1})$,

there follows the convergence for $n\longrightarrow\infty$ to an H-valued function E.

(ii) Equations (4.34) lead to the problem

$$D\,\gamma^{-1}DH_n=f+\operatorname{Im}(aE)+D\,\gamma^{-1}\operatorname{Im}(bH_{n-1}),$$

$$\operatorname{tr} H_n=g.$$

Theorem 4.4.3 yields

$$H_n=F_\Gamma g+T_G P_\gamma Dh+T_G Q_\gamma \gamma T_G(f+\operatorname{Im}(aE))+T_G Q \operatorname{Im}(bH_{n-1}).$$

Obviously

$$H_k-H_{k-1}=T_G Q_\gamma \operatorname{Im} b\,(H_{n-1}-H_{n-2})=(T_G Q_\gamma \operatorname{Im} b)^{k-1}H_1.$$

Because $\|Q\|_{L(L^\gamma_{2,H})}=1$, now the proof follows in the same way as before in (i). If for each n the function H_n satisfies tr $H_n=g$, then $\lim_{n\to\infty} H_n=\lim_{n\to\infty}\sum_{k=1}^{n}(H_k-H_{k-1})=H$ fulfils this condition, too. #

4.4.17 Remark

The proof of Theorem 4.4.16 immediately yields the following a-priori estimates for E and H:

$$\text{(i)} \quad \|E\|_{L_{2,H}} \leq \frac{\sqrt{\lambda_1}}{\sqrt{\lambda_1}-\|a\|_{C_R}} \|T_G f\|_{L_{2,H}},$$

$$\text{(ii)} \quad \|H\|_{L_{2,H}} \leq \frac{\sqrt{\lambda_1}}{\sqrt{\lambda_1}-\|b\|_{C_R}} \|F_\Gamma g + T_G P_\gamma D h + T_G Q_\gamma \gamma T_G (f + J m a E)\|_{L_{2,H}}.$$

4.5 STOKES' Equations

A. VALLI considered in [Val] STOKES system in the following general form:

$$-\Delta\hat{u} + \frac{1}{\eta}\,\mathrm{grad}\; p = \frac{\varrho}{\eta}\hat{f} \quad \text{in } G, \tag{4.42}$$

$$\mathrm{div}\;\hat{u} = f_0 \quad \text{in } G, \tag{4.43}$$

$$\hat{u} = g \quad \text{on } \Gamma. \tag{4.44}$$

With $\int_G f_0\, dG=\int_\Gamma \alpha\, g\, d\Gamma$ the necessary condition for the solvability is given. This system describes the stationary flow of a homogeneous viscous fluid for small REYNOLDS' numbers. $\hat{u}$ means the velocity of the fluid and p the hydrostatic pressure. The scalar function f_0 is a measure for the compressibility of the fluid. The boundary condition (4.44) means the adhesion at the boundary of the domain for $\hat{g}=\emptyset$. Furthermore ϱ denotes the density, η the viscosity and $\hat{f}$ the vector of the external forces. A detailed discussion of the references is given in [Val] and in the book of O.A. LADYZENSKAJA [Lad].

Put $u=\sum_{i=0}^{3} u_i e_i$, $f=\hat{f}$ and $g=\hat{g}$. STOKES' equations may be written in the following hypercomplex notation:

$$D\,D\,u + \frac{1}{\eta} D\,p = \frac{\varrho}{\eta} f \quad \text{in } G, \tag{4.45}$$

$$\mathrm{Re}\; D\,u = f_0 \quad \text{in } G, \tag{4.46}$$

$$u = g \quad \text{on } \Gamma, \tag{4.47}$$

where the scalar problem

$$-\Delta u_o = 0 \quad \text{in } G,$$
$$u_o = 0 \quad \text{on } \Gamma$$

has been added to system (4.42)-(4.43)-(4.44). First we deal with the special case of the system (4.45)-(4.46)-(4.47) if $f_o=0$ and $g=0$.

4.5.1 Proposition

Let $f \in L_{2,H}(G)$, $p \in W^1_{2,R}(G)$. Every solution of system (4.45)-(4.47) with $f_o=0$ and $g=0$ allows the representation

$$u = \frac{\varrho}{\eta} T_G Q T_G f - \frac{1}{\eta} T_G Q p . \tag{4.48}$$

In this sense, condition (4.46) may be written as

$$-\operatorname{Re} \frac{\varrho}{\eta} Q T_G f + \frac{1}{\eta} \operatorname{Re} Q p = 0 . \tag{4.49}$$

Proof

Applying the operator T_G^2 from the left to equation (4.45) we obtain by using VEKUA's theory (cf.2.4)

$$u = -\frac{1}{\eta} T_G^2 D p + \frac{\varrho}{\eta} T_G^2 f + T_G v_2 + v_1 \tag{4.50}$$

with $v_1, v_2 \in \ker D(G)$. Because $\operatorname{im} P_\Gamma \cap \operatorname{im} Q_\Gamma = \{0\}$,

$$\operatorname{tr}\left(-\frac{1}{\eta} T_G^2 D p + \frac{\varrho}{\eta} T_G^2 f + T_G v_2\right) \in \operatorname{im} Q_\Gamma \quad \text{and} \quad \operatorname{tr} v_1 \in \operatorname{im} P_\Gamma$$

follows from (4.47) $v_1=0$. BOREL-POMPEIU's formula now yields

$$u = -\frac{1}{\eta} T_G[p - F_\Gamma p] + \frac{\varrho}{\eta} T_G^2 f + T_G v_2$$

with

$$\operatorname{tr} v_2 = (\operatorname{tr} T_G F_\Gamma)^{-1}\left[-\operatorname{tr} \frac{\varrho}{\eta} T_G^2 f + \frac{1}{\eta} \operatorname{tr} T_G p - \frac{1}{\eta} (\operatorname{tr} T_G F_\Gamma) p\right],$$

and so

$$v_2 = F_\Gamma (\operatorname{tr} T_G F_\Gamma)^{-1}\left[-\operatorname{tr} \frac{\varrho}{\eta} T_G^2 f + \frac{1}{\eta} \operatorname{tr} T_G p - \frac{1}{\eta} (\operatorname{tr} T_G F_\Gamma) p\right].$$

Inserting this expression for v_2 in (4.50) we get

$$u = -\frac{1}{\eta} T_G[p - F_\Gamma (\operatorname{tr} T_G F_\Gamma)^{-1} \operatorname{tr} T_G p] +$$
$$+ \frac{\varrho}{\eta} T_G[T_G f - (F_\Gamma (\operatorname{tr} T_G F_\Gamma)^{-1} \operatorname{tr} T_G) T_G f].$$

By the aid of the orthoprojections P and Q which are defined in Proposition 4.1.6 we obtain the simplified form

$$u = \frac{\varrho}{\eta} T_G Q T_G f - \frac{1}{\eta} T_G Q p .$$

It is readily seen that the boundary condition $\operatorname{tr} u=0$ is satisfied. Applying D from the left to representation (4.48) relation (4.49) follows. #

4.5.2 Theorem (Equivalence)

Let $u \in \overset{\circ}{W}{}^1_{2,H}(G)$, $p \in L_{2,R}(G)$ be a solution of system (4.48)-(4.49). Then $u = \text{Im}\, u$ is a weak solution of system (4.45)-(4.47). If $\hat{u}$ is a weak solution of system (4.45)-(4.47), then there exists a function $p \in L_{2,R}(G)$, such that the pair $\{u,p\}$ with $u=\hat{u}$ solves system (4.48)-(4.49).

Proof

First we have to prove that for all $v \in \ker \operatorname{div} \overset{\circ}{W}{}^1_{2,H}(G)$

$$\frac{\eta}{\varrho} \sum_{i=1}^{3} (\text{grad } u_i, \text{grad } v_i) = (f,v), \qquad (4.51)$$

where (u,v) denotes the classical scalar product $(u,v) = \int_G \sum_{i=1}^{3} u_i v_i dG$. We choose a sequence $\{p^{(n)}\}_{n=1}^{\infty} \subset W^1_{2,R}(G)$ with the property $p^{(n)} \xrightarrow[n\to\infty]{} p$ in $L_{2,R}(G)$. Then we have

$$\sum_{i=1}^{3} (\text{grad}(T_G Qp)_i, \text{grad } v_i) = \lim_{n\to\infty} \sum_{i=1}^{3} (\text{grad}(T_G Qp^{(n)})_i, \text{grad } v_i) =$$

$$= \lim_{n\to\infty} (\triangle(T_G Qp^{(n)}), v) = \lim_{n\to\infty} (\text{grad } p^{(n)}, v) = 0,$$

because $(\text{grad } p, v) = -(p, \operatorname{div} v) = 0$ for all $p \in W^1_2(G)$ and $v \in \ker \operatorname{div} \cap \overset{\circ}{W}{}^1_{2,H}(G)$. It follows from Corollary 3.3 and Proposition 2.4.2

$$\frac{\eta}{\varrho} [(-\triangle(T_G QTf), v) + \frac{1}{\eta} \sum_{i=1}^{3} (\text{grad}(T_G Qp)_i, \text{grad } v_i)] = (f,v).$$

Partial integration yields

$$\frac{\eta}{\varrho} \sum_{i=1}^{3} (\text{grad}(T_G QT_G f)_i, \text{grad } v_i) = (f,v).$$

Conversely, let $\hat{u} \in \overset{\circ}{W}{}^1_{2,H}(G) \cap \ker \operatorname{div}$ a weak solution of system (4.45)-(4.47). That means that equation (4.51) is valid. It follows

$$\sum_{i=1}^{3} (\text{grad } u_i, \text{grad } v_i) + \frac{\varrho}{\eta} (\triangle(T_G QT_G f), v) +$$

$$+ \frac{1}{\eta} \sum_{i=1}^{3} (\text{grad}(T_G Qq)_i, \text{grad } v_i) = 0,$$

and therefore for $q \in L_{2,R}(G)$, $v \in \overset{\circ}{W}{}^1_{2,H}(G) \cap \ker \operatorname{div}$

$$\sum_{i=1}^{3} (\text{grad}(u_i - \frac{\varrho}{\eta} (T_G QT_G f)_i + \frac{1}{\eta} (T_G Qq)_i), \text{grad } v_i) = 0. \qquad (4.52)$$

We have proved in Theorem 4.5.5 (independently of these considerations) that the system

$$v + \frac{1}{\eta} T_G Q p = \frac{\varrho}{\eta} T_G Q T_G f,$$

$$\frac{1}{\eta} \operatorname{Re} Qp = \frac{\varrho}{\eta} \operatorname{Re} Q T_G f$$

has a solution $\{v,p\} \in \overset{\bullet}{W}{}^1_{2,H}(G) \times L_{2,R}(G)$. Substituting in (4.52) q by p we deduce

$$- \operatorname{Re} Du + \frac{\varrho}{\eta} \operatorname{Re} Q T_G f - \frac{1}{\eta} \operatorname{Re} Qp = 0, \tag{4.53}$$

thence

$$u - \frac{\varrho}{\eta} T_G Q T_G f + \frac{1}{\eta} T_G Q p \in \overset{\bullet}{W}{}^1_{2,H}(G) \cap \ker \operatorname{div} .$$

Choosing $w = u - \frac{\varrho}{\eta} T_G Q T_G f + \frac{1}{\eta} T_G Q p$ we get from (4.52)

$$\sum_{i=1}^{3} \| \operatorname{grad}(u_i - (\frac{\varrho}{\eta} T_G Q T_G f)_i + \frac{1}{\eta} (T_G Q p)_i \|_2^2 = 0,$$

and therefore

$$u - \frac{\varrho}{\eta} T_G Q T_G f + \frac{1}{\eta} T_G Q p = c , \; c \in H .$$

As $\operatorname{tr}(u - \frac{\varrho}{\eta} T_G Q T_G f + \frac{1}{\eta} T_G Q p) = 0$ we finally conclude $c=0$.

#

4.5.3 Theorem (A-priori-Estimate)

Let $\{u,p\} \in \overset{\circ}{W}{}^1_{2,H}(G) \times L_{2,R}(G)$ be a solution of system (4.48)-(4.49). Denote by λ_1 the smallest eigenvalue of the problem $\{-\Delta u = 0, \operatorname{tr} u = 0\}$. Then the following inequality holds:

$$\left(\frac{\lambda_1}{1+\lambda_1}\right)^{1/2} \|u\|_{W^1_{2,H}} + \frac{1}{\eta} \|Qp\|_{L_{2,H}} \leq 2^{1/2} \frac{\varrho}{\eta} \|T_G f\|_{L_{2,H}} . \tag{4.54}$$

Proof

Representation (4.48) leads to

$$Du + \frac{1}{\eta} Qp = \frac{\varrho}{\eta} Q T_G f. \tag{4.55}$$

It is clear that $u \in W^1_{2,H}(G)$, $Du \in \operatorname{im} Q$, $\operatorname{Re} Du = 0$ and $\operatorname{Im} p = 0$. Thence it follows by using the proof of Theorem 3.1

$$\operatorname{Re} (Du, Qp) = \operatorname{Re} (Du, p) - \operatorname{Re} (Du, Pp) = 0 \tag{4.56}$$

as $Du \in \operatorname{im} Q$ and $\operatorname{Re} Du = 0$. From (4.55) and (4.56) we deduce the identity

$$\|Du\|^2_{L_{2,H}} + \frac{1}{\eta} \|Qp\|^2_{L_{2,H}} = \frac{\varrho^2}{\eta^2} \|Q T_G f\|^2_{L_{2,H}} ,$$

and therefore

$$2^{-1/2} [\|Du\|_{L_{2,H}} + \frac{1}{\eta} \|Qp\|_{L_{2,H}}] \leq \frac{\varrho}{\eta} \|T_G f\|_{L_{2,H}}$$

Using Proposition 4.1.12 and Theorem 4.1.14 we obtain

$$\|Du\|_{L_{2,H}} \geq \left(\frac{\lambda_1}{\lambda_1 + 1}\right)^{1/2} \|u\|_{W^1_{2,H}}$$

and so the assertion. #

4.5.4 Proposition

The subspace $Q(M^1_{2,H}(G)) \subset L_{2,H}(G)$ allows the orthogonal decomposition

$$Q(M^1_{2,H}(G)) = \ker \operatorname{div} \ D(\overset{\circ}{W}{}^1_{2,H}(G)) \oplus_{Re} Q(L_{2,R}(G)),$$

where "$\oplus_{Re}$" denotes the direct sum with respect to the scalar product $(u,v) = Re \int_G \bar{u}v dG$. The space $M^1_{2,H}(G)$ was introduced in Definition 2.4.1.

Proof

Let $v \in Q(M^1_{2,H}(G))$. Then we get $v=QT_G h$. In (4.56) it is proved that $Re\ (Du,Qp)=0$ with $u \in \overset{\circ}{W}{}^1_{2,H}(G) \cap \ker \operatorname{div}$, $p \in L_{2,R}(G)$. It remains to be shown that the validity of the relations

$$Re\ (Du,QT_G h) = 0, \quad u \in \overset{\circ}{W}{}^1_{2,H}(G) \cap \ker \operatorname{div} \tag{4.57}$$

and

$$Re\ (Qp,QT_G h) = 0, \quad p \in L_{2,R}(G) \tag{4.58}$$

yields $h=0$. Let $h \in L_{2,H}(G)$. We deduce from (4.57) by using Proposition 2.4.2 and Corollary 3.3

$$Re\ (Du,QT_G h) = Re\ (u,DQT_G h) = Re\ (u,h) = 0.$$

This implies (see [Tem]) $h=\operatorname{grad} q=Dq$, with $q \in W^1_{2,R}(G)$. From relation (4.58) it follows for all $p \in L_{2,R}(G)$

$$Re\ (Qp,QT_G h) = Re\ (Qp,QT_G Dq) = Re\ (Qp,Q(q-F\ q)) =$$
$$= Re\ (Qp,Qq) = 0,$$

whence $Qq=0$ for $q=p$ and $h=DQq=0$. This completes our proof. #

4.5.5 Theorem (Existence and Uniqueness)

The system of STOKES' equations (4.46)-(4.48) has a unique solution $\{u,p\}$ which may be represented by

$$u + \frac{1}{\eta} T_G Qp = \frac{\varrho}{\eta} T_G QT_G f .$$

The hydrostatic pressure p is unique up to a real additive constant.

Proof

The orthogonal decomposition in accordance with Proposition 4.5.4 yields the existence of an H-valued function $u \in \overset{\circ}{W}{}^1_{2,H}(G) \cap \ker \operatorname{div}$ and a scalar function $p \in L_{2,R}(G)$ such that condition (4.56) is fulfilled. An application of Proposition 2.5.4 and (4.55) leads to

$$u + \frac{1}{\eta} T_G Qp = \frac{\varrho}{\eta} T_G QT_G f .$$

The uniqueness of u is a consequence of Theorem 4.5.3. Likewise, from the a-priori estimate (4.54), by supposition of two solutions p_1 and p_2 for the hydrostatic pressure, the equation $Q(p_1-p_2)=0$ follows. It follows from Theorem 3.1 $p_1-p_2 \in \ker D$, whence $p_1-p_2= c \in \mathbb{R}^1$. #

4.5.6 Proposition

Let $1<p<\infty$, $0<\beta<1$, $k\in\mathbb{N}$ and $k>0$. Then the operator

$$\operatorname{tr} T_G F: W^{k-1/p}_{p,H}(\Gamma) \cap \operatorname{im} P_\Gamma \cap \ker \operatorname{Re} \longrightarrow W^{k+1-1/p}_{p,H}(\Gamma) \cap \operatorname{im} Q_\Gamma \cap \ker \operatorname{tr} \operatorname{div},$$

$$\operatorname{tr} T_G F: C^{0,\beta}_H(\Gamma) \cap \operatorname{im} P_\Gamma \cap \ker \operatorname{Re} \longrightarrow C^{1,\beta}_H(\Gamma) \cap \operatorname{im} Q_\Gamma \cap \ker \operatorname{tr} \operatorname{div},$$

respectively, is an isomorphism.

Proof

The smoothness relations follow from Proposition 4.1.5. Let $v \in W^{k-1/p}_{p,H}(\Gamma) \cap \operatorname{im} P_\Gamma \cap \ker \operatorname{Re}$. Suppose $\operatorname{tr} T_G F_\Gamma v=0$. As in the proof of Proposition 4.1.5 we obtain $v=0$. Let $w \in W^{k+1-1/p}_{p,H}(\Gamma) \cap \operatorname{im} Q_\Gamma \cap \ker \operatorname{tr} \operatorname{div}$. Because $w \in \operatorname{im} Q_\Gamma$, there exists an H-valued function $u \in \ker \Delta(G)$ with $w=\operatorname{tr} u$. BOREL-POMPEIU's formula yields

$$u = F_\Gamma w + T_G Du = T_G v$$

with $v=Du \in \ker D(G)$. It is clear that $u=T_G F_\Gamma z$ with $z=\operatorname{tr} v \in \operatorname{im} P_\Gamma$ and $w=\operatorname{tr} T_G F_\Gamma z$. For the assumption it follows that $v \in W^k_{p,H}(G)$ and $T_G v \in W^{k+1}_{p,H}(G)$. Therefore

$$\operatorname{div} T_G v = -\operatorname{Re} D T_G v = -\operatorname{Re} v$$

and

$$\operatorname{tr} \operatorname{div} T_G v = -\operatorname{tr} \operatorname{Re} v = 0,$$

whence $\operatorname{Re} z=0$. The proof for HOELDER-continuous functions is similar. #

4.5.7 Theorem (Representation and Regularity)

Let $f \in W^k_{p,H}(G)$, $k\geq 0$, $1<p<\infty$. Then the only solution of $\{u,p\}$ of (4.48)-(4.49) has the representation

$$u = \frac{\varrho}{\eta} T_G \operatorname{Im} T_G f - \frac{\varrho}{\eta} T_G \operatorname{Im} F_\Gamma (\operatorname{tr} T_G \operatorname{Im} F_\Gamma)^{-1} \operatorname{tr} T_G \operatorname{Im} T_G f, \qquad (4.59)$$

$$p = \varrho \operatorname{Re} T_G f - \varrho \operatorname{Re} F_\Gamma (\operatorname{tr} T_G \operatorname{Im} F_\Gamma)^{-1} \operatorname{tr} T_G \operatorname{Im} T_G f. \qquad (4.60)$$

Consequently, we have $u \in W^{k+2}_{p,H}(G) \cap \overset{\circ}{W}{}^1_{2,H}(G)$ and $p \in W^{k+1}_{p,R}(G)$.

Proof

It is easy to prove that $\operatorname{tr} T_G \operatorname{Im} T_G f \in \operatorname{im} Q_\Gamma$. Furthermore we have

$$\operatorname{div} T_G \operatorname{Im} T_G f = -\operatorname{Re} DT_G \operatorname{Im} T_G f = -\operatorname{Re} \operatorname{Im} T_G f = 0,$$

whence

$$\operatorname{tr} T_G \operatorname{Im} T_G f \in \ker \operatorname{tr} \operatorname{div}.$$

Let $\operatorname{Re} z=0$. If for $\operatorname{tr} T_G F z \in \ker \operatorname{tr} \operatorname{div}$, then immediately $\operatorname{tr} \operatorname{Re} F_\Gamma z=0$. $\operatorname{Re} F_\Gamma z$ is a harmonic function and therefore $\operatorname{Re} F_\Gamma z=0$. The operator $\operatorname{tr} T_G F_\Gamma$ may be written in the form $\operatorname{tr} T_G \operatorname{Im} F_\Gamma$. Now we prove that u and p defined in (4.59) and (4.60), respectively, satisfy the equation

$$u + \frac{1}{\eta} T_G Q p = \frac{\varrho}{\eta} T_G Q T_G f.$$

Substituting (4.59) and (4.60) on the left-hand side we get

$$T_G \operatorname{Im} T_G f - T_G \operatorname{Im} F_\Gamma (\operatorname{tr} T_G \operatorname{Im} F_\Gamma)^{-1} \operatorname{tr} T_G \operatorname{Im} T_G f + T_G Q[\operatorname{Re} T_G f -$$

$$-\operatorname{Re} F_\Gamma (\operatorname{tr} T_G \operatorname{Im} F_\Gamma)^{-1} \operatorname{tr} T_G \operatorname{Im} T_G f] = T_G \operatorname{Im} T_G f -$$

$$-T_G \operatorname{Im} F_\Gamma (\operatorname{tr} T_G \operatorname{Im} F_\Gamma)^{-1} \operatorname{tr} T_G \operatorname{Im} T_G f + T_G Q T_G f - T_G Q \operatorname{Im} T_G f -$$

$$-T_G Q F_\Gamma (\operatorname{tr} T_G \operatorname{Im} F_\Gamma)^{-1} \operatorname{tr} T_G \operatorname{Im} T_G f +$$

$$+T_G Q \operatorname{Im} F_\Gamma (\operatorname{tr} T_G \operatorname{Im} F_\Gamma)^{-1} \operatorname{tr} T_G \operatorname{Im} T_G f.$$

It is clear that $T_G Q F_\Gamma (\operatorname{tr} T_G \operatorname{Im} F_\Gamma)^{-1} \operatorname{tr} T_G \operatorname{Im} T_G f=0$. Using the special structure of the projection P (Proposition 4.1.6) we apparently get

$$T_G P \operatorname{Im} T_G f - T_G P \operatorname{Im} F_\Gamma (\operatorname{tr} T_G \operatorname{Im} F_\Gamma)^{-1} \operatorname{tr} T_G \operatorname{Im} T_G f = 0. \qquad (4.61)$$

Equation (4.50) remains to be considered. From (4.61) it follows

$$P[\operatorname{Im} T_G f - \operatorname{Im} F_\Gamma (\operatorname{tr} T_G \operatorname{Im} F_\Gamma)^{-1} \operatorname{tr} T_G \operatorname{Im} T_G f] = 0,$$

whence

$$\operatorname{Im} T_G f - \operatorname{Im} F_\Gamma (\operatorname{tr} T_G \operatorname{Im} F_\Gamma)^{-1} \operatorname{tr} T_G \operatorname{Im} T_G f \in \operatorname{im} Q.$$

Therefore

$$\operatorname{Re} Q(\operatorname{Im} T_G f - \operatorname{Im} F_\Gamma (\operatorname{tr} T_G \operatorname{Im} F_\Gamma)^{-1} \operatorname{tr} T_G \operatorname{Im} T_G f) = 0$$

and by consideration of (4.60)

$$\operatorname{Re} Q T_G f = \operatorname{Re} Q[\operatorname{Re} T_G f - \operatorname{Re} F_\Gamma (\operatorname{tr} T_G \operatorname{Im} F_\Gamma)^{-1} \operatorname{tr} T_G \operatorname{Im} T_G f] =$$

$$= \frac{1}{\varrho} \operatorname{Re} Q p.$$

We have shown that (4.59)-(4.60) is a correctly defined solution of problem (4.49)-(4.50). The statements which relate to the smoothness may be obtained without difficulties by the aid of Theorem 2.3.6, Theorem 2.5.5 and Proposition

4.5.6 from representations (4.59) and (4.60). #

4.5.8 Theorem (Local Regularity)

Let $f \in L_{2,H}(G) \cap W^{k,loc}_{p,H}(G)$, $k \in \mathbb{N}$, $1<p<\infty$. Then

$$u \in W^{k+2,loc}_{p,H}(G),\quad p \in W^{k+1,loc}_{p,R}(G).$$

Proof

Because of $\operatorname{Im} p=0$ we have $\operatorname{Im} Pp=-\operatorname{Im} Qp \in \ker \Delta(G)$, and so $\operatorname{Im} Qp \in C^{\infty}_{R}(G')$ for any G' with $\overline{G'} \subset G$. Applying Theorem 4.1.1 from $\operatorname{Re} Qp= \operatorname{Re} QT_G f$ it follows $\operatorname{Re} Qp \in W^{k+1,loc}_{p,R}(G)$ and therefore $Qp \in W^{k+1,loc}_{p,H}(G)$. Relation (4.55) now gives $Du \in W^{k+1,loc}_{p,H}(G)$ and finally BOREL-POMPEIU's formula $u=T_G Du \in W^{k+2,loc}_{p,H}(G)$. Besides it follows from $Qp \in W^{k+1,loc}_{p,H}(G)$, using Corollary 3.3 $Dp \in W^{k,loc}_{p,H}(G)$. A new application of BOREL-POMPEIU's formula leads to $p=T_G Dp+F_\Gamma p \in W^{k+1,loc}_{p,H}(G)$ $(F_\Gamma p \in \ker \Delta(G))$. #

4.5.9 Remark

It is possible to weaken the requirement $f \in L_{2,H}(G)$. In the framework which we present here (we look for strong solutions of (4.49)-(4.50)) it is necessary to suppose $f \in L_{p,H}(G)$, $p>6/5$ for $T_G f \in L_{2,H}(G)$. Only in these cases we can use the properties of the orthoprojection Q. In the case of $f \in H^{-1}(G)$ we refer to [Tem] and [LM]. Similar results in regard to statements of local regularity are also given in generalized situations in [Nau]. Our approach does not seem to be so complicated as the others.

4.5.10 Remark

In the general situation (4.42)-(4.44) it is impossible to prescribe the functions $f_\emptyset$ and g independently of each other. We always have to satisfy the necessary condition (4.45). If we fix f and $f_\emptyset$, the following system arises:

$$DDu + \frac{1}{\eta} Dp = \frac{\varrho}{\eta} f \quad \text{in } G, \tag{4.42}$$

$$\operatorname{Re} Du = f_\emptyset \quad \text{in } G, \tag{4.43}$$

$$u = \operatorname{tr} T_G f_\emptyset \quad \text{on } \Gamma. \tag{4.62}$$

The only solution in the sense of Theorem 4.5.5 may be constructed starting from the solution $\{u,p\}$ expressed by (4.59)-(4.60). So system (4.42)-(4.43)-(4.62) is solved by

$\{u',p'\}$ with

$$u' = u + T_G f_\emptyset ,$$

$$p' = p - \eta f_\emptyset .$$

In the case of given f and g we have to consider the system

$$\begin{aligned} DDu + \frac{1}{\eta} Dp &= \frac{\varrho}{\eta} f && \text{in } G , \\ \mathrm{Re}\, Du &= \mathrm{Re}\, PDh && \text{in } G , \\ u &= g && \text{on } \Gamma , \end{aligned}$$

where h denotes a corresponding smooth extension of g into the domain G. The only solution $\{u',p'\}$ in the sense of Theorem 4.5.5 is given by

$$u' = u + F_\Gamma g + T_G PDh ,$$

$$p' = p ,$$

where $\{u,p\}$ is defined by (4.59)-(4.60).

4.6 NAVIER-STOKES Equations

The aim of the following considerations is the foundation of an iterative method for solving the time-independent NAVIER-STOKES equations

$$-\Delta\hat{u} + \frac{\varrho}{\eta}(\hat{u}\cdot\mathrm{grad})\hat{u} + \frac{1}{\eta}\,\mathrm{grad}\; p = \frac{\varrho}{\eta}\hat{f} \quad \text{in } G, \qquad (4.63)$$

$$\mathrm{div}\;\hat{u} = 0 \quad \text{in } G, \qquad (4.64)$$

$$\hat{u} = 0 \quad \text{on } \Gamma. \qquad (4.65)$$

$\hat{u}$ means the velocity of the fluid, and p is the hydrostatic pressure. Furthermore ϱ denotes the density, η the toughness and $\hat{f}$ the vector of the external forces. A discussion of several methods to solve this system is given in Remark 5.6.13. OSEEN showed in 1928 that in the case of a ball approximative solutions of good quality may be obtained if the convection term $C(\hat{u}) = \frac{\varrho}{\eta}(\hat{u}\cdot\mathrm{grad})\hat{u}$ is replaced by $\frac{\varrho}{\eta}(\hat{v}\cdot\mathrm{grad})\hat{u}$, where $\hat{v}$ denotes the solution of the corresponding STOKES problem. Based on this idea we intend to solve NAVIER-STOKES equations by reduction to a sequence of STOKES problems.

Let $f=\sum_{i=1}^{3} f_i e_i$, $u=\sum_{i=0}^{3} u_i e_i$. Then system (4.63)-(4.64)-(4.65) allows the hypercomplex notation

$$DDu + \frac{\varrho}{\eta}M(u) + \frac{1}{\eta}Dp = 0 \quad \text{in } G, \qquad (4.66)$$

$$\mathrm{Re}\; Du = 0 \quad \text{in } G, \qquad (4.67)$$

$$u = 0 \quad \text{on } \Gamma \qquad (4.68)$$

with $M(u)=M^*(u)-f=[\mathrm{Re}\;(u\;D)]u-f$,

where DIRICHLET's problem

$$-\Delta u_o = 0 \quad \text{in } G,$$

$$u_o = 0 \quad \text{on } \Gamma$$

has been added. Note that $\mathrm{Re}\;[u_o\;D]=0$.

4.6.1 Proposition

Let $f \in L_{2,H}(G)$, $p \in W^1_{2,R}(G)$. Every solution of system (4.64)-(4.65)-(4.66) allows the representation

$$u = -\frac{\varrho}{\eta}T_G Q T_G M(u) - \frac{1}{\eta}T_G Q p \qquad (4.69)$$

and

$$\mathrm{Re}\;\frac{\varrho}{\eta}QT_G M(u) + \frac{1}{\eta}\mathrm{Re}\;Qp = 0. \qquad (4.70)$$

Proof

Replacing f by $M(u)$ in the proof of Proposition 4.5.1 we verify the assertion.

#

4.6.2 Theorem (Equivalence)

Let $u \in \overset{\circ}{W}{}^1_{2,H}(G)$, $p \in L_{2,R}(G)$ be a solution of (4.69)-(4.70). Then $\hat{u}=\mathrm{Im}\ u$ is a weak solution of system (4.66)-(4.67)-(4.68).

If $\hat{u}$ is a weak solution of (4.66)-(4.67)-(4.68), then there exists a function $p \in L_{2,R}(G)$ such that the pair $\{u,p\}$ with $u=\hat{u}$ solves system (4.69)-(4.70).

Proof

First we verify that for all $v \in \ker \mathrm{div} \cap \overset{\circ}{W}{}^1_{2,H}(G)$ we have the identity

$$\sum_{i=1}^{3} \frac{\eta}{\varrho}(\mathrm{grad}\ u_i, \mathrm{grad}\ v_i) + \sum_{i=1}^{3}(u_i \partial_i u, v) = (f,v), \qquad (4.71)$$

where (u,v) denotes the scalar product $(u,v)=\sum_{i=1}^{3}\int_G u_i v_i dG$.

Following the proof of Theorem 4.5.2 we also find

$$\sum_{i=1}^{3}(\mathrm{grad}\ (T_G Qp)_i, \mathrm{grad}\ v_i) = 0.$$

It follows from Corollary 3.3 and Proposition 2.4.2 that

$$\frac{\varrho}{\eta}\left[(\triangle T_G Q T_G M(u), v) + \frac{1}{\eta}\sum_{i=1}^{3}(\mathrm{grad}(T_G Qp)_i, \mathrm{grad}\ v_i)\right] +$$

$$+ \sum_{i=1}^{3}(u_i \partial_i u, v) = (f,v) .$$

Partial integration yields

$$\frac{\varrho}{\eta}\left[\sum_{i=1}^{3}(-\mathrm{grad}(T_G Q T_G M(u))_i, \mathrm{grad}\ v_i) +\right.$$

$$\left. + \frac{1}{\eta}\sum_{i=1}^{3}(\mathrm{grad}(T_G Qp)_i, \mathrm{grad}\ v_i)\right] + \sum_{i=1}^{3}(u_i \partial_i u, v) = (f,v).$$

From (4.69) we get (4.71).

Now let $\hat{u} \in \overset{\circ}{W}{}^1_{2,H}(G) \cap \ker \mathrm{div}$ be a weak solution of (4.66)-(4.67)-(4.68). From (4.71) it follows, similarly to the proof of Theorem 4.5.2,

$$\sum_{i=1}^{3}(\mathrm{grad}\ u_i, \mathrm{grad}\ v_i) - \frac{\varrho}{\eta}(\triangle(T_G Q T_G M(u), v) +$$

$$+ \frac{1}{\eta}\sum_{i=1}^{3}(\mathrm{grad}(T_G Qq)_i, \mathrm{grad}\ v_i) = 0,$$

and so for any $q \in L_{2,R}(G)$ and $v \in \overset{\circ}{W}{}^1_{2,H}(G) \cap \ker \mathrm{div}$

$$\sum_{i=1}^{3}(\mathrm{grad}(u_i + \frac{\varrho}{\eta}(T_G Q T_G M(u))_i + \frac{1}{\eta}(T_G Qq)_i), \mathrm{grad}\ v_i) = 0 . \qquad (4.72)$$

An application of Theorem 4.5.5 gives us that the system

$$v + \frac{1}{\eta} T_G Qp = - \frac{\varrho}{\eta} T_G Q T_G M(u), \qquad (4.73)$$

$$\frac{1}{\eta} \operatorname{Re} Qp = - \frac{\varrho}{\eta} \operatorname{Re} Q T_G M(u) \qquad (4.74)$$

has a solution $\{v,p\} \in \mathring{W}^1_{2,H}(G) \times L_{2,R}(G)$. Replacing q by p in (4.72), it follows

$$- \operatorname{Re} Du - \frac{\varrho}{\eta} \operatorname{Re} Q T_G M(u) - \frac{1}{\eta} \operatorname{Re} Qp = 0,$$

thence

$$u + \frac{\varrho}{\eta} T_G Q T_G M(u) + \frac{1}{\eta} T_G Qp \in \mathring{W}^1_{2,H}(G) \cap \ker \operatorname{div}.$$

Set $v = u + \frac{\varrho}{\eta} T_G Q T_G M(u) + \frac{1}{\eta} T_G Qp$. One get now

$$\sum_{i=1}^{3} \| \operatorname{grad}(u_i + \frac{\varrho}{\eta} (T_G Q T_G M(u))_i + \frac{1}{\eta} (T_G Qp)_i \|^2 = 0$$

and therefore, similarly to Theorem 4.5.2,

$$u + \frac{\varrho}{\eta} T_G Q T_G M(u) + \frac{1}{\eta} T_G Qp = 0.$$

#

4.6.3 Theorem

Let $u \in \mathring{W}^1_{2,H}(G) \cap \ker \operatorname{div}$, $p \in L_{2,R}(G)$ be a solution of system (4.69)-(4.70). The inequality

$$\left(\frac{\lambda_1}{1+\lambda_1}\right)^{1/2} \| u \|_{W^1_{2,H}} + \frac{1}{\eta} \| Qp \|_{L_{2,H}} \leq 2^{1/2} \frac{\varrho}{\eta} \| T_G M(u) \|_{L_{2,H}} \qquad (4.75)$$

is valid.

Proof

Similar to that of Theorem 4.5.3

#

4.6.4 Remark

Inequality (4.75) has the same structure as the a-priori estimate (4.54) for the solutions of STOKES' equations. Indeed, it is not an a-priori estimate for the solutions of NAVIER-STOKES equations, but it is without doubt very important with respect to further considerations. Making use of (4.75) one may estimate the term of the hydrostatic pressure Qp by the velocity u and the right-hand side f.

4.6.5 Proposition

Let $u \in \mathring{W}^1_{2,H}(G)$, $1<p<3/2$. Then

$$\| M^*(u) \|_{L_{p,H}} \leq C_1 \| u \|^2_{W^1_{2,H}} .$$

Proof

We have

$$\| M^*(u) \|^p_{L_{p,H}} \leq \sum_{i,j=1}^{3} \int_G | u_i \partial_i u_j |^p dG \leq \sum_{i,j=1}^{3} \| u_i \|^p_{L_{q,R}} \| \partial_i u_j \|^p_{L_{2,R}} \leq$$

$$\leq \sum_{i,j=1}^{3} C^p \| u_i \|^p_{W^1_{2,H}} \| u_j \|^p_{W^1_{2,H}} \leq 9C^p \| u \|^{2p}_{W^1_{2,H}} ,$$

where $q<6$, $p=\frac{2q}{2+q}$.
The last estimate is easy to verify by using HOELDER's inequality. C denotes the embedding constant from $\overset{\circ}{W}{}^1_{2,R}(G)$ to $L_{q,R}(G)$. Obviously, it holds $C_1=8^{1/p}C$.

#

4.6.6 Corollary

The constant C_1 of Proposition 4.6.5 may be estimated by $\|T_G\|_{[L_{q,H},L_{2,H}]}$. More exactly it holds

$$C_1 \leq 8^{1/p}\|T_G\|_{[L_{2,H},L_{q,H}]},$$

where $q<6$ and $p=\frac{2q}{2+q}$.

Proof

Let $u\in \overset{\circ}{W}{}^1_{2,H}(G)$. BOREL-POMPEIU's formula leads to

$$\|u\|_{L_q} = \|T_G Du\|_{L_q} \leq \|T_G\|_{[L_{2,H},L_{q,H}]}\|Du\|_{L_{2,H}} \leq \|T_G\|_{[L_{2,H},L_{q,H}]}\|u\|_{W^1_{2,H}}.$$

Together with the proof of Proposition 4.6.5 the wanted estimate follows.

#

4.6.7 Remark

In the case of a star shaped domain with reference to a certain ball in [Gue3] an estimate is given for the embedding constant C, which only depends on the volume of G. Other estimates for C may be obtained by using methods which are presented in [Mi3]. In this paper a connection between the constant C and the smallest eigenvalue of $\{-\Delta,tr\}$ is given.

4.6.8 Theorem (Iteration Method)

System (4.69)-(4.70) has a unique solution
$\{u,p\}\in(\overset{\circ}{W}{}^1_{2,H}(G)\cap \ker \operatorname{div})\times L_{2,R}(G)$
(p is unique up to a real constant) if the right-hand side f satisfies the condition

$$\frac{\varrho}{\eta}\|f\|_{L_{p,H}} \leq (16K^2C_1)^{-1} \tag{4.76}$$

with $K=\frac{\varrho}{\eta}\|T_G\|_{[L_{2,H}\cap \operatorname{im} Q,W^1_{2,H}]}\|T_G\|_{[L_{p,H},L_{2,H}]}$.

For any function $u_\emptyset\in \overset{\circ}{W}{}^1_{2,H}(G)\cap \ker \operatorname{div}$ with

$$\|u_\emptyset\| \leq \min(R,1/(4KC_1)+W) \tag{4.77}$$

$(R=(2KC_1)^{-1},\ W=[(4KC_1)^{-2}-\varrho\|f\|_{L_{p,H}}/(\eta C_1)]^{1/2})$

the iteration method

$$u_n = - \quad T_G Q T_G M(u_{n-1}) - \frac{1}{\eta} T_G Q p_n , \quad n=1,2,\ldots, \tag{4.78}$$

$$\frac{1}{\eta} \operatorname{Re} Q p_n = - \cdot \quad \operatorname{Re} Q T_G M(u_{n-1}), \tag{4.79}$$

$$u_\emptyset \in \mathring{W}^1_{2,H}(G) \cap \ker \operatorname{div}$$

converges in $W^1_{2,H}(G) \times L_{2,R}(G)$.

<u>Proof</u>

Theorem 4.5.5 ensures the unique solvability of the STOKES problems which we have to solve at each step. We have to prove the convergence of the procedure. Throughout the whole proof we settle that all norms without footnotes are $W^1_{2,H}(G)$-norms. Using inequality (4.75) we get

$$\|u_n - u_{n-1}\| \le \|T_G Q T_G (M(u_{n-1}) - M(u_{n-2}))\| +$$

$$+ \frac{1}{\eta} \|T_G Q(p_n - p_{n-1})\| \le$$

$$\le 2K \|M(u_{n-1}) - M(u_{n-2})\|_{L_{p,H}} \qquad (\|Q\|_{L(L_{2,H})} = 1\ !) \quad .$$

As in the proof of Proposition 4.6.5, we find

$$\|M(u_{n-1}) - M(u_{n-2})\|_{L_{p,H}} = \|M^*(u_{n-1}) - M^*(u_{n-2})\|_{L_{p,H}} \le$$

$$\le \|((u_{n-1}-u_{n-2})\cdot \operatorname{grad})u_{n-1}\|_{L_{p,H}} + \|(u_{n-2}\cdot \operatorname{grad})(u_{n-1}-u_{n-2})\|_{L_{p,H}} \le$$

$$\le C_1 \|u_{n-1}-u_{n-2}\|(\|u_{n-1}\|+\|u_{n-2}\|) \ .$$

We set for short $L_n = 2KC_1(\|u_{n-1}\|+\|u_{n-2}\|)$. Then one get

$$\|u_n - u_{n-1}\| \le L_n \|u_{n-1} - u_{n-2}\| \ .$$

Now assume $\|u_{n-1}\| \le (4KC_1)^{-1} - W$. Then it follows

$$\|u_n\| \le \quad \|T_G Q T_G M(u_{n-1})\| + \frac{1}{\eta} \|T_G Q p_n\| \le$$

$$\le 2KC_1 \|u_{n-1}\|^2 + 2K\frac{\xi}{\eta} \|f\|_{L_{p,H}} \le$$

$$\le 2KC_1((4KC_1)^{-2} - W/(2KC_1) + W^2) + 2K\frac{\xi}{\eta} \|f\|_{L_{p,H}} \le$$

$$\le (4KC_1)^{-1} - W.$$

In a similar way we get from

$$(4KC_1)^{-1} - W \le \|u_{n-1}\| \le (4KC_1)^{-1} + W$$

the inequality

$$\|u_n\| \le \|u_{n-1}\| \ .$$

A necessary condition for all these considerations is, of

course, the assumption

$$\frac{\varrho}{\eta} \|f\|_{L_{p,H}} \leq (16K^2C_1)^{-1}. \tag{4.80}$$

Using the inequality $\|u_{n-1}\| \leq (4KC_1)^{-1} - W$ we may find the estimate

$$L_n \leq L = 1 - 4KC_1W < 1. \tag{4.81}$$

Let $R=(2KC_1)^{-1}$ (obviously $R \geq (4KC_1)^{-1} - W$). Then we conclude from $\|u_{n-1}\| < R$ immediately $L_n < 1 - \varepsilon$, $\varepsilon > 0$ a certain constant. If we additionally suppose

$$\|f\|_{L_{p,H}} > 0.25\ (4KC_1)^{-2} C_1 \frac{\eta}{\varrho},$$

then we even have $R > (4KC_1)^{-1} + W$. The proof is finished by making use of BANACH's fixed-point theorem. #

<u>4.6.9 Corollary</u>

Under the suppositions of Theorem 4.6.8 we have the a-priori estimate

$$\|u\| \leq (4KC_1)^{-1} - W. \tag{4.82}$$

An a-priori estimate for the term $\|Qp\|_{L_{2,H}}$ is easy to find.

<u>Proof</u>

It follows from $\|u_{n-1}\| \leq (4KC_1)^{-1} - W$. #

<u>4.6.10 Corollary</u>

There holds the error estimate

$$\|u_n - u\| \leq L^n \|u_0 - u\|.$$

In the case of $u_0 = 0$ we have

$$\|u_n - u\| \leq L^n[(4KC_1)^{-1} - W].$$

<u>Proof</u>

It follows from the proof of Theorem 4.6.8. #

<u>4.6.11 Theorem (Regularity)</u>

Let $f \in W^k_{q,H}(G)$, $q > 6/5$. Then the solution $\{u,p\}$ of the system (4.69)-(4.70) belongs to $W^{k+2}_{q,H}(G) \cap \overset{\circ}{W}{}^1_{2,H}(G) \times W^{k+1}_{q,R}(G)$.

<u>Proof</u>

We confine our considerations to the case $f \in L_{q,H}(G)$. In the general case the proof can be given by the same technique. First we consider the STOKES problem

$$v + \frac{1}{\eta} T_G Qg = \frac{\varrho}{\eta} T_G Q T_G M(u), \tag{4.83}$$

$$\frac{1}{\eta} \text{ Re } Qg = - \frac{\varrho}{\eta} \text{ Re } QT_G M(u). \tag{4.84}$$

Theorem 4.5.7, Theorem 4.6.8 and Remark 4.5.9 yield $v=u \in W^2_{s,H}(G)$ and $g=p \in W^1_{s,R}(G)$ for $s<3/2$. By the aid of HOELDER's inequality and embedding theorems we get $M^*(u) \in L_{t,H}(G)$ for all $t<3$. Now, let $q<3$, $M(u) \in L_{t,H}(G)$. Then, by a new study of problem (4.83)-(4.84) we find $u \in W^2_{t,H}(G)$, $p \in W^1_{t,R}(G)$. Renewing this procedure over and over again obtain we may obtain $M^*(u) \in L_{r,H}(G)$ for $r<\infty$. Therefore we have $M(u) \in L_{q,H}(G)$. A new investigation of STOKES problem (4.83)-(4.84) leads to the wanted result. #

4.6.12 Remark

The foregoing considerations show that hypercomplex methods may be used successfully for the treatment of non-linear problems. In the characteristic case of the time-independent NAVIER-STOKES equations we have demonstrated that a unified approach with elementary means can be used to solve not only the analytical problems of existence, uniqueness, regularity and a-priori estimates but also approximative methods may be established. Such special properties of this differential operator as monotony were not used. It is necessary to stress that the sequence of approximative solutions for u in the space $\overset{\circ}{W}{}^1_{2,H}(G)$ converges strongly . Other methods (see [Zei]) only lead to weak convergent sequences (or subsequences) of approximative solutions. Frequently weak convergence is not sufficient for the necessary evaluation of the quality of the approximative solutions.

4.6.13 Remark

In most of the papers which deal with NAVIER-STOKES equations the convection term has been replaced by several linear expressions or non-linear terms of a simplified form. Furthermore, it has been tried to separate the calculation of pressure from the calculation of velocity.

Essential methods in this direction are the UZAWA-algorithm and the ARROW-HURWITZ algorithm. The numerical solution of these linear or nonlinear problems with variable coefficients (dependence on the position and on the solution at the preceding step of iteration) requires comprehensive computations. Difficulties occur for the satisfaction of condition (4.64). In recent years boundary collocation methods have gained a growing importance for the solution of elliptical boundary value problems. Successful examples are contained in [GST], [Mue] and [Ke]. Both the boundary collocation methods and the boundary approximation methods proposed by some authors (cf. G.BURGESS/E.MAHAJERIN [BM]) assume the explicit knowledge of proper systems of basis functions which must be constructed by the aid of the fundamental solution of the differential operator. Thence these methods seem to be effective only for equations with constant coefficients. Our approach to solving NAVIER-STOKES equations makes it possible to apply a boundary collocation method.

4.7 Stream Problems with Free Convection

In this chapter it is planned to deal with such problems by our method which combine NAVIER-STOKES equations and the equations of heat conduction. More exactly, the following problem shall be considered:

$$-\Delta u + \frac{\varrho}{\eta}(u\cdot\mathrm{grad})u + \frac{1}{\eta}\,\mathrm{grad}\ p + \frac{\gamma}{\eta}\, g\, w = - f \quad \text{in } G, \qquad (4.85)$$

$$\mathrm{div}\ u = 0 \quad \text{in } G, \qquad (4.86)$$

$$-\Delta w + \frac{m}{\varkappa}(u\cdot\mathrm{grad}\ w) = \frac{1}{\varkappa} h \quad \text{in } G, \qquad (4.87)$$

$$u = 0,\ w = 0 \quad \text{on } \Gamma. \qquad (4.88)$$

Here w denotes the temperature, γ the GRASSHOF number, m the PRANDTL number, $\varkappa$ the number of temperature conductivity and g the vector $(0,0,-1)^T$. By similar considerations as in the case of NAVIER-STOKES equations we get the following eqivalent problem:

$$u = - \frac{\varrho}{\eta} T_G Q T_G [M(u) + \frac{\gamma}{\eta} e_3 w] - \frac{1}{\eta} T_G Q p \quad \text{in } G, \qquad (4.89)$$

$$\mathrm{Re}\ \frac{\varrho}{\eta} Q T_G [M(u) + \frac{\gamma}{\eta} e_3 w] + \frac{1}{\varkappa}\ \mathrm{Re}\ Qp = 0 \quad \text{in } G, \qquad (4.90)$$

$$w = - \frac{m}{\varkappa} T_G Q T_G (u\cdot\mathrm{grad}\ w) + \frac{1}{\varkappa} T_G Q T_G h \quad \text{in } G. \qquad (4.91)$$

We propose the iteration method

$$u_n = - \frac{\varrho}{\eta} T_G Q T_G [M(u_{n-1} + \frac{\gamma}{\eta} e_3 w_{n-1}] - \frac{1}{\eta} T_G Q p_n, \qquad (4.92)$$

$$\mathrm{Re}\ \frac{\varrho}{\eta} Q T_G [M(u_{n-1}) + \frac{\gamma}{\eta} e_3 w_{n-1}] + \frac{1}{\eta}\ \mathrm{Re}\ Qp_n = 0, \qquad (4.93)$$

$$w_n = - \frac{m}{\eta} T_G Q T_G (u_n\cdot\mathrm{grad}\ w_n) + \frac{1}{\eta} T_G Q T_G h. \qquad (4.94)$$

By equations (4.92)-(4.93) certain STOKES problems are given which are solvable by our statements in Chapter 4.5. Therefore we have to study the solvability of equation (4.94) and to consider the properties of its solution. For this purpose we investigate the following "inner" iteration:

$$w_n^j = -\frac{m}{\eta} T_G Q T_G (u_n \cdot \mathrm{grad}\, w_n^{j-1}) + \frac{1}{\eta} T_G Q T_G h. \tag{4.95}$$

4.7.1 Theorem

Let $u_n \in \overset{\circ}{W}{}^1_{2,H}(G)$. Then problem (4.94) has a solution. Under the assumptions

$$\text{(i)}\quad 1 - \frac{m}{\varkappa} KC_1 \|u_n\|_{W^1_{2,H}} > 0, \qquad \text{(ii)}\quad m < 4\varkappa$$

the sequence defined by (4.95) converges in the space $\overset{\circ}{W}{}^1_{2,H}(G)$ to a unique solution of (4.94), and we have the a-priori estimate

$$\|w_n\|_{W^1_{2,H}} \le \frac{4K}{4\varkappa - m}\|h\|_p \tag{4.96}$$

Proof

From the assumption (i) it follows for each $j \in \mathbb{N}$ that

$$\|w_n^j\|_{W^1_{2,H}} \le K\|h\|_p/(\varkappa - mKC_1\|u_n\|_{W^1_{2,H}}).$$

Consequently, the sequence $\{w_n^j\}_{j\in\mathbb{N}}$ is bounded in $W^1_{2,H}(G)$. Choosing a weakly convergent subsequence $\{w_n^{j'}\}$ there exists for $j \longrightarrow \infty$ a solution w_n of equation (4.94). Using Proposition 4.6.5 an easy calculation leads to

$$\|w_n^j - w_n^{j-1}\|_{W^1_{2,H}} = \|\frac{m}{\varkappa} T_G Q T_G (u_n \cdot \mathrm{grad}(w_n^{j-1} - w_n^{j-2}))\|_{W^1_{2,H}} \le$$

$$\le \frac{m}{\varkappa} KC_1 \|u_n\|_{W^1_{2,H}} \|w_n^{j-1} - w_n^{j-2}\|_{W^1_{2,H}} .$$

Obviously, we have for $\frac{m}{\varkappa}KC_1\|u_n\|_{W^1_{2,H}} < 1$ convergence of the sequence $\{w_n^j\}_{j\in\mathbb{N}}$ in $\overset{\circ}{W}{}^1_{2,H}(G)$. From relation (4.92) we obtain, in virtue of the a-priori estimate (4.75),

$$\|u_n\|_{W^1_{2,H}} \le (4KC_1)^{-1} - ((4KC_1)^{-2} - \frac{9}{\eta C}\|f\|_{L_p} - \frac{\gamma}{\eta C}\|w_{n-1}\|_{L_p})^{1/2} \le (4KC_1)^{-1}.$$

This implies that the condition $m < 4\varkappa$ is a sufficient convergence criterion.

#

4.7.2 Remark

If it can be ensured that condition (4.96) is fulfilled, then assumption (i) can be left out. This requires additional suppositions relative to $\|h\|_p$. Furthermore it is possible to weaken assumption (ii) if a refined estimate is used for $\|u_n\|_{W^1_{2,H}}$.

4.7.3 Remark

The investigation of equation (4.91) has shown that by the help of hypercomplex means equations with variable coefficients which do not occur in the main part of the differential operator may also be treated.

4.7.4 Theorem

Let $f \in L_{p,H}(G)$, $h \in L_{p,H}(G)$, $p>6/5$. Furthermore assume the conditions

$$\text{(i)} \quad \frac{\varrho}{\eta C_1}\|f\|_{L_p} + 4\gamma K(\|h\|_{L_p}/(\eta C_1(4\varkappa - m))) \le (4KC_1)^{-2},$$

$$\text{(ii)} \quad \|h\|_{L_p} < (1 - 2^{(1/p)-1})\,\eta\,(4\varkappa - m)^2/(32K_1^3 C_1 m),$$

$$\text{(iii)} \quad m < 4\varkappa.$$

Then the boundary value problem (4.89)-(4.90)-(4.91) has a solution $\{u,w,p\} \in \overset{\circ}{W}{}^1_{2,H}(G) X \overset{\circ}{W}{}^1_{2,R}(G) X L_{2,R}(G)$, where u and w are uniquely defined and p uniquely up to an additive real constant. The iteration method (4.92)-(4.93)-(4.94) converges in $\overset{\circ}{W}{}^1_{2,H}(G) X \overset{\circ}{W}{}^1_{2,R}(G) X L_{2,R}(G)$ to the solution of (4.89)-(4.90)-(4.91).

Proof

The repetition of the consideration which is made in the proof of Theorem 4.6.8 in order to get condition (4.80) leads to the necessary condition (i) for the boundedness of the sequence $\{u_n\}_{n \in N}$ in $\overset{\circ}{W}{}^1_{2,H}(G)$. With the a-priori estimate (4.96) we conclude from (i)

$$\frac{\varrho}{\eta C_1}\|f\|_{L_p} + 4\gamma K\|h\|_{L_p}/[\eta C_1(4\varkappa - m)] \le (4KC_1)^{-2}.$$

Similarly as in Section 4.6 one now obtains the following estimates:

$$\|u_n\|_{W^1_{2,H}} \le (4KC_1)^{-1} - [(4KC_1)^{-2} - \frac{\varrho}{\eta C_1}\|f\|_{L_p} - \frac{\gamma}{\eta C_1}4K\|h\|_{L_p}/(4\varkappa - m)]^{1/2}.$$

So the sequences $\{u_n\}_{n \in N}$ and $\{w_n\}_{n \in N}$ are bounded. We have

$$\|w_n - w_{n-1}\|_{W^1_{2,H}} \le 16K^2 C_1 m(4\varkappa - m)^{-2}\|h\|_p \|u_n - u_{n-1}\|_{W^1_{2,H}}$$

and

$$\|u_n - u_{n-1}\|_{W^1_{2,H}} \le L_n\|u_{n-1} - u_{n-2}\|_{W^1_{2,H}}$$

with

$$L_n=2KC_1(\|u_{n-1}\|_{W^1_{2,H}}+\|u_{n-2}\|_{W^1_{2,H}})+32K^3C_1\gamma m\|h\|_p/(\eta(4\varkappa-m)^2).$$

It follows, that for $L_n<1-\varepsilon$, $\varepsilon>0$ the additional condition (ii) must be fulfilled. #

4.7.5 Remark

Similarly to Section 4.6, sharper statements of regularity may be obtained.

4.7.6 Remark

At present S.BERNSTEIN (Freiberg) considers boundary value problems over unbounded domains G with a compact, sufficiently smooth boundary $\partial G=\Gamma$. The following formula of BOREL-POMPEIU's type.

$$-F_\Gamma u + T_G Du = \begin{cases} u & \text{in } G, \\ 0 & \text{in } R^3\setminus\bar{G} \end{cases}$$

is also valid.

If $G=R^3$, then we obtain $T_G Du=u$. By the help of three examples we will characterize the different situations in case of unbounded domains.

Example 1:

Let $G=R^3$, $u=\frac{1}{2}\sum_{i=1}^{3} x_i(1+|x|^2)^{-3/4}e_i$.

Then u is a continuous and bounded H-valued function, but

$T_G u=T_G D(1+|x|^2)^{1/4}=(1+|x|^2)^{1/4}$

is an unbounded H-valued function.

Example 2:

Let $G=R^3$, $u = -\sum_{i=1}^{3} x_i(1+|x|^2)^{-3/2}e_i$.

Then $u\in L_{2,H}(G)$. On the other hand it is easy to see that

$$T_G u = T_G D(1+|x|^2)^{-1/2} = (1+|x|^2)^{-1/2}\notin L_{2,H}(G).$$

Example 3:

$$\text{Let } G=R^3,\ v_a=\begin{cases} \prod_{i=1}^{3}\sin^3(\pi x_i/a) & \text{within a cube of the edge length } a, \\ 0 & \text{otherwise.} \end{cases}$$

Setting $u_a=Dv_a$, then $u_a\in \overset{\circ}{W}_{2,H}(G)$. Regarding the action of the T_G-operator on u_a, we obtain

$$T_G u_a = T_G D v_a = v_a,$$

$$\|u_a\|^2_{L_{2,H}} = \|Dv_a\|^2_{L_{2,H}} = (-\Delta v_a, v_a)_{L_{2,H}} = c' a,$$

and

$$\|T_G u_a\|^2_{L_{2,H}} = \|T_G D V_a\|^2_{L_{2,H}} = \|v_a\|^2_{L_{2,H}} = c'' a^3 .$$

Finally, we get

$$\|T_G u_a\|^2_{L_{2,H}} / \|u_a\|^2_{L_{2,H}} = ca^2 \longrightarrow \infty \quad \text{for} \quad a \longrightarrow \infty .$$

Hence the operator T_G, as a mapping from the subspace $\overset{\circ}{W}{}^1_{2,H}(G) \subset L_{2,H}(G)$ into $L_{2,H}(G)$, is not continuous in the norm of $L_{2,H}(G)$.

To preserve great parts of our theory it is necessary to use weighted spaces.

Let

$$W^{1,\delta}_{2,H}(G) = \{u \in \mathcal{D}'_H(G) : p_\delta u \in L_{2,H}(G) \text{ and } D_i u \in L_{2,H}(G),\ i=1,2,3\},$$

where $p_\delta = (1+|x|^2)^{\delta/2}$, $\delta \in R$. $W^{k,\delta}_{2,H}(G)$ is analogously defined.

Then we have

$$\|u\|^2_{W^{1,\delta}_{2,H}} = \|p_\delta u\|^2_{L_{2,H}} + \sum_{i=1}^{3} \|D_i u\|^2_{L_{2,H}} .$$

$\overset{\circ}{W}{}^{1,\delta}_{2,H}(G)$ denotes the $W^{1,\delta}_{2,H}(G)$-closure of $C^\infty_{\emptyset,H}(G)$. Besides the space

$$L^\delta_{2,H}(G) = \{u \in \mathcal{D}'_H(G) : p_\delta u \in L_{2,H}(G)\}$$

is needed.

First results are devoted to the outer stationary STOKES problem. Let

$u = \sum_{i=1}^{3} u_i e_i$ be the velocity of the fluid, p be the hydrostatic pressure, and

$f = \sum_{i=1}^{3} f_i e_i$ the external forces.

The STOKES problem in a weak formulation has the following form:

$$\sum_{i=1}^{3}(Du_i, Dv_i)_{L_{2,H}} = \frac{\varrho}{\eta}(f,v)_{L_{2,H}} \qquad (4.97)$$

(where $f \in L^{\gamma}_{2,H}(G)$, $\gamma > 1$, ϱ the density and η the viscosity) for all H-valued functions $v \in V = \{w \in \overset{\circ}{W}{}^{1,\delta}_{2,H}(G),\ \mathrm{div}\ w = 0,\ \delta < -1\}$.

It can be proved that there is a well-defined connection to the following problem:

$$u + \frac{1}{\eta} T_G Q p = \frac{\varrho}{\eta} T_G Q T_G f,$$
$$\frac{1}{\eta} \operatorname{Re} Qp = \frac{\varrho}{\eta} \operatorname{Re} QT_G f. \qquad (4.98)$$

Theorem 1':

Let $u \in \overset{\circ}{W}{}^{1,\delta}_{2,H}(G) \cap \ker \mathrm{div}$, $\delta < -1$ be a solution of problem (4.97). Then there exists a scalar function $p \in L_{2,R}(G)$, such that the pair $\{u,p\}$ fulfils the equations (4.98). Conversely, if $u \in \overset{\circ}{W}{}^{1,\delta}_{2,H}(G) \cap \ker \mathrm{div}$, $\delta < -1$ and $p \in L_{2,R}(G)$ satisfy equations (4.98), then Im u realizes the weak formulation (4.97).

#

Theorem 2':

The solution of problem (4.98) is uniquely defined and allows the following a-priori estimate:

$$\left(1 + \frac{1}{2(|\delta|-1)}\right)^{-1} \|u\|^2_{W^{k+1,\delta}_{2,H}} + \frac{1}{\eta^2}\|Qp\|^2_{W^k_{2,H}} \le \frac{\varrho^2}{\eta^2}\|T_G f\|^2 . \qquad \#$$

Theorem 3':

Let $f \in W^{k,\gamma}_{2,H}(G)$, $\gamma > 1$, $k \geq 1$. Then problem (4.98) is equivalent to the STOKES problem in the strong sense

$$-\Delta u + \frac{1}{\eta}\operatorname{grad} p = \frac{\varrho}{\eta} f \quad \text{in } G,$$
$$\mathrm{div}\ u = 0 \quad \text{in } G,$$
$$\mathrm{tr}\ u = 0 \quad \text{on } \Gamma.$$

#

4.8 Approximation of STOKES Equations by Boundary Value Problems of Linear Elasticity

It is well-known that the solutions of STOKES' problem can be approximated arbitrarily exactly by a sequence of suitably chosen boundary value problems of linear elasticity. By the help of hypercomplex methods, this connection may be demonstrated very effectively. Besides, we have succeeded in expressing the quality of approximation by known data, which is an essential assumption for eventual numerical application. In this way we obtain explicit representations for the velocity u and the pressure p given in Theorem 4.5.7. This result means the complete separation of the computation of velocity and hydrostatic pressure. Thence it is possible to execute the iteration procedure for the solution of NAVIER-STOKES equations treated in Theorem 4.6.8 without calculating the pressure up to the break off.

4.8.1 Theorem (A-priori Estimate)

The solution of the boundary value problem (4.28)-(4.29) with $g=0$ will be denoted by u_m. Let $f \in L_{2,H}(G)$. Then we have

$$\|u_m\|_{W^1_{2,H}} \leq (1+1/\lambda_1)^{1/2}\|T_G\|_{[L_{2,H},L_{2,H}]}\|f\|_{L_{2,H}} .$$

Proof

Obviously, the function $u_m=T_G Q_m M T_G f$ solves problem (4.28)-(4.29). From that immediately follows

$$\|u_m\|_{W^1_{2,H}} \leq (1+1/\lambda_1)^{1/2}\|Q_m M T_G f\|_{L_{2,H}} .$$

Recalling the notation $\|w\|_m=[w,w]$ (cf.4.30) it is easy to prove $\|w\|_{L_{2,H}} \leq \|w\|_m$. Of course, for the orthoprojection Q_m onto the subspace $D(\overset{\circ}{W}{}^1_{2,H}(G))$ we have $\|Q_m\|_m=1$. Therefore we get

$$\|Q_m M T_G f\|_{L_{2,H}} \leq \|M T_G f\|_m = \|T_G f\|_{L_{2,H}} \leq \|T_G\|_{[L_{2,H},L_{2,H}]}\|f\|_{L_{2,H}},$$

and the assertion follows. #

4.8.2 Theorem

The solutions u_m of the boundary value problem (4.28)-(4.29) of linear elasticity converge for $m \longrightarrow 2$, $m>2$ in the $W^1_{2,H}(G)$-norm to $\frac{\eta}{\varrho}u$, where u solves STOKES' problem (4.48)-(4.49). Besides, we have the estimate

$$\| u_m - \frac{\eta}{\varrho} u \|_{W^1_{2,H}} \leq \frac{1}{\varrho} (1+1/\lambda_1)^{1/2} \frac{m-2}{2m-2} \| p \|_{L_{2,R}} .$$

Proof

There holds

$$D[u_m - \frac{\eta}{\varrho} u] = Q_m M T_G f - D(\frac{\eta}{\varrho} u) = Q_m M Q T_G f - D(\frac{\eta}{\varrho} u) =$$

$$= \frac{\eta}{\varrho} Q_m M D u - D(\frac{\eta}{\varrho} u) + \frac{1}{\varrho} Q_m M Q p .$$

Note that $Q_m MDu = Du$. Indeed, there holds

$$MDu = \frac{m-2}{2m-2} \operatorname{Re} Du + \operatorname{Im} Du = \operatorname{Im} Du \in \operatorname{im} Q_m$$

for all $m \geq 0$, $m \neq 1$. Thus we obtain

$$Du_m - D(\frac{\eta}{\varrho} u) = \frac{1}{\varrho} Q_m M Q p = \frac{1}{\varrho} Q_m M p ,$$

and so

$$u_m - \frac{\eta}{\varrho} u = T_G D u_m - T_G D(\frac{\eta}{\varrho} u) = \frac{1}{\varrho} T_G Q_m M p .$$

Theorem 4.1.14 yields

$$\| u_m - \frac{\eta}{\varrho} u \|_{W^1_{2,H}} \leq \frac{1}{\varrho} (1+1/\lambda_1)^{1/2} \frac{m-2}{2m-2} \| p \|_{L_{2,R}} ,$$

and our proof is finished. #

4.8.3 Remark

Using representation (4.60) the norm of the pressure $\| p \|_{L_{2,R}}$ can be estimated by the $L_{2,H}(G)$-norm of the right-hand side f.

4.8.4 Remark

The assumption of Theorem 4.8.2 is also important for the derivation of numerical methods. In the proof of Proposition 4.3.5 a possibility was explained to construct a solution of the boundary value problem $\{DM^{-1}Du=0,\ \operatorname{tr} u=g\}$ on the basis of the solution of DIRICHLET's problem $\{-\Delta u=0,\ \operatorname{tr} u=g\}$. Besides, we find that $DM^{-1}Du$ tends to $-\Delta u$ if $m \longrightarrow 0$. Theorem 4.8.2 states that a solution of STOKES equations may be approximated by solutions of boundary value problems of linear elasticity (with exact error bounds). The iteration principle for solving NAVIER-STOKES equations based on the fixed-point theorem allows that STOKES' problems may be solved at each step even with a certain error (stability) if the domain of convergence is not left (this domain is explicitly known). If taken in the iteration procedure as solutions of STOKES problems, the corresponding solutions of the equations of linear elasticity nevertheless convergence or at least weak convergence can be proved and in both cases an error estimate can be obtained. Therefore it could be sufficient to work out an effective numerical algorithm to solve the first boundary value problem of linear elasticity (or the LAPLACE equation) to master the classical boundary value problems of mathematical physics from the numerical point of view. We shall deal with additional requirements of realizing a numerical method in Chapter 5.

4.8.5 Remark

The considerations leading to the proof of Theorem 8.4.2 pointed out that it is really possible to compute the velocity $u=\frac{\varrho}{\eta}\lim_{m\to 2} u_m$ and pressure p (with $Qp=-\eta Du+\varrho T_G Q T_G f$) in the solution of STOKES' problem by separating calculations. The question of using the limit

$$\lim_{m\to 2} u_m = \lim_{m\to 2} T_G Q_m M T_G f$$

to obtain the results of Theorem 4.5.7 arises. Indeed, we have

$$u=\frac{\varrho}{\eta}\lim_{m\to 2} u_m =$$

$$=\frac{\varrho}{\eta} T_G \mathrm{Im}\, T_G f - \frac{\varrho}{\eta} T_G \mathrm{Im}\, F_\Gamma (\mathrm{tr}\, T_G \mathrm{Im}\, F_\Gamma)^{-1} \mathrm{tr}\, T_G \mathrm{Im}\, T_G f \quad . \qquad (4.99)$$

It is clear that $\mathrm{tr}\ u=0$ and $\mathrm{div}\ \hat{u}=-\mathrm{Re}\ Du=0$.
Now we consider the representation $Qp=-\eta Du+\varrho T_G Q T_G f$.
Inserting (4.99) in the latter formula we get a scalar term

$$p=\varrho\, \mathrm{Re}\, T_G f - \varrho\, \mathrm{Re}\, F_\Gamma (\mathrm{tr}\, T_G \mathrm{Im}\, F_\Gamma)^{-1} \mathrm{tr}\, T_G \mathrm{Im} T_G f \quad . \qquad (4.100)$$

These constructions may also be applied in the case of NAVIER-STOKES equations. However, a complete separation of u and p is not possible. Note that the representation of u (a non-linear equation) can be investigated independently of p.

5. H-REGULAR BOUNDARY COLLOCATION METHODS

5.1. Complete Systems of H-regular Functions

To get statements about H-completeness of certain systems of H-valued functions it is necessary to use some essential theorems of classical functional analysis in quaternionic calculus. The technique of the proofs will be characterized by HAHN-BANACH's theorem. Other proofs are left to the reader.

5.1.1 Proposition (HAHN-BANACH's Theorem)

Let X be a normed right-vector space over H, $X_\emptyset \subset X$ be a closed subspace and f be a bounded right-linear H-valued functional. Then there exists a right-linear extension F of f bounded on X.

Proof

Let $X_{\emptyset,R}$ be a subset considered as a vector space over the field of real numbers. Denote by $f_\emptyset$ an R-valued bounded linear functional which is defined on $X_\emptyset$. This may be extended onto X_R and will be denoted by $F_\emptyset$. We define an H-valued functional F by

$$F(x) = F_\emptyset(x)e_\emptyset - F_\emptyset(xe_1)e_1 - F_\emptyset(xe_2)e_2 - F_\emptyset(xe_3)e_3 . \qquad (5.1)$$

This is the announced extension as will now be proved.

On account of $|F(x)| \leq 4\|F_\emptyset\|\|x\|$ the functional F is bounded. Obviously F appears additive. A single consideration of the terms $F_\emptyset(x)$, $F_\emptyset(xe_1)$, $F_\emptyset(xe_2)$, $F_\emptyset(xe_3)$ leads to the verification of the homogeneity of F. Thus F is a bounded linear functional. Similarly to the proof in [KA] for complex-valued functionals, we can show that $\|F\| = \|f\|$. We only have to replace $\text{sgn } f(z_\emptyset)$ by $\overline{F(x_\emptyset)}|F(x_\emptyset)|^{-1}$ and $z_1 = \text{sgn } \overline{f(z_\emptyset)}z_\emptyset$ by $x_1 = x_\emptyset\overline{F(x_\emptyset)}|F(x_\emptyset)|^{-1}$. To prove that every bounded linear H-valued functional can be written in the form (5.1), we start from the representation

$$f(x) = f_\emptyset(x)e_\emptyset + f_1(x)e_1 + f_2(x)e_2 + f_3(x)e_3 .$$

Using the relations $f(xe_i) = f(x)e_i$, $i=1,2,3$, we find

$$f_\emptyset(x) = f_1(xe_1) = f_2(xe_2) = f_3(xe_3),$$

$$f_1(x) = -f_\emptyset(xe_1) = f_3(xe_2) = -f_2(xe_3),$$

$$f_2(x) = -f_3(xe_1) = -f_\emptyset(xe_2) = f_1(xe_3),$$

$f_3(x) = f_2(xe_1) = -f_1(xe_2) = -f_\emptyset(xe_3)$.

Hence follows $f_i(x) = -f_\emptyset(xe_i)$, $i=1,2,3$. The above constructed functional F is an extension of f. #

5.1.2 Remark

The proof is included in [Gue1]. A corresponding theorem for A-valued locally convex spaces where A denotes a CLIFFORD algebra is deduced in [BDS]. G.A.SUCHOMLINOV also proved a HAHN-BANACH-type theorem in quaternionic spaces [Such].

5.1.3 Definition

Let X be a normed right-vector space over H. A system of points $\{x^{(i)}\}_{i \in N} \subset X$ is called H-complete if and only if every element $x \in X$ may be approximated arbitrarily closely by finite right-linear-combinations of the elements $\{x^{(i)}\}_{i \in N}$. $\{x^{(i)}\}_{i \in N}$ is called closed in X if for every bounded right-linear functional F over X with values in H it follows from $F(x^{(i)}) = \emptyset$, $i \in N$, $F = \emptyset$.

5.1.4 Corollary

Let X be a normed right-vector space over H. The system $\{x^{(i)}\}_{i \in N} \subset X$ is closed if and only if it is H-complete in X.

Proof

The proof is left to the reader. #

5.1.5 Proposition (RIESZ' Theorem)

Every bounded right-linear H-valued functional F over $L_{p,H}(G)$ allows the representation

$$F(u) = \int_G \bar{f}u \, dG \quad , \; u \in L_{p,H}(G)$$

with $f \in L_{q,H}(G)$, $\frac{1}{p} + \frac{1}{q} = 1$.

Proof

For the proof, see [BDS]. #

5.1.6 Theorem

Let G and G_a be star-shaped domains with sufficiently smooth boundary. Suppose $G_a \supset \bar{G}$. Set $\Gamma_a = \partial G_a$. Let $\{x^{(i)}\}_{i \in N}$ be a dense subset of Γ_a, $\varphi_i(x) = \sum_{k=1}^{3} (x_k - x_k^{(i)}) |x - x_i^{(i)}|^{-3} e_k$.

Then the system $\{\varphi_i\}_{i\in N} \subset L_{p,H}(G) \cap A_H(G)$ is H-complete in $L_{p,H}(G) \cap A_H(G)$.

Proof

Considering Corollary 5.1.4 it is sufficient to verify that $\{\varphi_i\}_{i\in N}$ is closed. Let $F \in (L_{p,H}(G) \cap A_H(G))'$ be such that $F(\varphi_i) = 0$, $\forall i \in N$. Proposition 5.1.1 yields that F can be extended onto the whole space $L_{p,H}(G)$, maintaining the norm. Denote this extension by H. Using Proposition 5.1.5 we have

$$H(\varphi) = \int_G \bar{h}\varphi\, dG\ ,\ \varphi \in L_{p,H}(G)\ ,\ \bar{h} \in L_{q,H}(G)\ ,\ \frac{1}{p} + \frac{1}{q} = 1,$$

$$H(\varphi) = F(\varphi)\quad ,\quad \varphi \in L_{p,H}(G) \cap A_H(G).$$

Next, let G' be a star-shaped domain with sufficiently smooth boundary such that $G \subset G' \subset G_a$. Furthermore let $\eta \in C^{\infty}_{0,R}(G')$ with $\eta(x) = e_0$ for $x \in G$. We have

$$H(\varphi_i) = \int_G \overline{h(y)} \sum_{k=1}^{3} \frac{y_k - x_k^{(i)}}{|y - x^{(i)}|^3} e_k\, dy .$$

$$\text{Hence}\qquad \int_G \sum_{k=1}^{3} \frac{x_k^{(i)} - y_k}{|y - x^{(i)}|^3} e_k h(y)\, dy = -(T_G h)(x^{(i)}) = 0 .$$

It is clear that it immediately follows $(T_G h)(x) = 0$ for all $x \in co\,\bar{G}$.

Let $\varphi \in L_{p,H}(G') \cap A_H(G') \subset L_{p,H}(G) \cap A_H(G)$. Then

$$H(\varphi) = H(\varphi\eta) = (H*\delta)(\varphi\eta) = \frac{1}{4\pi}\Big[H*\sum_{k=1}^{3} \frac{y_k}{|y|^3} e_k\Big] D(\varphi\eta) =$$

$$= \int_G \Big[\int_G \bar{h}(y) \frac{1}{4\pi} \sum_{k=1}^{3} \frac{x_k^{(i)} - y_k}{|x^{(i)} - y|^3} e_k\, dy\Big] D(\varphi\eta)(x)\, dx +$$

$$+ \int_{co\,G} \Big[\int_G \bar{h}(y) \frac{1}{4\pi} \sum_{k=1}^{3} \frac{x_k^{(i)} - y_k}{|x^{(i)} - y|^3} e_k\, dy\Big] D(\varphi\eta)(x)\, dx .$$

Indeed, the first integral vanishes because $\varphi \in A_H(G)$, and the second integral becomes zero since $(T_G h)(x) = 0$ in $co\,\bar{G}$.

Let $\{G_n\}_{n\in N}$ be a sequence of domains with the following properties, namely $\bar{G} \subset G_{n+1}$, $\overline{G_{n+1}} \subset G_n$, $\bar{G}_n \subset G_a$, $n \in N$, and $mes(G_n \setminus G) \to 0$, $mes(\partial G_n) \to mes\,\partial G$ for $n \to \infty$. Renewing the previous considerations then it follows

$$H(\varphi) = 0\quad ,\ \varphi \in \bigcup_{i=1}^{\infty} L_{p,H}(G_i) \cap A_H(G_i).$$

Because of the star-shapedness of the domains G and G_a we finally have

$H(\varphi) = 0 \quad , \varphi \in L_{p,H}(G) \cap A_H(G)$

and so $H=0$. #

5.1.7 Remark

The supposition of the star-shapedness of the domains G and G_a may by omitted if it is ensured that the closure in $L_{p,H}(G)$ of $\bigcup_{i=1}^{\infty} L_{p,H}(G_i) \cap A_H(G_i)$ coincides with $L_{p,H}(G) \cap A_H(G)$.

5.1.8 Theorem

Let $\{x^{(i)}\}_{i \in \mathbb{N}} \subset \Gamma$ be adense subset on Γ_a, $\varphi_i(x) = \sum_{k=1}^{3} \frac{x_k^{(i)} - x_k}{|x^{(i)} - x|^3} e_k$.

The system

$\{tr\ \varphi_i\}_{i \in \mathbb{N}}$ is H-complete in $L_{2,H}(\Gamma)$-clos[im $P_\Gamma \cap C_H^{0,\beta}(\Gamma)$]

with the scalar product

$(u,v)_\Gamma = \int_\Gamma \bar{u}v\, d\Gamma \quad , u,v \in L_{2,H}(\Gamma)$.

Proof

Define the function $U = \overline{n\ u}$. First of all we prove

(i) $(\bar{U}, \varphi_i) = 0$, $i \in \mathbb{N}$, if and only if

$u \in L_{2,H}(\Gamma)$ - clos[im $P_\Gamma \cap C_H^{0,\beta}(\Gamma)$] .

Let $(\bar{U}, \varphi_i) = 0$, $i \in \mathbb{N}$, whence $\int_\Gamma \overline{n\ u}\ \varphi_i d\Gamma = 0$, $i \in \mathbb{N}$,

and so

$$\int_\Gamma \varphi_i nu\, d\Gamma = \int_\Gamma \sum_{k=1}^{3} \frac{x_k^{(i)} - y_k}{|x^{(i)} - y|^3} e_k nu\, d\Gamma = (F_\Gamma u)(x^{(i)}) = 0 \quad \forall i \in \mathbb{N}.$$

We get $(F_\Gamma u)(x) = 0$ in $co\ \bar{G}$. Using PLEMELJ-SOCHOTZKIJ's formula we obtain

$$\lim_{\substack{x \to x_0 \in \Gamma \\ x \in co\ \bar{G}}} (F_\Gamma u)(x) = \frac{1}{2} [(S_\Gamma u)(x_0) - u(x_0)], \tag{5.2}$$

which implies $u \in L_{2,H}(\Gamma)$ - clos[im $P_\Gamma \cap C_H^{0,\beta}(\Gamma)$] .

Conversely , let $u \in L_{2,H}(\Gamma)$ - clos[im $P_\Gamma \cap C_H^{0,\beta}(\Gamma)$] then holds $S_\Gamma u = u$, and with (5.2) it follows that $\lim_{\substack{x \to x_0 \\ x \in co\ \bar{G}}} (F_\Gamma u)(x) = 0$ almost everywhere on Γ .

Because of $\lim_{|x| \to \infty} (F_\Gamma u)(x) = 0$ and $F_\Gamma u \in A_H(co\ \bar{G})$ it immediately follows that $(F_\Gamma u)(x) = 0$ in $co\ \bar{G}$. In particular, we have $(F_\Gamma u)(x^{(i)}) = 0$, $i \in \mathbb{N}$, and so $(\bar{U}, \varphi_i) = 0$, $i \in \mathbb{N}$. By this (i) is shown. It remains to check that $(\bar{U}, \varphi) = \emptyset$, $\forall \varphi$, $u \in L_{2,H}(\Gamma)$ - clos[im $P_\Gamma \cap C_H^{\emptyset,\beta}(\Gamma)$]. For this purpose let $(G_i)_{i \in N}$ be a monotonically decreasing sequence of sets furnished with the properties given in the proof of Theorem 5.1.6. Now let $\varphi \in L_{2,H}(G_i) \cap A_H(G_i)$, then

$$\varphi(x) = \lim_{k \to \infty} \sum_{i=1}^{k} \varphi_i(x) c_{ik} \quad \text{in } L_{2,H}(G_i)$$

since the system $\{\varphi_i\}_{i \in N}$ is H-complete in $L_{2,H}(G_i) \cap A_H(G_i)$ (Theorem 5.1.6) , which implies

$$\varphi(x) = \lim_{k \to \infty} \sum_{i=1}^{k} \varphi_i(x) c_{ik} \quad \text{in } L_{2,H}(\Gamma),$$

and therefore

$$(\bar{U}, \varphi) = \lim_{k \to \infty} \sum_{i=1}^{k} (\bar{U}, \varphi_i) c_{ik} = \emptyset .$$

For every $L_{2,H}(\Gamma)$-clos[im $P_\Gamma \cap C_H^{\emptyset,\beta}(\Gamma)$] there exists a sequence $\{\psi_n\}_{n \in N} \subset$ im $P_\Gamma \cap C_H^{\emptyset,\beta}(\Gamma)$ with $\psi_n \to \varphi$ in $L_{2,H}(\Gamma)$ for n tends to infinity. It is clear that $F_\Gamma \psi_n \in C_H(\bar{G}) \cap A_H(G)$. There exists a sequence of H-valued functions $a_{nj} \in C_H(\bar{G}_j) \cap A_H(G)$ with $a_{nj} \to F_\Gamma \psi_n$ in $C_H(\bar{G})$ as $j \longrightarrow \infty$ and tr a_{nj} converges to $\mathrm{tr} F_\Gamma \psi_n = \psi_n$ in $L_{2,H}(\Gamma)$.
Thence

$$(\bar{U}, \varphi) = \lim_{n \to \infty} (\bar{U}, \psi_n) = \lim_{n \to \infty} \lim_{j \to \infty} (\bar{U}, a_{nj}) = \emptyset .$$

Now we conclude $u = \emptyset$, and the proof is finished. #

5.1.9 Theorem

Let G, G_i be bounded star-shaped domains with the boundaries Γ, Γ_i , respectively. Let $\{y^{(i)}\}_{i \in N}$ be a dense subset of Γ_i and set

$$\psi_i(x) = \sum_{k=1}^{3} (x_k - y_k^{(i)}) | x - y^{(i)} |^{-3} e_k .$$

Then the system $\{\mathrm{tr}\, \psi_i\}_{i \in N}$ is H-complete in $L_{2,H}(\Gamma)$ - clos[im $Q_\Gamma \cap C_H^{\emptyset,\beta}(\Gamma)$].

Proof

This may be proved in the same way as Theorem 5.1.8. #

5.1.10 Theorem

Let G, G_i, G_a, Γ_i, Γ_a, ψ_i, φ_i be defined as in Theorem 5.1.8 and Theorem 5.1.9. Then the system $\{\mathrm{tr}\, \varphi_i\}_{i \in N} \cup \{\mathrm{tr}\, \psi_i\}_{i \in N}$ is H-complete in $L_{2,H}(\Gamma)$.

Proof

Any H-valued function $u \in L_{2,H}(\Gamma)$ can be written in the form

$u = \frac{1}{2}(I+S_\Gamma)u + \frac{1}{2}(I-S_\Gamma)u$,

where $(I+S_\Gamma)u \in L_{2,H}(\Gamma) - \text{clos}[\text{im } P_\Gamma \cap C_H^{\emptyset,\beta}(\Gamma)]$

and $(I-S_\Gamma)u \in L_{2,H}(\Gamma) - \text{clos}[\text{im } Q_\Gamma \cap C_H^{\emptyset,\beta}(\Gamma)]$.

Using Theorem 5.1.8 and Theorem 5.1.9 we obtain the proof. #

5.1.11 Remark

Theorem 5.1.11 is also true in the case of piecewise-smooth LIAPUNOV surface.

5.1.12 Proposition

Let $G \subset R^3$ be a bounded domain with sufficiently smooth boundary, which is star-shaped with respect to zero. Let $G_u \subset R^3$ be a bounded domain with $G \subset G_u$. If a function system $\{h_i\}_{i \in N} \subset \subset W^k_{2,H}(G) \cap L_{2,H}(G_u) \cap A_H(G_u)$ is H-complete in $L_{2,H}(G_u) \cap A_H(G_u)$, then $\{h_i\}_{i \in N}$ is H-complete in $W^k_{2,H}(G) \cap A_H(G)$.

Proof

Let $G_t = \{x \in R^3\colon tx \in G ,\ t \in (t_\emptyset,1)\}$, $g_t(x) = f(tx)$, where $t_\emptyset = \inf \{t\colon G_t \subset G_u\}$ and $f \in W^k_{2,H}(G) \cap A_H(G)$.

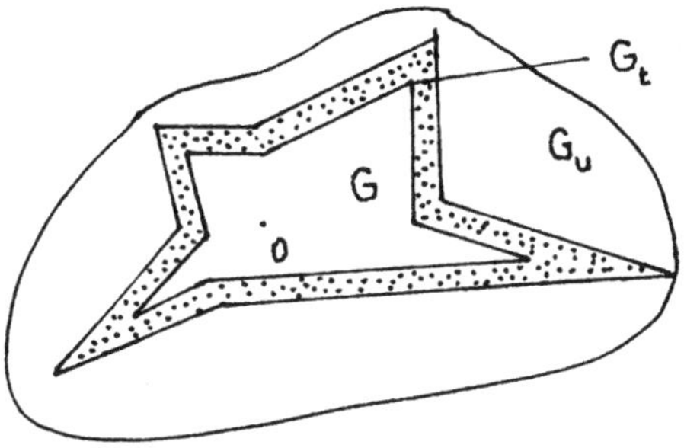

Figure 6:

For the proof it is advantageous to verify the identity

$$W^k_{2,H}(G) - \text{clos}[\bigcup_{t \in (t_\emptyset,1)} W^k_{2,H}(G_t) \cap A_H(G_t)] = W^k_{2,H}(G) \cap A_H(G), \qquad (5.3)$$

First, let $f \in W^k_{2,H}(G) \cap A_H(G)$, then it follows $g_t \in W^k_{2,H}(G_t) \cap A_H(G_t)$. Besides it is easy to see

$$\lim_{t \to 1} \|g_t\|_{W^k_{2,H}(G)} = \|f\|_{W^k_{2,H}(G)}.$$

Therefore it is possible to choose a subsequence $(g_{t'}) \subset (g_t)$ such that $g_{t'}$ converges weakly to a certain H-valued function g if t' tends to 1, and so $\lim_{t'\to 1} g_{t'}(x)=g(x)$ for any $x \in G$.

Obviously

$$\lim_{t\to 1} g_t(x) = \lim_{t\to 1} f(tx) = f(x),$$

which means that $g(x) = f(x)$ for all $x \in G$. Hence

$$W^k_{2,H}(G) - \lim_{t'\to 1} g_{t'} = f, \qquad (5.4)$$

and the identity (5.3) is shown.

By the help of the techniques which are used for the proof of Theorem 5.1.6 we can prove that the system $\{h_i\}_{i\in\mathbf{N}}$ is also H-complete in $L_{2,H}(G_t) \cap A_H(G_t)$ for $t \in (t_\emptyset, 1)$. By the aid of HARNACK s theorem we have

$$tr_G g_t \in C^k_H(\bar{G}) - \underset{H}{clos}[span\{h_i\}_{i\in\mathbf{N}}],$$

whence

$$tr_G g_t \in W^k_{2,H}(\bar{G}) - \underset{H}{clos}[span\{h_i\}_{i\in\mathbf{N}}].$$

From (5.4) now it follows $f \in W^k_{2,H}(G) - \underset{H}{clos}[span\{h_i\}_{i\in\mathbf{N}}]$. #

5.1.13 Theorem

Let G, G_a, Γ, Γ_a, $\{x^{(i)}\}_{i\in\mathbf{N}}$, $\{\varphi_i\}_{i\in\mathbf{N}}$ be defined as in Theorem 5.1.6. Then the system $\{\varphi_i\}_{i\in\mathbf{N}} \subset W^k_{2,H}(G) \cap A_H(G)$ is H-complete in $W^k_{2,H}(G) \cap A_H(G)$ for $k \in \mathbf{N}$.

Proof

Choose G_u in such a way that $G \subset G_u \subset G_a$. Applying Theorem 5.1.6. and Proposition 5.1.12 we find the assertion. #

5.1.14 Theorem

Let G, G_a, Γ, Γ_a, $\{x^{(i)}\}_{i\in\mathbf{N}}$, $\{\varphi_i\}_{i\in\mathbf{N}}$ be defined as in Theorem 5.1.6. Then the function system $\{tr\, \varphi_i\}_{i\in\mathbf{N}}$ is complete in $W^{k-1/2}_{2,H}(\Gamma) \cap \mathrm{im}\, P_\Gamma$ for each $k \in \mathbf{N}$.

Proof

Making use of the trace theorems from [LM] and Theorem 5.1.13 we obtain our statement. #

5.1.15 Remark

The significance of Theorem 5.1.13 and Theorem 5.1.14 consists in the possibility of applying the same basis of approximation or interpolation independently of the smoothness of the given function.

5.1.16 Theorem

Let G, G_i be bounded star-shaped domains with the boundaries Γ, Γ_i, respectively. Let $\{y^{(i)}\}_{i\in N}$ be a dense subset of Γ_i and set

$$\psi_i(x) = \sum_{k=1}^{3} (x_k - y_k^{(i)})\, |x - y^{(i)}|^{-3} e_k .$$

Then the system $tr\{\psi_i\}_{i\in N}$ is H-complete in $W_{2,H}^{k-1/2}(\Gamma) \cap \operatorname{im} Q_\Gamma$ for every $k \in N$.

Proof

Applying trace theorems and the proof of Theorem 5.1.13 we obtain our assertion. #

5.2 Numerical Properties of H-complete Systems of H-regular Functions

A successful application of the above formulated results considering the H-completeness makes it necessary to carry out further investigations of these approximative systems. For this end we introduce and sketch the concept of the H-minimality of a given system of H-valued basis functions. This appears as a slight generalization of the minimal property in MICHLIN's sense [Mi], which is an essential characteristic of the numerical realization of interpolation and approximation procedures. Finally, connections between the solutions of the interpolation problem (1.16) - (1.17) (Theorem 1.2.9) and the best approximation will be explained in an appropriate sense.

5.2.1 Definition

Let $\{h_i\}_{i\in N}$ be a function system in a normed right-vector space of H-valued functions X. If for each h_j we have

$$h_j \notin X - \operatorname{clos}[\operatorname{span}_H\{h_i\}_{i\in N, i\neq j}] ,$$

then the system $\{h_i\}_{i\in N}$ is called H-minimal in X.

5.2.2 Remark

The function system $\{\varphi_i\}_{i\in N}$ with $\varphi_i(x) = \sum_{k=1}^{3} \frac{x_k - x_k^{(i)}}{|x - x^{(i)}|^3} e_k$ is not H-minimal, because a finite number of points $x^{(i)}$

may be left out with the remainder still being closed on Γ_a. Therefore the H-completeness of $\{\varphi_i\}_{i\in N}$ remains maintained.

5.2.3 Definition

Let X be a normed right-vector space of H-valued functions, $\{y^{(i)}\}_{i=1}^{n} \subset G$, $\{h_i\}_{i=1}^{n} \subset X$. The system $\{h_i\}_{i=1}^{n}$ is named H-unisolvent with respect to $\{y^{(i)}\}_{i=1}^{n}$ if the algebraic polynomial

$$\sum_{i=1}^{n} h_i(x)c_i \quad \text{for any } c_i \in H \text{ with } \sum_{i=1}^{n} |c_i|^2 > 0$$

on $\{y^{(i)}\}_{i=1}^{n}$ possesses n-1 zeros at most.

5.2.4 Algorithm

The numerical implementation of interpolation and approximation procedures on the basis of a non-H-minimal system leads to algebraic systems of equations which are of badly conditioned. By the help of an orthogonalization method one can transform a given system into an H-minimal system. But this procedure requires the numerical realization of scalar products over 3-dimensional domains which is connected with considerable expense. Therefore we will write down another algorithm to obtain an H-minimal system, where the shape of the domain G does not play such an essential role as for the orthogonalization method.

Let $\{h_i\}_{i\in N} \subset X$, $\{y^{(i)}\}_{i\in N} \subset G$. Consider the following transformation principle:

$$\begin{aligned} g_1 &= h_1 \\ g_2 &= h_2 - g_1 a_{12} \\ &\cdots\cdots \\ g_n &= h_n - g_1 a_{1n} - \dots - g_{n-1} a_{n-1,n} , \end{aligned} \tag{5.5}$$

where $a_{ik} \in H$ for $i<k$, $k \in N$, may be computed by n systems of algebraic equations, namely

$$g_1(y^{(1)})a_{12} = h_2(y^{(1)}) \quad \text{for } g_2$$

(5.6)

$$\cdots\cdots$$

$$\begin{bmatrix} g_1(y^{(1)}) & & \mathbb{0} & \\ g_1(y^{(2)}) & g_2(y^{(2)}) & & \\ \cdots\cdots & & & \\ g_1(y^{(n)}) & g_2(y^{(n)}) & \dots & g_n(y^{(n)}) \end{bmatrix} \begin{bmatrix} a_{1,n+1} \\ a_{2,n+1} \\ \\ a_{n,n+1} \end{bmatrix} = \begin{bmatrix} h_{n+1}(y^{(1)}) \\ h_{n+1}(y^{(2)}) \\ \\ h_{n+1}(y^{(n)}) \end{bmatrix}$$

for g_{n+1}.

By the aid of the relations (5.5) - (5.6) the transformed system $\{g_i\}_{i\in\mathbb{N}}$ is H-minimal. More exactly we have

__5.2.5 Theorem__

Let $X=L_{2,H}(G)\cap A_H(G)$. Further let $\{h_i\}_{i\in\mathbb{N}}\subset X$ be H-complete in X, $\{y^{(i)}\}_{i\in\mathbb{N}}\subset G$ and $\{h_i\}_{i\in\mathbb{N}}$ be H-unisolvent on $\{y^{(i)}\}_{i=1}^n, n\in\mathbb{N}$. Then the system $\{g_i\}_{i\in\mathbb{N}}\subset X$ generated by (5.5) - (5.6) is H-complete and H-minimal in X.

__Proof__

To carry out the algorithm described in 5.2.4 it is necessary that $g_i(y^{(i)})\neq 0$ for all $i\leq n$, $n\in\mathbb{N}$. If there exists an index $j\in\mathbb{N}$ such that $g_j(y^{(i)})=0$, then the function g_j, which may be represented in form of a right-linear combination of the functions $h_1,\dots,h_j$ has j zeros on the point-system $\{y^{(i)}\}_{i=1}^j$. This is in contradiction to the supposition of the H-unisolvence. Now let $f\in X$. Assume $(g_i,f)_X = 0$, $i\in\mathbb{N}$, then it follows, due to (5.5), $(h_i,f)_X = 0$, and so $f=0$. Hence the H-completeness is proved.

The system $\{g_i\}_{i\in\mathbb{N}}$ is characterized by the properties $g_i(y^{(j)})=0$ for $j<i$, $i\in\mathbb{N}$, and $g_i(y^{(i)})\neq 0$ for $i\in\mathbb{N}$.

We prove the H-minimality of the system $\{g_i\}_{i\in\mathbb{N}}$, by contradiction. Let $\{g_i\}_{i\in\mathbb{N}}$ be non-H-minimal. There exists an index $i_0\in\mathbb{N}$ with

$$g_{i_0}\in X - \operatorname{clos}[\operatorname{span}_H\{g_i\}_{i\in\mathbb{N}, i\neq i_0}],$$

whence it follows

$$g_{i_0} = \lim_{n\to\infty} \sum_{\substack{i=1\\ i\neq i_0}}^{n} g_i a_{in} \quad \text{in } X .$$

Using HARNACK's theorem we get

$$g_{i_0}(x) = \lim_{n\to\infty} \sum_{\substack{i=1\\ i\neq i_0}}^{n} g_i(x) a_{in} \quad \text{for } x\in G .$$

Our construction now yields $g_{i_0}(y^{(1)}) = 0$ and

$$g_{i_o}(y^{(1)}) = \lim_{n\to\infty} \sum_{\substack{i=1\\ i\neq i_o}}^{n} g_i(y^{(1)})a_{in} = \lim_{n\to\infty} g_1(y^{(1)})a_{1n} = 0.$$

Then we have

$$\lim_{n\to\infty} a_{1n} = 0 \quad . \tag{5.7}$$

In a similar way we can verify

$$\lim_{k\to\infty} a_{jk} = 0 \quad , \; j < i_0 \quad . \tag{5.8}$$

This implies

$$g_{i_o}(y^{(i_o)}) = \lim_{n\to\infty} \sum_{\substack{i=1\\ i\neq i_o}}^{n} g_i(y^{(i_o)})a_{in} = \lim_{n\to\infty} \sum_{i=1}^{i_o-1} g_i(y^{(i_o)})a_{in},$$

and because of (5.7) and (5.8) $g_{i_o}(y^{(i_o)})=0$. This result contradicts the construction of g_{i_o}, whence the assertion follows. #

<u>5.2.6 Remark</u>

The accomplishment of the transformation (5.5) - (5.6) only requires the resolution of a triangular system of algebraic equations. The structure of the system matrix solely requires a simple computation of function values. Numerical integration with integrals over 3-dimensional domains is not necessary. Using the minimal system $\{g_i\}_{i=1}^{n}$ as basis of interpolation, the system of equations which is to be solved will again be triangular by reason of the properties of the zeros of the functions g_i.

<u>5.2.7</u> In the following we will study an interesting connection between the best approximation of f in $X_n = \operatorname{span}_H\{g_1,\dots,g_n\}$ and the solution of the generalized collocation problem

$$\sum_{i=1}^{n} g_i(y^{(j)})c_i = f(y^{(j)}), \; j=1,\dots,n \; , \; f\in X \; . \tag{5.9}$$

Introducing the algebraic interpolation function

$$(\Pi_n f)(x) = \sum_{i=1}^{n} g_i(x)c_i \; ,$$

(5.9) may be written as

$$(\Pi_n f)(y^{(j)}) = f(y^{(j)}), \quad j=1,\dots,n. \tag{5.10}$$

5.2.8 Theorem

Let $\{g_i\}_{i\in\mathbb{N}} \subset X=L_{2,H}(G)\cap A_H(G)$ be an H-complete system in X with $g_i(y^{(j)})=0$ for $j<i$, $i\in\mathbb{N}$, and $g_i(y^{(i)})\neq 0$ for $i\in\mathbb{N}$.

Let $P_n f = \sum_{i=1}^{n} g_i a_{in}$ be the best $L_{2,H}$-approximation of $f\in X$ in X_n. Then we have

$\lim_{n\to\infty} a_{in} = c_i \in H$ for all $i\in\mathbb{N}$,

where $(c_1,\dots,c_n)\in H^n_-$ fulfils the collocation conditions (5.10).

Proof

The completeness of the system $\{g_i\}_{i\in\mathbb{N}}\subset X$ yields

$f(x) = \lim_{n\to\infty} \sum_{i=1}^{n} g_i(x)a_{in}$. Setting x in turn $y^{(1)},y^{(2)},\dots,y^{(n)}$, we obtain

$$f(y^{(1)}) = \lim_{n\to\infty} g_1(y^{(1)})a_{1n} = g_1(y^{(1)})c_1,$$

$$f(y^{(2)}) = \lim_{n\to\infty} (g_1(y^{(2)})a_{1n}+g_2(y^{(2)})a_{2n}) = g_1(y^{(2)})c_1+g_2(y^{(2)})c_2,$$

$$\dots\dots\dots\dots\dots\dots\dots\dots$$

$$f(y^{(n)}) = \lim_{n\to\infty} \sum_{k=1}^{n} g_k(y^{(n)})a_{kn} = g_1(y^{(n)})c_1+\dots+g_n(y^{(n)})c_n. \qquad \#$$

5.2.9 Remark

Theorem 5.2.8 remains true if $P_n f$ will be replaced by another sequence of approximations of f in X_n, which converges to f. Theorem 5.2.8 also expressed for H-minimal systems a close connection between the sequence of best approximations and the solutions of certain collocation problems.

5.2.10 Theorem

Let $X=L_{2,H}(G)\cap A_H(G)$, $\{h_i\}_{i\in\mathbb{N}}$ be H-complete orthonormal system. Let $\{y^{(i)}\}_{i\in\mathbb{N}}\subset G$ be a set of collocation points in such a way that Proposition 1.5.11 is fulfilled. Furthermore let $\{h_i\}_{i=1}^n$ be H-unisolvent with respect to $\{y^{(i)}\}_{i=1}^n$, $n\in\mathbb{N}$. Then we obtain the following connection between the best $L_{2,H}$-approximation $P_n f$ and the solutions of the collocation problem (5.10)

$$\Pi_n f = P_n f + \sum_{k=n+1}^{\infty} \Pi_n h_k (h_k,f)_X \quad , \tag{5.11}$$

where $\Pi_n f = \sum_{k=1}^{n} h_k c_{kn}$, $c_{kn} \in H$.

Proof

Considering the right-linearity of X f allows the generalized FOURIER expansion

$$f = \sum_{k=1}^{\infty} h_k(h_k,f)_X \quad \text{for every } f \in X \quad (L_{2,H}\text{-convergence}) .$$

Using HARNACK's theorem we get

$$f(x) = \sum_{k=1}^{\infty} h_k(x)(h_k,f)_X , \text{ for each } f \in X , x \in G .$$

This implies

$$\Pi_n f = \sum_{k=1}^{\infty} \Pi_n h_k(h_k,f)_X ,$$

since $\Pi_n \in L(L_{2,H},L_{2,H})$ and by virtue of the H-unisolvence of $\{h_i\}_{i=1}^n$

$$\Pi_n f = \sum_{k=1}^{n} \Pi_n h_k(h_k,f)_X + \sum_{k=n+1} \Pi_n h_k(h_k,f)_X$$

$$= \sum_{k=1}^{n} h_k(h_k,f)_X + \sum_{k=n+1} \Pi_n h_k(h_k,f)_X .$$

Hence the assertion follows. #

5.2.11 Corollary

The conditions $\Pi_n h_k=0$ for $k>n$, $n \in \mathbb{N}$, are necessary and sufficient so that it holds

$$\Pi_n f = P_n f, \quad n \in \mathbb{N} . \tag{5.12}$$

5.2.12 Remark

Obviously $\Pi_n h_k=0$ for $k>n$, $n \in \mathbb{N}$ means $h_k(y^{(j)})=0$ for $j<k$ and $k \in \mathbb{N}$. Moreover notice that the conditions for the accordance of the generalized interpolation polynomial $\Pi_n f$ and the best $L_{2,H}$-approximation $P_n f$ for every n correspond with the demand for the position of the zeros of the basis functions in the algorithm (5.5) - (5.6).

5.2.13 Proposition

Let G, G_a, Γ , Γ_a, $\{x^{(i)}\}_{i \in \mathbb{N}}$, $\{\varphi_i\}_{i \in \mathbb{N}}$ be defined as in Theorem 5.1.6. Then the function system $\{\varphi_i\}_{i \in \mathbb{N}}$ is linearly independent in G, and the system of restrictions on the boundary $\Gamma = \partial G$ $\{tr\, \varphi_i\}_{i \in \mathbb{N}}$ is linearly independent on Γ.

Proof

It is clear that $\{\varphi_i\}_{i\in\mathbb{N}}$ is right-linear-independent in G.

Suppose $\sum_{i=1}^{n} \operatorname{tr} \varphi_i(x)c_i = 0$, $x\in\Gamma$, $c_i\in H$, $\sum_{i=1}^{n} |c_i|^2 \neq 0$.

Using the maximum modulus theorem we get $\sum_{i=1}^{n} \varphi_i(x)c_i = 0$, $x\in G$, whence it clearly follows $c_i = 0$, $i=1,\dots,n$. #

5.2.14 Proposition

Let $\{x^{(i)}\}_{i\in\mathbb{N}}\subset\Gamma_a$, $\{y^{(i)}\}_{i\in\mathbb{N}}\subset G$ ($\{y^{(i)}\}_{i\in\mathbb{N}}\subset\Gamma$) and $\{\varphi_i\}_{i=1}^{n}$ ($\{\operatorname{tr}\varphi_i\}_{i=1}^{n}$) be H-unisolvent with respect to $\{y^{(i)}\}_{i=1}^{n}$. Then there exists a point $y^{(n+1)}\in G$ ($y^{(n+1)}\in\Gamma$) such that $\{\varphi_i\}_{i=1}^{n+1}$ ($\{\operatorname{tr}\varphi_i\}_{i=1}^{n+1}$) is H-unisolvent with respect to $\{y^{(i)}\}_{i=1}^{n+1}$.

Proof

Making use of the linear independence it follows easily. #

5.2.15 Remark

Proposition 5.2.14 is interesting for the collocation on Γ . Similar statements are obtained by ALEKZIDSE [Ale] for some other functions . It is known that for polynomials of several variables in dependence on the degree n there exist surfaces such that an arbitrary choice of n points on these polynomials never may be unisolvent. Such situations can be excluded by the help of Proposition 5.2.14.

5.2.16 Proposition

Let $G\subset G_a$, $\{x^{(i)}\}_{i=1}^{n}\subset\Gamma_a$, $\{y^{(i)}\}_{i=1}^{n}\subset\Gamma$, $n^{(i)}$ the unit vector of the outer normal on Γ at the point $y^{(i)}$, $x^{(i)} = y^{(i)} + tn^{(i)}$, $i=1,\dots,n$, $t>0$. Then there exists a real number $\delta>0$ with the property that the system $\{\varphi_i\}_{i=1}^{n}$ with $\varphi_i = \sum_{k=1}^{3}\frac{x_k-x_k^{(i)}}{|x-x^{(i)}|^3}e_k$ is H-unisolvent with respect to $\{y^{(i)}\}_{i=1}^{n}$ for an arbitrary $t\in(0,\delta)$.

Proof

We have to investigate the following system of equations:

$$\left[\varphi_i(y^{(j)})\right]_{\substack{j=1,\dots,n\\ i=1,\dots,n}}\left[c_i\right]_{i=1,\dots,n} = 0 ,$$

$(c_1, c_2, \ldots, c_n) \in \mathbb{H}^n$.

It may be transformed into

$$\left[\frac{y^{(j)} - y^{(i)} - t n^{(i)}}{|y^{(j)} - y^{(i)} - t n^{(i)}|^3}\right]_{j,i=1,2\ldots n} [c_i]_{i=1,\ldots,n} = \emptyset.$$

For non-diagonal elements we have

$$\left|\sum_{j=1}^{n} \frac{\overline{y^{(i)} - y^{(j)} - t n^{(j)}}}{|y^{(i)} - y^{(j)} - t n^{(j)}|^3}\right| \leq \sum_{\substack{j=1 \\ j \neq i}}^{n} |y^{(i)} - y^{(j)} - t n^{(j)}|^{-2} \xrightarrow[t \to \emptyset]{} \sum_{\substack{j=1 \\ j \neq i}}^{n} |y^{(i)} - y^{(j)}|^{-2} < \infty.$$

For diagonal elements in the i-th row we have

$$\left|\frac{y^{(i)} - y^{(i)} - t n^{(i)}}{|y^{(i)} - y^{(i)} - t n^{(i)}|^3}\right| = t^{-2} \xrightarrow[t \to \emptyset]{} \infty.$$

Letting $t \longrightarrow \emptyset$ the diagonal elements dominate, and therefore the system of equations has a unique solution. Hence $c_i = \emptyset$, $i=1,\ldots,n$, and the H-unisolvence is proved. #

5.2.17 Remark

The algorithm to construct H-minimal systems described by (5.5)-(5.6) may be modified in virtue of Proposition 5.2.16 that its success is always guaranteed. One has to pay for the diagonal dominance by worse convergence. There are papers in which hints are given for the optimal choice of the distance (cf. [J]).

5.2.18 Definition

Let $\{g_i\}_{i \in \mathbb{N}}$ be a H-complete system in $L_{2,\mathbb{H}}(G) \cap A_{\mathbb{H}}(G)$. Let $\{y^{(i)}\}_{i \in \mathbb{N}} \subset G$ and $X_n = \operatorname{span}_{\mathbb{H}}\{g_1, \ldots, g_n\}$. If for all $f \in X$ and every $n \in \mathbb{N}$ the generalized interpolation function

$$\Pi_n f = \sum_{i=1}^{n} g_i a_{in} \quad , a_{in} \in \mathbb{H}, \tag{5.13}$$

with $(\Pi_n f)(y^{(i)}) = f(y^{(i)})$, $i=1,\ldots,n$, and the best $L_{2,\mathbb{H}}$-approximation of f in X_n coincide, then $\{g_i\}_{i \in \mathbb{N}}$ is called an <u>optimal basis of interpolation</u> with respect to $\{y^{(i)}\}_{i \in \mathbb{N}}$. Further a set $\{y^{(i)}\}_{i \in \mathbb{N}} \subset G$ of collocation points is called <u>total</u> , if for arbitrary $u \in A_{\mathbb{H}}(G)$ the condition $u(y^{(i)}) = \emptyset$ for all $i=1,2,\ldots$ implies $u = \emptyset$.

5.2.19 Theorem

For all $\{y^{(j)}\}_{j\in N} \subset G$ there exists an orthonormal system $\{g_i\}_{i\in N} \subset L_{2,H}(G) \cap A_H(G)$ with $g_i(y^{(j)})=0$ for all $j<i$, $i=2,3,\ldots$, and $g_i(y^{(i)})\neq 0$ for all $i=1,2,3,\ldots$. If $\{y^{(j)}\}_{j\in N}$ is chosen total in G, then the system $\{g_i\}_{i\in N}$ is H-complete in $L_{2,H}(G) \cap A_H(G)$.

Proof

For the proof we refer to [Gue 5] . #

5.2.20 Theorem

Let $\{g_i\}_{i\in N}$ be an H-complete orthonormal system in $L_{2,H}(G) \cap A_H(G)$, let $\{y^{(j)}\}_{j\in N} \subset G$ a total set, and let Π_n be defined by (5.13). Then it holds for all $f \in L_{2,H}(G) \cap A_H(G)$

$$\lim_{n\to\infty} \| \Pi_n f - f \|_{L_{2,H}(G)} = 0 ,$$

and the convergence is monotonic .

Proof

The proof is carried out in [Gue 5] . #

5.2.21 Remark

Several possibilities for the construction of optimal systems and the connection with optimal interpolation in the sense of A. SARD are discussed in detail in [Gue5] and in a somewhat other connection in [Ta] .

For certain distributions of the collocation points and of the singularities local-uniformly convergence of the interpolation procedure can be obtained. We only succeed in proving these statements in the case of H-regular functions in the unit ball $B_1(0)$ of R^3 and the best approximation of their boundary values in $L_{2,H}(S_1)$.

5.2.22 Theorem

Let $\{y^{(j)}\}_{j\in N} \subset \text{co}\, \overline{B_1(0)}$, $\tilde{y}^{(j)} = -\dfrac{y^{(j)}}{|y^{(j)}|^2}$ and

$$(\Pi_n f)(x) = \sum_{i=1}^{n} e(x-y^{(j)})a_i \quad , \text{ with } a_i \in H .$$

Then

$$(f-\Pi_n f, \overline{e(x-y^{(j)})})_{L_{2,H}(S_1)} = 0 \text{ for all } j\leq n \quad \text{if and only if}$$

$$f(\tilde{y}^{(j)}) = (\Pi_n f)(\tilde{y}^{(j)}) \quad \text{for all } j\leq n .$$

Proof

Suppose that $(f-\Pi_n f,\overline{e(x-y^{(j)})})_{L_{2,H}(S_1)}=0$, $j\leq n$. Then it follows that

$$(e(x-y^{(j)}),f-\Pi_n f)_{L_{2,H}(S_1)}=0 \quad , j\leq n,$$

and

$$\left(\frac{1-y^{(j)}\bar{x}}{|1-y^{(j)}\bar{x}|^3}\, e(x),f-\Pi_n f\right)_{L_{2,H}(S_1)}=0 \quad , j\leq n .$$

Since $x\in B_1(0)$, $|x|=1$, we conclude

$$\left(\frac{\bar{y}^{(j)-1}-\bar{x}}{|\bar{y}^{(j)-1}-\bar{x}|^3}\frac{x}{|x|} ,f-L_n f\right)_{L_{2,H}(S_1)}=0 \quad , j\leq n .$$

Thence we achieve because of

$$F_\Gamma(f-\Pi_n f)(\bar{y}^{(j)-1})=0 \quad , j\leq n .$$

Using $F_\Gamma u = u$ for $u\in A_H(G)$ we finally get to

$$f(\bar{y}^{(j)-1}) - (\Pi_n f)(\bar{y}^{(j)-1}) = 0 , j\leq n .$$

Reverse considerations provide the second part of the proof. #

5.2.23 Remark

Theorem 5.2.22 generalizes results of the complex function theory concerning the approximation and collocation with rational functions [Gai] , [Wal] .

5.2.24 Theorem

Under the hypothesis of Theorem 5.2.22 we have

$\Pi_n f \underset{n\to\infty}{\rightrightarrows} f$ in $G'\subset B_1(0)$ with $\bar{G}'\subset B_1$,

which have boundary values in $L_{2,H}(S_1)$ if $\{\bar{y}^{(j)(-1)}\}\subset G'$.

Proof

The statement of Theorem 5.2.22 yields that

$tr\, \Pi_n f \to tr\, f$ in $L_{2,H}(S_1)$.

The required result follows from the representation

$(\Pi_n f - f)(x) = [F_\Gamma\, tr(\Pi_n f - f)](x)$. #

5.2.25 Remark

The statement of Theorem 5.2.22 can be transformed into the case of the ball $B_R(0)$, if we use the singularities $y^{(i)}$, $|y^{(i)}| > R$ and the collocation points $y^{(i)-1}R^2$. Theorem 5.2.24 also permits a formulation in a generalized situation, namely if $G\subset R^3$ is a bounded domain , $B_R(x^*)$ the largest ball contained in G, $\{x^{(i)}\}_{i\in N}\subset B_R(x^*)$, $y^{(i)} = \frac{\overline{(x^{(i)}-x^*)}^{-1}}{R^2}+x^*$

then

$$\Pi_n f - f \underset{n\to\infty}{\rightrightarrows} 0 \quad \text{for all} \quad f \in A_H(G) \text{ in all } B_\varrho(x^*) \text{ with } \varrho < R.$$

5.3 Foundation of a Collocation Method with H-regular Functions for Several Elliptic Boundary Value Problems

Let L be an elliptic differential operator of second order with constant coefficients. The following boundary value problem is to be solved approximately :

$$Lu = 0 \quad \text{in } G,$$
$$Ru = g \quad \text{on } \Gamma. \tag{5.14}$$

Suppose that a continuous mapping M_L with

$$M_L : A_H(G) \times A_H(G) \xrightarrow{\text{onto}} \ker L$$

is known in suitably chosen spaces. On account of the continuity then it follows that the system $\{M_L(h_i,h_j)\}_{i,j\in\mathbf{N}} \subset \ker L$ is H-complete in $\ker L$ if the system $\{h_i\}_{i\in\mathbf{N}} \subset A_H(G)$ is H-complete in $A_H(G)$.

If we succeed in describing functions $M_L\{h_i,h_j\}$ in simple analytic terms, all constructions carried out in Subsection 5.1 and Subsection 5.2 may be repeated in $\ker L$, $R(\ker L)$, respectively. Mappings of the required manner are given by the representations of the solutions of the equations $Lu = 0$ which are obtained in Section 4. Let X be a right-vector space. We look for an approximative solution of the boundary value problem (5.14) in the form

$$u_n(x) = \sum_{i=1}^{n} h_i(x)a^{(i)} ,$$

where $h_i \in X \cap \ker L$ and $a^{(i)} \in H$, $i=1,\dots,n$. The coefficients $a^{(i)}$ shall be defined by the equations

$$Ru_n(z^{(i)}) = g(z^{(i)}), \quad i=1,\dots,n ,$$

where $\{z^{(i)}$, $i=1,\dots,n\}$ denotes the set of the collocation points which are lying on Γ.

It is a matter of decisive importance to find an effective description of the system $\{T_G h_i\}_{i\in\mathbf{N}}$ for special interesting systems $\{h_i\}_{i\in\mathbf{N}}$.

5.3.1 Theorem

Let $G \subset \mathbf{R}^3$ a bounded domain with sufficient smooth boundary, $\{x^{(i)}\}_{i\in\mathbf{N}} \subset \Gamma_a$, $\Gamma_a \subset \operatorname{co} \bar{G}$, $x \in G$. Then

$$T_G\left(\sum_{k=1}^{3} \frac{x_k - x_k^{(i)}}{|x - x^{(i)}|^3} e_k\right) = -\frac{1}{|x - x^{(i)}|} + \phi_G , \quad \phi_G \in A_H(G), \tag{5.15}$$

$$T_G\left(\frac{x_k^{(i)} - x_k}{|x^{(i)} - x|^3} e_k\right) = \frac{1}{2}\frac{1}{|x^{(i)} - x|} e_0 - \frac{1}{2}\sum_{k=1}^{3} \frac{x_k^{(i)} - x_k}{|x^{(i)} - x|^3}(x_k^{(i)} - x_k) e_k + \psi_{G,k} , \psi_{G,k} \in A_H . \tag{5.16}$$

<u>Proof</u>

Letting the differential operator D act on the left on the relations (5.15) and (5.16), the proof follows by a straightforward computation.

<u>5.3.2 Remark</u>

Owing to the right-linearity of the operator T_G one may at once deduce expressions for $T_G(\frac{x_k^{(i)} - x_k}{|x^{(i)} - x|^3} e_j)$ with $k \neq j$.

<u>5.3.3 Theorem</u>

Let $G \subset R^3$ be a bounded star-shaped domain with sufficiently smooth boundary , $G_a \subset R^3$ with $G \Subset G_a$, $\Gamma_a = \partial G_a$, $\{x^{(i)}\} \subset \Gamma_a$ a dense subset on Γ_a. Then the system

$$\left\{\sum_{k=1}^{3} \frac{x_k^{(i)} - x_k}{|x^{(i)} - x|^3} e_k\right\}_{i \in N} \cup \left\{\frac{1}{|x^{(i)} - x|} e_0\right\}$$

is H-complete in $L_{2,H}(G) \cap \ker \Delta(G)$.

<u>Proof</u>

Let $u \in L_{2,H}(G) \cap \ker \Delta(G)$, $\varepsilon > 0$ be arbitrarily chosen and $k > 2$. Then there exist H-valued functions $u_n \in W^k_{2,H}(G) \cap \ker \Delta(G)$ with the property

$u_n \xrightarrow[n \to \infty]{} u$ in $L_{2,H}(G)$ (see Proposition 5.1.12) .

There exists an index $n_0 \in N$ with $\|u_n - u\|_{L_{2,H}} \leq \varepsilon/2$

By virtue of Theorem 4.1.11 , u_n allows the representation

$u_{n_0} = \phi_1 + T_G \phi_2$.

Using Proposition 5.1.12 there exist $n_1 \in N$ and appropriate quaternions $(a_{i,n})_{i=1}^{n}$ with

$$\left\| \phi_2 - \sum_{i=1}^{n_1} \sum_{k=1}^{3} \frac{x_k^{(i)} - x_k}{|x^{(i)} - x|^3} e_k a_{i,n_1} \right\|_{W^k_{2,H}} \leq \frac{\varepsilon}{4 \|T_G\|_{[W^{k-1}_{2,H}, L_{2,H}]}} ,$$

whence it follows

$$\left\| T_G \phi_2 - T_G \sum_{i=1}^{n_1} \sum_{k=1}^{3} \frac{x_k^{(i)} - x_k}{|x^{(i)} - x|^3} e_k a_{i,n_1} \right\|_{L_{2,H}} \leq \frac{\varepsilon}{4} ,$$

$$T_G\left(-\sum_{i=1}^{n_1}\sum_{k=1}^{3}\frac{x_k^{(i)}-x_k}{|x^{(i)}-x|^3}e_k a_{i,n_1}\right)=\sum_{i=1}^{n_1}\left(-\frac{1}{|x^{(i)}-x|}e_\emptyset a_{i,n_1}\right)+\sum_{j=1}^{n_1}\phi_{j,G}a_{j,n_1},$$

where $\phi_{j,G}\in A_H(G)$ for $j\le n_1$.

Denote by $\tilde{\phi}_1$ the expression $\phi_1+\sum_{j=1}^{n_1}\phi_{j,G}a_{j,n_1}$.

Using Theorem 5.1.6 there exist $n_2\in N$ and $b_{i,n_2}\in H$ with

$$\left\|\tilde{\phi}_1-\sum_{i=1}^{n_2}\sum_{k=1}^{3}\frac{x_k-x_k^{(i)}}{|x-x^{(i)}|^3}e_k b_{i,n_2}\right\|_{L_{2,H}}\le\frac{\varepsilon}{4}.$$

Hence

$$\left\|u_{n_0}-\sum_{i=1}^{n_1}\left(-\frac{1}{|x^{(i)}-x|}e_\emptyset a_{i,n_1}\right)-\sum_{i=1}^{n_2}\sum_{k=1}^{3}\frac{x_k-x_k^{(i)}}{|x-x^{(i)}|^3}e_k b_{i,n_2}\right\|\le$$

$$\le\left\|\tilde{\phi}_1-\sum_{i=1}^{n_2}\sum_{k=1}^{3}\frac{x_k^{(i)}-x_k}{|x^{(i)}-x|^3}e_k b_{i,n_2}\right\|+\left\|T_G\phi_2-\sum_{j=1}^{n_1}\phi_{j,G}a_{j,n_1}-\right.$$

$$\left.-\sum_{j=1}^{n_1}\left(-\frac{1}{|x^{(j)}-x|}e_\emptyset a_{j,n_1}\right)\right\|\le\frac{\varepsilon}{4}+\left\|T_G\left(\phi_2-\sum_{i=1}^{n_1}\sum_{k=1}^{3}\frac{x_k-x_k^{(i)}}{|x-x^{(i)}|^3}e_k a_{i,n_1}\right)\right\|\le\frac{\varepsilon}{2}.$$

This provides, together with $\|u_{n_\emptyset}-u\|<\frac{\varepsilon}{2}$, our statement. #

5.3.4 Theorem

Under the suppositions of Theorem 5.3.3 the system

$$\left\{\sum_{k=1}^{3}\frac{x_k-x_k^{(i)}}{|x-x^{(i)}|^3}e_k\right\}_{i\in N}\cup\left\{\frac{1}{|x-x^{(i)}|}e_\emptyset\right\}_{i\in N}$$

is H-complete in $W_{2,H}^k(G)\cap\ker\Delta(G)$ for arbitrary $k\in N$.

Proof

Let $U_\varepsilon(G)=\{x\in R^3:\ \mathrm{dist}(x,\Gamma)<\varepsilon\}$. Then the system

$$\left\{\sum_{k=1}^{3}\frac{x_k-x_k^{(i)}}{|x-x^{(i)}|^3}e_k\right\}_{j\in N}\cup\left\{\frac{1}{|x^{(i)}-x|}e_\emptyset\right\}_{i\in N}$$

is H-complete in $L_{2,H}(U_\varepsilon(G))\cap\ker\Delta(U_\varepsilon(G))$ for sufficiently small $\varepsilon>0$. Similarly to Proposition 5.1.12, the assertion follows now.

#

5.3.5 Theorem

Under the suppositions of Theorem 5.3.3 the system

$$\left\{tr\sum_{k=1}^{3}\frac{x_k-x_k^{(i)}}{|x-x^{(i)}|^3}e_k\right\}_{i\in N}\cup\left\{\frac{1}{|x^{(i)}-x|}e_\emptyset\right\}_{i\in N}$$

is H-complete in $W_{2,H}^k(\Gamma)$, $k>\frac{3}{2}$, and the system

$$\left\{ tr \sum_{k=1}^{3} \frac{\partial}{\partial n} \frac{x_k - x_k^{(i)}}{|x - x^{(i)}|^3} e_k \right\}_{i \in N} \cup \left\{ tr \frac{\partial}{\partial n} \frac{1}{|x - x^{(i)}|} e_\emptyset \right\}_{i \in N}$$

is H-complete in $W^k_{2,H}(\Gamma)/\underset{H}{\text{span}}\{1\}$ for $k > \frac{3}{2}$.

Proof

This follows by using Theorem 5.3.4 and the corresponding trace theorems. #

5.3.6 Remark

Theorem 5.3.5 ensures the possibility of constructing H-minimal systems in the spaces $L_{2,H}(G) \cap \ker \Delta(G)$, $W^k_{2,H}(G) \cap \ker \Delta(G)$, $W^k_{2,H}(\Gamma)$ and $W^k_{2,H}(\Gamma)/\underset{H}{\text{span}}\{1\}$, $k > \frac{3}{2}$, respectively.

The equations of the stationary linear elasticity are written down by the quaternionic equation (4.28) in Subsection 4.3 .

5.3.7 Theorem

Under the assumptions of Theorem 5.3.3 every element $u \in \ker DM^{-1}D \cap W^k_{2,H}(G)$ may be approximated in $W^k_{2,H}(G)$ arbitrarily closely by the expressions

$$u_{n_1 n_2} = \sum_{i=1}^{n_1} \sum_{k=1}^{3} \frac{x_k - x_k^{(i)}}{|x - x^{(i)}|^3} e_k a_i + \sum_{i=1}^{n_2} \left[\frac{1}{|x - x^{(i)}|} e_\emptyset b_{\emptyset,i} + \sum_{k=1}^{3} \left(\frac{3m-4}{4m-4} \frac{1}{|x^{(i)} - x|} e_k + \right.\right.$$

$$\left.\left. + \frac{m}{4m-4} (x_k^{(i)} - x_k) \sum_{l=1}^{3} \frac{x_l - x_l^{(i)}}{|x - x^{(i)}|^3} e_l \right) b_{k,i} \right] \qquad (5.17)$$

with suitably chosen coefficients $a_i, b_{k,i} \in H$.

Proof

Using the representation $u = \Phi_1 + T_G M \Phi_2$, where $\Phi_1, \Phi_2 \in A_H(G)$ (Theorem 4.3.11) , it can be proved similarly to Theorems 5.3.3 and 5.3.4 . #

5.3.8 Remark

The operator $DM^{-1}D$ is not right-linear with respect to the quaternionic multiplication. This property causes real factors and real multiplications to occur in the representation (5.17).

5.3.9 Theorem

Under the assumptions of Theorem 5.3.3 every element $u \in \mathrm{tr}\ \ker DMD(G) \cap W^{k+1/2}_{2,H}(\Gamma)$, $k > \frac{3}{2}$ can be approximated in $W^{k}_{2,H}(\Gamma)$ arbitrarily closely by

$$v_{n_1,n_2} = \mathrm{tr}\ u_{n_1,n_2}. \tag{5.18}$$

Let $R = \mathrm{Im}(\frac{\partial}{\partial\alpha} - \alpha BD) + (\mathrm{Re\ tr})e_0$, $\alpha = \sum_{i=1}^{3}\alpha_i e_i$ and $(\alpha_1, \alpha_2, \alpha_3)$ is the unit vector of the outer normal, and B is defined by $Bu = -\frac{u_0}{m-2}e_0 + \frac{1}{2}\sum_{i=1}^{3} u_i e_i$. Every element $u \in R(\ker DM^{-1}D(G) \cap W^{k}_{2,H}(G))$, $k > \frac{3}{2}$ may be approximated in $W^{k}_{2,H}(\Gamma)$ arbitrarily closely by

$$w_{n_1,n_2} = R(u_{n_1,n_2}) . \tag{5.19}$$

The coefficients occurring in (5.18), (5.19), respectively, are to be determined separately in each case.

Proof

The proof follows from Theorem 5.3.7 and the corresponding trace theorems. #

5.3.10 Remark

Note that the boundary operator R is not right-linear with respect to the quaternionic multiplication.

5.3.11 Remark

For constructing H-complete systems in $(\ker DM^{-1}D)(G)$ it is also possible to start from the representation

$$u = v - \frac{m}{2m-2} T_G(\mathrm{Re}\ Dv) \tag{5.20}$$

for $u \in \ker DM^{-1}D$ with $v \in \ker \Delta$ and use Theorems 5.3.3 and 5.3.4 .

5.3.12 Proposition

For the solution of the equations of linear elasticity the NEUBER-PAPKOVIC statements

$$\hat{u} = \hat{w} - \frac{m}{4m-4}\,\mathrm{grad}(x\hat{w}) - \frac{m}{4m-4}\,\mathrm{grad}\ w_0 \tag{5.21}$$

are well-known, where $\hat{w} = (w_1, w_2, w_3)$, $w_i : R^3 \to R^1$, $i = 0,1,2,3$, $w_i \in (\ker \Delta)(G)$. Between the statements (5.20) and (5.21) there exists the following relation:

$v = \mathrm{Im}\ w + \phi$

with $\phi \in A_H(G)$, $w = w_\emptyset e_\emptyset + \hat{w}$, $\hat{w} = \sum_{i=1}^{3} w_i e_i$.

Proof

Set $u = \sum_{i=\emptyset}^{3} u_i e_i$. The representation (5.21) may be transformed into

$$\operatorname{Im} u = \operatorname{Im} w - \frac{m}{4m-4} \operatorname{Im} D(x \cdot \operatorname{Im} w) - \frac{m}{4m-4} \operatorname{Im} Dw .$$

A straightforward computation yields

$$D[\operatorname{Im} w - \frac{m}{2m-2} T_G \operatorname{Re} D \cdot \operatorname{Im} w] = D[\operatorname{Im} w - \frac{m}{4m-4} D(x \cdot \operatorname{Im} w) - \frac{m}{4m-4} D \operatorname{Re} w] .$$

Hence follows the existence of an H-regular function ϕ with

$$\operatorname{Im} w - \frac{m}{2m-2} T_G \operatorname{Re} D \cdot \operatorname{Im} w + \phi = \operatorname{Im} w - \frac{m}{4m-4} D(x \cdot \operatorname{Im} w) - \frac{m}{4m-4} D \operatorname{Re} w$$

and finally

$$\operatorname{Im}(w+\phi) - \frac{m}{2m-2} T_G \operatorname{Re} D \cdot \operatorname{Im}(w+\phi) = \operatorname{Im} w - \frac{m}{4m-4} D(x \cdot \operatorname{Im} w) - D \operatorname{Re} w ,$$

whence the relation

$$v = \operatorname{Im} w + \phi \quad , \ \phi \in \ker D(G) ,$$

follows. #

As a further example, STOKES' equations will be treated. We look for approximate solutions of the problem (4.42)-(4.43)-(4.44) under the necessary condition (4.45).

5.3.13 Proposition

The boundary value problem (4.42)-(4.43)-(4.44) may be reduced to the problem

$$-\Delta v + \frac{1}{\eta} Dq = \emptyset \quad \text{in } G, \qquad (5.22)$$

$$\operatorname{div} v = \emptyset \quad \text{in } G, \qquad (5.23)$$

$$\operatorname{tr} v = H \quad \text{on } \Gamma \qquad (5.24)$$

with $H = g - \operatorname{tr} T_G(\operatorname{Im} T_G \frac{\varrho}{\eta} f - f_\emptyset)$, where the condition (4.45) is converted into

$$\int_\Gamma H \cdot \alpha \, d\Gamma = \emptyset . \qquad (5.25)$$

Proof

In accordance with representation (4.51) generalized VEKUA's theory yields

$$u = -\frac{1}{\eta} T_G p + \frac{\varrho}{\eta} T_G^2 f + T_G \phi_1 + \phi_2 \quad , \ \phi_i \in A_H(G) . \qquad (5.26)$$

Setting (5.26) in (4.43) we obtain

$$f_\emptyset = \operatorname{div} u = -\operatorname{Re} Du = \frac{1}{\eta} p - \frac{\varrho}{\eta} \operatorname{Re} T_G f - \operatorname{Re} \Phi_1 =$$

$$= \frac{1}{\eta} p - \operatorname{Re}(\Phi_1 + \frac{\varrho}{\eta} T_G f) .$$

Therefore

$$p = \eta f_\emptyset + \operatorname{Re}(\eta \Phi_1 + \varrho T_G f). \tag{5.27}$$

Substituting the expression (5.27) in (5.26) we get

$$u = - T_G f_\emptyset + T_G \operatorname{Im} \Phi_1 + T_G \frac{\varrho}{\eta} \operatorname{Im} T_G f + \Phi_2 , \quad \Phi_i \in A_H(G) .$$

Hence , a special solution (u_s, p_s) is given by

$$(u_s, p_s) = (- T_G f_\emptyset + T_G \frac{\varrho}{\eta} \operatorname{Im} T_G f , \eta f_\emptyset + \operatorname{Re} \varrho T_G f) .$$

Setting $v = u - u_s$ and $q = p - p_s$ we achieve the problem (5.22)-(5.23)-(5.24) . Using GAUSS' formula the condition (5.25) is easy to check. #

5.3.14 Theorem

An approximate solution of STOKES' problem (5.22)-(5.23)-(5.24) is given by

$$v_{n_1 n_2} = T_G \operatorname{Im} \Phi_{1n_1} + \sum_{j=1}^{n_2} \sum_{k=1}^{3} \frac{x_k - x_k^{(j)}}{|x - x^{(j)}|^3} e_k c_j ,$$

$$q_{n_1} = - \operatorname{Re} \Phi_{1n_1}$$

with $\Phi_{1n_1} \in \operatorname{Ker} D$, $c_j \in \mathbb{H}$.

The quaternionic constants occurring may be calculated by the conditions

(i) $$\sum_{j=1}^{n_1} \sum_{k=1}^{3} \frac{z_k^{(j)} - x_k^{(j)}}{|z^{(j)} - x^{(j)}|^3} e_k c_j = (F_\Gamma g)(z^{(j)}) , \; j=1,\dots,4n_1,$$

(ii) $$(\operatorname{tr} T_G \operatorname{Im} \Phi_{1n_1})(t^{(j)}) = (Q_\Gamma H)(t^{(j)}) , \; j=1,\dots,4n_2 ,$$

where $z^{(j)}$, $j=1,\dots,4n_1$, and $t^{(j)}$, $j=1,\dots,4n_2$, are the corresponding collocation points lying on Γ .

Proof

Starting from the system

$$q = - \operatorname{Re} \Phi_1 , \tag{5.28}$$

$$v = T_G \operatorname{Im} \Phi_1 + \Phi_2 \tag{5.29}$$

we approximate Φ_1 by $$\Phi_{1n_1} = \sum_{j=1}^{n_1} \sum_{k=1}^{3} \frac{x_k - x_k^{(j)}}{|x - x^{(j)}|^3} e_k a_j$$

and Φ_2 by $$\Phi_{2n_2} = \sum_{j=1}^{n_2} \sum_{k=1}^{3} \frac{x_k - x_k^{(j)}}{|x - x^{(j)}|^3} e_k c_j ,$$

where a_j , $j=1,\dots,n_1$, and c_j , $j=1,\dots,n_2$, are quaternionic constants. Letting the operator F_Γ act from the left on equation (5.29) we obtain

$F_\Gamma v = F_\Gamma H = \Phi_2$,

whence condition (i) follows. Besides we get

$v - F_\Gamma H = T_G \operatorname{Im} \Phi_1$.

Finally , PLEMELJ-SOKHOTZKIJ's formulas yield

$Q_\Gamma H = tr(T_G \operatorname{Im} \Phi_1)$,

and (ii) may be easily verified. #

5.3.15 Remark

More exactly, we have

$$T_G \operatorname{Im} \Phi_{1n_1} = \sum_{i=1}^{n_1} [(T_G r_1 e_1)a_{i0} + (T_G r_2 e_1)a_{i3} - (T_G r_3 e_1)a_{i2} +$$
$$+(T_G r_2 e_2)a_{i0} + (T_G r_3 e_2)a_{i1} - (T_G r_1 e_2)a_{i3} +$$
$$+(T_G r_3 e_3)a_{i0} + (T_G r_1 e_3)a_{i2} - (T_G r_2 e_3)a_{i1}] ,$$

where $r_k = \frac{x_k - x_k^{(i)}}{|x - x^{(i)}|^3}$, $k=1,2,3$, $a_i = \sum_{k=0}^{3} a_{ik} e_k$. The expressions

$(T_G r_j e_k)$ may be calculated by the help of Theorem 5.3.1 .

5.3.16 Conclusion

The results of Chapter 5 were completely applied to boundary approximation methods, see also in [Mue2] , [Ke] , [Te] . By the explicit construction of basis functions the fundamentals of boundary collocation methods were established. But an application of this method is only proved for some special cases, see for example Theorem 5.2.22. Then we have very nice numerical properties. A realization of the general case is still an open problem. Boundary collocation methods differ by their simplicity , fast numerical realization and good judgement of the quality of the approximate solutions through the user. Some examples are given to emphasize our statements.

5.4 Numerical Examples

Now we demonstrate the nice numerical properties of the

collocation method by means of some examples of two-dimensional DIRICHLET problems for the LAPLACE equation and of some three-dimensional boundary value problems of linear elasticity.

The boundary value problems for the LAPLACE equation shall be used to show numerical effects connected with certain properties of the solutions of the boundary value problem. Using the maximum principle for the LAPLACE equation the $C(\bar{G})$-error in the approximation u_n (see 5.3) to the solution is bounded by $\|u_n-u\|_{C(\Gamma)}$. In the following, the norm of $C(\Gamma)$ is approximated by a discrete $C(\Gamma)$-seminorm with respect to 5000 nodes on Γ.

The disk $\{(x,y)\in R^2 : x^2+y^2<1\}$ is denoted by K.

<u>Examples 1-3:</u>

$$\Delta u=0 \quad \text{in } K, \; u=g_i \quad \text{on } \Gamma=\partial K, \; i=1,2,3,$$

with $g_1(x,y)=e^x\cos y - x^2 + y^2$,

$g_2(x,x)=2x^2 - 5y^2 + 3$,

$g_3(x,y)=|x|$.

The singularities of the basis functions are located on $\{(x,y): x^2+y^2=2\}$ in each case, n denotes the number of used basis functions, δ_i is the relative error of the approximative solution in $C(\Gamma)$ for the i-th example and def_i is the absolute error of the approximative solution in the collocation points.

n	δ_1	δ_2	δ_3	def_1	def_2	def_3
20	9.5E-06	2.5E-06	5.8E-02	E-15	E-15	E-15
40	7.6E-12	9.3E-08	3.0E-02	E-15	E-15	E-12
60	7.5E-15	5.3E-08	2.0E-02	E-15	E-14	E-09
80	5.8E-15	8.2E-08	1.5E-02	E-15	E-12	E-06

These examples show that the rate of convergence and the stability depend upon the smoothness of g_i.

If the distance between the singularities and Γ is changed in dependence on n then we can obtain some corrections.

Example 4:

$$\Delta u=0 \quad \text{in} \quad K, \; u=g_3 \quad \text{on} \quad \Gamma = \partial K .$$

Let h be the minimal distance of adjoining collocation points on Γ. The singularities are chosen on $\{(x,y):x^2+y^2=1+d,\ d>0\}$.

n	d/h	δ	def
20	0.639	3.6E-02	E-15
40	0.637	1.7E-02	E-15
60	0.762	1.3E-02	E-15
80	0.769	1.0E-02	E-14
100	0.793	7.7E-03	E-14

Improvements of the rate of convergence and of the stability can be seen in comparison with example 3. However, the quality of the approximative solution can depend upon d/h only a little (in a certain range of d/h). For this reason, we consider the same example 4 in case of n=80.

d/h	δ	def
0.256	2.1E-01	E-14
0.513	2.1E-02	E-14
0.769	1.0E-02	E-14
1.282	1.3E-02	E-15
2.564	1.4E-02	E-14
6.410	1.4E-02	E-11
12.820	1.5E-02	E-06

The question of an optimal choice of d/h is discussed, for instance, in [Fra].

Example 5:

$\Delta u=0$ in $K\backslash(K_1\cup K_2\cup K_3)$,

$u=2$ on Γ, $u=1$ on $\Gamma_1\cup\Gamma_2\cup\Gamma_3$,

where

$K_1=\{(x,y):16x^2+(4y-2)^2<1\}$,

$K_2=\{(x,y):(4x+2)^2+(4y+2)^2<1\}$,

$K_3=\{(x,y):(4x-2)^2+(4y+2)^2<1\}$,

and $\Gamma_i=\partial K_i$. Now n/4 collocation points are located on every circle , Γ_1, Γ_2, Γ_3. The singularities are arranged on $\{(x,y):x^2+y^2=2\}$ and on circles within K_1, K_2, and K_3, respectively.

n	δ	def
40	8.1E-02	E-15
60	4.3E-02	E-15
80	2.5E-02	E-15
100	1.6E-02	E-14
120	9.4E-03	E-14

The example demonstrates the practical applicability of our collocation method for a non-convex domain with non-connected boundary.

Example 6:

The last example in the plane shows a comparison between the collocation method (CM) and the GALERKIN method (GM) with respect to exactness, stability, computing time (T) and main storage requirement (MSR). The integrals necessary in using the GALERKIN method are calculated by means of the mid-point rule.

n	δ_{CM}	δ_{GM}	def_{CM}	def_{GM}	MSR_{CM}	MSR_{GM}	T_{GM}/T_{CM}
20	1.4E-07	1.4E-07	E-15	E-13	88	82	1.5
40	5.8E-13	9.8E-10	E-15	E-09	94	82	3.3
60	4.7E-15	2.6E+10	E-15	E+10	110	98	5.6
80	6.1E-15	4.0E+10	E-14	E+10	132	108	7.8
100	7.3E-15	6.5E+15	E-14	E+10	160	124	11.5
					\K Bytes/		

Now, let us consider some three-dimensional problems. The aim of the following examples is only to present a "practical proof" of the possibility to use collocation methods also in such situations. For this reason we shall omit the details and give only the boundary value problem and the results.

Example 7:

$$DMDu=0 \quad \text{in} \quad G=\{s\in R^3: |s|<1\},$$

$$t_n(\hat{u})=\frac{\partial\hat{u}}{\partial n}+n\,\frac{\operatorname{div}\hat{u}}{m-2}+\frac{1}{2}n\times\operatorname{rot}\hat{u}=(0,x,y,z)^T \quad \text{on} \quad \Gamma.$$

n	14	22	32	44	58	74
δ_C	0.64	0.32	0.18	0.11	0.06	0.03

Examples 8-9:

$$DMDu=0 \quad \text{in} \quad G=\{s=(x,y,z)^T: |z|<1, x^2+y^2<1\},$$
$$t_n(\hat{u})=g_i \quad \text{on} \quad \Gamma.$$

$$g_1=\begin{cases} 0 & \text{for} \quad |z|=1, \\ (0,x,y,0)^T & \text{for} \quad |z|<1,\ x^2+y^2=1, \end{cases}$$

$$g_2=\begin{cases} 1 & z=1, \\ -1 & z=-1, \\ 0 & |z|<1. \end{cases}$$

n	$\delta_C^{(8)}$	$\delta_C^{(9)}$
21	0.19	0.49
30	0.15	0.43
40	0.07	0.20
50	0.03	0.09
60	0.01	0.04

Example 10:

$$DMDu=0 \quad \text{in} \quad G=\{s \in R^3 : |s|<1\},$$
$$u=g_i \quad \text{on} \quad \Gamma,\ i=1,2,$$

$$g_1(x,y)=(0,x,y,z)^T,$$

$$g_2(x,y)=(0,x^3+2x^2y-3z^2,x-5y+3z^3,xyz-4x^2y^2z^2-y^2z)^T.$$

n	$\delta_C^{(1)}$	$\delta_C^{(2)}$
14	1.03	1.20
22	0.65	0.59
32	0.35	0.27
44	0.18	0.45
58	0.092	0.18
78	0.046	0.04

Example 11:

$$\Delta u=0 \quad \text{in} \quad G=\{s \in R^3 : |s|<1\},$$

$$\frac{\partial u}{\partial n}=(45xz^2 \cdot 15x^3,3y^3-9x^2y,3xy^2-3x^3+15z^3-45x^2z)^T \quad \text{on } \Gamma.$$

n	6	14	22	44	58	63
δ_C	0.66	0.59	0.71	0.42	0.22	0.15

Example 12:

$$\Delta u=\mathrm{grad}|x|^2 \quad \text{in} \quad G=\{s \in R^3 : |s|<1\},$$
$$\frac{\partial u}{\partial n}=0 \quad \text{on } \Gamma.$$

n	8	14	22	44	83
C	1.09	0.89	0.66	0.24	0.06

The last problem was solved by constructing a special solution of the inhomogeneous differential equation and transformation to a homogeneous equation which was treated using collocation methods.
We finish the discussion of numerical examples and refer for further information to the papers [GL],[GSS],[Gue5] and [S].
In [GL] a boundary value problem for the biharmonic equation was investigated. The papers [GSS] and [S] contain a lot of examples which show that the successful use of boundary collocation method is possible in case of a parabolic problem. Especially in [S] theoretical difficulties in the procedure of solving the linear equations of the interpolation problem are discussed.

6. DISCRETE QUATERNIONIC FUNCTION THEORY

A decisive disadvantage of the numerical implementation of continuous models seems to be that essential analytical and algebraic properties which are characteristic of the continuous theory have to be dropped out if we change over to a corresponding discrete problem. As far as the theory of quaternionic analysis is concerned, such properties are, for instance, algebraical relations of the continuous operators, orthogonality, statements of invertibility and the behaviour of H-regular functions in the neighbourhood of singular points. In order to change this state, in several papers [Z], [DD], [Hay], [DL], [Du], [Fe] attempts were made to develop a discrete analogue to the complex function theory. For functions over lattices which satisfy a well-defined difference equation similar to the CAUCHY-RIEMANN system it is possible to transfer some essential elements of the classical function theory, for instance, CAUCHY's theorem, idea of the complex curve integral and power series expansions. Up to the present time a discrete BOREL-POMPEIU formula and the restriction to bounded domains have been wanted. Statements in this direction also contain the theory of finite difference methods. A.A. SAMARSKIJ [Sam] considers discrete harmonic functions and shows that these functions fulfil a maximum priciple. It has been tried to maintain several properties such as monotonicity, sourcelessness, theorems of conservation of the solutions of boundary value problems in the discrete case.
New researches to construct a discrete function theory on lattices were done by P. BECHER, H. JOOS and M. GÖCKELER in [BJ],[GJ]. Our investigations in Section 6 are dedicated to the development of a discrete quaternionic calculus.

6.1 Fundamental Solutions of the Discrete Laplacian

Considerations in the continuous case point out that an important fact in the definition of the operators T_G and F_Γ consists in the explicit knowledge of the fundamental solution of $-\Delta$ and its properties Therefore we shall attempt to define a corresponding idea of the fundamental solution of the Laplacian. For bounded domains (square stones) similar constructions were made by A.A. SAMARSKIJ and his collaborators. In [Sam] discrete GREEN's functions are given. These investigations are not sufficient for our purposes. A careful study of the proof of the algebraic properties of the operators T_G and F_Γ shows that the H-regularity of the fundamental solution of the operator D in $R^3\setminus\{0\}$ is used. By the aid of discrete functions as carried out by A.A.SAMARSKIJ, H-regularity cannot be guaranteed. For further considerations we define the equidistant lattice

$$R_h^3 = \{(ih,jh,kh)^T,\ h>0 \text{ real},\ i,j,k \quad \text{integer numbers}\}$$

and briefly write

$$u(ih,jh,kh) = u_{i,j,k}\ .$$

The operator $-\Delta_h$ is given by

$$-\Delta_h u=(6u_{i,j,k}-u_{i-1,j,k}-u_{i+1,j,k}-u_{i,j-1,k}-u_{i,j+1,k} - u_{i,j,k-1}-u_{i,j,k+1})/h^2. \qquad (6.1)$$

First of all we look for a function $E_h(S)$ which fulfils the equation

$$(-\Delta_h E_h)(s) = \begin{cases} 1/h & s=0 \\ 0 & s\neq 0 \end{cases} \quad s\in R_h^3\ . \qquad (6.2)$$

The following notations shall be settled.

$$Q_l = ([-l,l]X[-l,l]X[-l,l])\cap R_h^3,$$

$$\mathring{Q}_l = \text{int}\ Q_l,\ l=nh,\ n\in \mathbf{N},$$

$$K_h(s) = \begin{cases} 1/h & s=0 \\ 0 & s\neq 0 \end{cases} \quad ,\quad s\in R_h^3\ ,$$

$$G_h = G\cap R_h^3,\ G\subset R^3,\ \mathring{G}_h = \text{int}\ G_h\ .$$

For functions defined on G_h with values in H we introduce the inner product

$$\langle f,g\rangle = \sum_{s\in G_h} \overline{f(s)}g(s)h^3 \quad . \qquad (6.3)$$

The inner product (6.3) induces the norm $\|\ \|_{2,h,H}$. Obviously, every function defined on G_h with finite values belongs to $L_{2,h}(G_h)$ if G_h is bounded and h the fixed meshwidth. This notation is different from the notation in [SLM]. To prove the existence of a discrete fundamental solution E_h we start from the results obtained by A.A. SAMARSKIJ.

6.1.1 Proposition [Sam]

The boundary value problem

$$
\begin{aligned}
-\Delta_h u &= K_h \quad \text{in} \quad \text{int } Q_l, \\
u &= 0 \quad \text{on} \quad \partial Q_l
\end{aligned}
\tag{6.4}
$$

has a unique solution $E_{h,l}$, and it holds

$$E_{h,l}(s) \geq 0 \ , \ s \in Q_l. \tag{6.5}$$

Proof

The proof is based on the discrete maximum principle. #

6.1.2 Remark

In case of $s \in \text{int } Q_l \setminus \{0\}$ we even have $E_{h,l} > 0$. $E_{h,l}(0)$ will be investigated later.

6.1.3 Proposition

Let $l < L$ and $E_{h,l}$, $E_{h,L}$ are solutions of the boundary value problem (6.4) in Q_l, Q_L, respectively, which are to be extended by zero into co Q_l, co Q_L, respectively. Then the inequality

$$E_{h,l}(s) \leq E_{h,L}(s), \quad s \in R_h^3, \tag{6.6}$$

holds.

Proof

We have in Q_l $K_l = K_L$. On the bondary ∂Q_l it holds $E_{h,L} \geq E_{h,l}$. Using the discrete maximum principle, we obtain the assertion in Q_l. By the help of Proposition 6.1.1 follows our statement in $Q_L \setminus Q_l$. #

6.1.4 Proposition

It holds

$$E_{h,l}(0) \geq E_{h,l}(s) \ , \quad s \in Q_l \ .$$

Proof

Assume that there exists a lattice point $s_o \in Q_l \setminus \{0\}$ with the property

$$E_{h,l}(s_o) \geq E_{h,l}(s) \ , \quad s \in Q_l \ .$$

Then there exists a connected (in the sense of [Sam]) set $Q' \subset Q_l$ with $s_o \in Q'$, $0 \notin Q'$, $Q' \cap \partial Q_l \neq \phi$ and

$$
\begin{aligned}
-\Delta_h E_{h,l} &= 0 \quad \text{in} \quad Q', \\
E_{h,l} &\geq 0 \quad \text{on} \quad \partial Q' \qquad (E_{h,l}(s) = 0 \quad \text{on} \quad Q' \cap \partial Q_l).
\end{aligned}
$$

From the assumption and by the help of the discrete maximum principle it follows

$$E_{h,l} = \text{const} \quad \text{in} \quad Q'.$$

Because $Q' \cap \partial Q_l \neq \phi$, we have $E_{h,l}=0$ in Q'. This contradicts to $E_{h,l}(s_0)>0$ and proves the proposition. #

Now let us consider the limit procedure $l \longrightarrow \infty$. We already know that the sequence $\{E_{h,l}\}_l$ is monotonically increasing with respect to l and has a lower bound. Thence, for the proof of the existence, we have to use some other properties, too. For that reason we expand the function $E_{h,l}$ in Q_l in terms of eigenfunctions of $\{-\Delta_h, tr\}$.

__6.1.5 Proposition__

The functions

$$v_{ijk}^{(h,l)}(x,y,z) = \\ = l^{-3/2}\cos\left(\frac{2i+1}{2l}\pi x\right)\cos\left(\frac{2j+1}{2l}\pi y\right)\cos\left(\frac{2k+1}{2l}\pi z\right),$$

$$i,j,k=0,1,\ldots,n-1 \;;\; (x,y,z)^T \in Q_l,$$

are normed eigenfunctions of $\{-\Delta_h, tr\}$ in Q_l which correspond to the eigenvalues

$$\lambda_{ijk}^{(h,l)}=4h^{-2}\left\{\sin^2\left(\frac{2i+1}{4l}\pi h\right)+\sin^2\left(\frac{2j+1}{4l}\pi h\right)+\sin^2\left(\frac{2k+1}{4l}\pi h\right)\right\}$$

with $i,j,k=0,1,\ldots,n-1$.

__Proof__

For the proof we refer to [Sam]. #

By this we obtain the following representation of $E_{h,l}$:

__6.1.6 Proposition__

$$E_{h,l}(x,y,z) = \sum_{i,j,k=0}^{n-1} l^{-3/2}\left(\lambda_{ijk}^{(h,l)}\right)^{-1} v_{ijk}^{(h,l)}(x,y,z) \cdot \qquad (6.8)$$

__Proof__

Representation (6.8) is the discrete FOURIER series expansion of $E_{h,l}$. #

To estimate $E_{h,l}$ by the help of (6.8), it is useful to have an estimate of sums of the form

$$\sum_{i,j,k=0}^{n-1} \left((2i+1)^2 + (2j+1)^2 + (2k+1)^2\right)^{-p}$$

for $p>0$.

__6.1.7 Proposition__

Let $n=l/h$, h the meshwidth, $0<p<3$. Then

$$\sum_{i,j,k=0}^{n-1}\left((2i+1)^2+(2j+1)^2+(2k+1)^2\right)^{-p/2} \leq C_1 h^{p-3} l^{3-p}/(3-p) + C_2 \qquad (6.9)$$

with positive constants C_i, $i=1,2$, independently of h and l.

Proof

Set

$V_{ijk}=[(i-\frac{1}{2})h,(i+\frac{1}{2})h]\times[(j-\frac{1}{2})h,(j+\frac{1}{2})h]\times[(k-\frac{1}{2})h,(k+\frac{1}{2})h]$,

$r=(x^2+y^2+z^2)^{1/2}$, $r_{ijk}=(i^2+j^2+k^2)^{1/2}h$.

Then, under the supposition $i^2+j^2+k^2>m_o$, we get

$$\left|\int_{V_{ijk}} r^{-p}dG - r_{ijk}^{-p}h^3\right| \le \frac{p}{2}\sup\{(|x|+|y|+|z|)r^{-p-2};V_{ijk}\}h^4 \le$$

$$\le 3^{1/2}p/2\ \sup\{r^{-p-1};V_{ijk}\}h^4 \le 3^{1/2}p/2\ h^4(r_{ijk}-3^{1/2}h/2)^{-p-1}$$

$$\le 3^{1/2}p/2\ C\ h^3 r_{ijk}^{-p}\ ,\quad C<2p^{-1}\ 3^{-1/2}\ .$$

Furthermore we immediately obtain

$$r_{ijk}^{-p}h^3 \le \int_{V_{ijk}} r^{-p}dG + 3^{1/2}(p/2)\cdot C\cdot h^3 r_{ijk}^{-p}\ .$$

Now we are able to estimate the left-hand side of (6.9) as follows:

$$\sum_{i,j,k=0}^{n-1}((2i+1)^2+(2j+1)^2+(2k+1)^2)^{-p/2} \le$$

$$\le h^{p-3}\sum_{i,j,k=1}^{n}[(i^2+j^2+k^2)h^2]^{-p/2}h^3 =$$

$$= h^{p-3}\sum_{0<i^2+j^2+k^2<m_0}[(i^2+j^2+k^2)h^2]^{-p/2}h^3 +$$

$$+ h^{p-3}\sum_{i^2+j^2+k^2\ge m_0}[(i^2+j^2+k^2)h^2]^{-p/2}h^3 \le$$

$$\le C_2 + h^{p-3}\sum_{i^2+j^2+k^2\ge m_0} r_{ijk}^{-p}h^3 \le C_2 + h^{p-3}(1-3^{1/2}Cp/2)^{-1}\int_{V_{ijk}} r^{-p}dG$$

$$\le C_2 + h^{p-3}(1-3^{1/2}Cp/2)^{-1}\ 3^{(3-p)/2}(3-p)^{-1}l^{3-p} =$$

$$= C_2 + h^{p-3}(3-p)^{-1}C_1 l^{3-p}.$$

#

Using the inequality (6.9), the existence of the discrete fundamental solution E_h may be shown.

6.1.8 Theorem

For each fixed meshwidth $h>0$ there exists a function E_h which satisfies the discrete boundary value problem (6.2) We have

$$E_h(s) = \lim_{l\to\infty} E_{h,l}(s), \quad s\in R_h^3, \tag{6.10}$$

The convergence is uniform with respect to $s\in R_h^3$.

Proof

For the validity of Proposition 6.1.3 and Proposition 6.1.4 it is sufficient to verify the boundedness of the set $\{E_{h,l}(0)\}_l$. Then pointwise convergence follows. Making use of (6.8) we conclude

$$|E_{h,l}(0)| \le \sum_{i,j,k=0}^{n-1} l^{-3/2}(\lambda_{ijk}^{(h,l)})^{-1} v_{ijk}^{(h,l)}(0) \le$$

$$\le 0.25h^2 l^{-3} \sum_{i,j,k=0}^{n-1} \{\sin^2(\frac{(2i+1)\pi h}{4l})+\sin^2(\frac{(2j+1)\pi h}{4l})+$$

$$+\sin^2(\frac{(2k+1)\pi h}{4l})\}^{-1} \le l^{-1} \sum_{i,j,k=0}^{n-1} \{(2i+1)^2+(2j+1)^2+(2k+1)^2\}^{-1} \le$$

$$\le l^{-1}\{C_2+C_1 l\} = C_2/l + C_1 \le K \quad \text{for } l \longrightarrow \infty.$$

If l tends to infinity, it follows that $|E_{h,l}(0)|\le K$.

The boundedness of $\{E_{h,l}\}_l$ and monotonic convergence lead to the uniform convergence $E_{h,l} \xrightarrow{\longrightarrow} E_h$ for an arbitrary bounded subset of R_h^3. In Theorem 6.1.14 it is proved by other means that $|E_h(s)|\le\frac{C}{|s|}$, whence uniform convergence in R_h^3 can be obtained at once. It is obvious that $E_h(s)$ fulfils (6.2).

#

The above-deduced representation of E_h by the help of a limit procedure is not sufficient for further investigations. Putting a connection between a discrete and continuous theory we have to explain in which sense the convergence of E_h for $h\longrightarrow 0$ can be understood.

6.1.9 Proposition

Let $n=1/h$, $h>0$. Then

$$\sum_{i,j,k=1}^{n} (i^2+j^2+k^2)^{-3/2} \le C_1 + C_2(h) + C_3\ln(l) \tag{6.11}$$

is valid.

Proof

Similarly to Proposition 6.1.7, it may be proved that

$$\left| \int_{V_{ijk}} r^{-3} dG - h^3 r_{ijk}^{-3} \right| \le 3^{3/2} C h^3 r_{ijk}^{-3}$$

if $r_{ijk} > (1+2C\, 3^{1/2}) h C^{-1}$.

If $C < 3^{-3/2}$, we get with $r > m_0 h$ a sufficient condition and therefore

$$\sum_{i,j,k=1}^{n} [(i^2+j^2+k^2)h^2]^{-3/2} h^3 = \sum_{i,j,k=1}^{n} r_{ijk}^{-3} h^3 = C_1 + \sum_{r_{ijk} > m_0 h} r_{ijk}^{-3} h^3 \le$$

$$\le C_1^* + (1-3^{3/2}C)^{-1} \int_{Q_l \setminus Q_{m_0 h}} r^{-3} dG =$$

$$= C_1^* + (1-3^{3/2}C)^{-1} 4\pi [\ln(l) + \ln 3^{1/2} - \ln(m_1 h)] =$$

$$= C_1 + C_2(h) + C_3 \ln(l).$$ #

<u>6.1.10 Theorem</u>

Let $\chi_{\pi/h}$ denote the characteristic function of the cube $[-\pi/h, \pi/h]^3$. Let the function f_h be given by

$$f_h(x,y,z) = 0.25h^2(\sin^2(hx/2)+\sin^2(hy/2)+\sin^2(hz/2))^{-1} \chi_{\pi/h}$$

and $\hat{F}$ the FOURIER transform in R^3. Then the discrete fundamental solution of $-\Delta_h$ has the following representation:

$$E_h(x,y,z) = (2\pi)^{-3} \hat{F}[f_h](x,y,z). \qquad (6.12)$$

<u>Proof</u>

Introduce the following notations:

$$\alpha_i = \frac{2i+1}{2l}, \quad \beta_j = \frac{2j+1}{2l}, \quad \gamma_k = \frac{2k+1}{2l},$$

$$V_{ijk} = [\alpha_i - \frac{\pi}{2l}, \alpha_i + \frac{\pi}{2l}] X [\beta_j - \frac{\pi}{2l}, \beta_j + \frac{\pi}{2l}] X [\gamma_k - \frac{\pi}{2l}, \gamma_k + \frac{\pi}{2l}],$$

$$S(\alpha,\beta,\gamma) := \sin^2(h\alpha/2)+\sin^2(h\beta/2)+\sin^2(h\gamma/2).$$

Then the series (6.8) can be written in the form

$$E_{h,l}(x,y,z) = h^2/(4\pi^3) \sum_{i,j,k=0}^{n-1} \cos(\alpha_i x)\cos(\beta_j y)\cos(\gamma_k z) S^{-1}(\alpha_i,\beta_j,\gamma_k) \frac{\pi^3}{l^3}. \qquad (6.13)$$

If we regard (6.13) as a quadratur formula, we obtain

$$E_{h,l}(x,y,z) = h^2/(4\pi^3) \{ \int_0^{\frac{\pi}{h}} \int_0^{\frac{\pi}{h}} \int_0^{\frac{\pi}{h}} \frac{\cos(\alpha x)\cos(\beta y)\cos(\gamma z)}{S(\alpha,\beta,\gamma)} d\alpha d\beta d\gamma +$$

$$+ R_{h,l}(x,y,z)\}, \qquad (6.14)$$

where $R_{h,l}$ denotes the error of quadrature. A generous estimate of $R_{h,l}$ yields in V_{ijk}

$$R_{h,l}|V_{ijk} \leq \sup\{(|x|+|y|+|z|)\pi^2 h^{-2} r^{-2} + 3^{1/2}\pi^3 h^{-2} r^{-3}; V_{ijk}\}\frac{\pi}{l},$$

with $r^2=(\alpha^2+\beta^2+\gamma^2)$. Altogether we find out

$$|E_{h,l}(x,y,z)-h^2/(4\pi^3)\iiint_{V_{ijk}} \frac{\cos(\alpha x)\cos(\beta y)\cos(\gamma z)}{S(\alpha,\beta,\gamma)}\,d\alpha d\beta d\gamma| \leq$$

$$\leq |\sum_{i^2+j^2+k^2<3} h^2/(4\pi^3)\cos(\alpha_i x)\cos(\beta_j y)\cos(\gamma_k z)S^{-1}(\alpha_i,\beta_j,\gamma_k)\pi^3 l^{-3} -$$

$$- h^2/(4\pi^3)\iiint_{V_{ijk}} \frac{\cos(\alpha x)\cos(\beta y)\cos(\gamma z)}{S(\alpha,\beta,\gamma)}\,d\alpha d\beta d\gamma| +$$

$$+ |\sum_{i^2+j^2+k^2\geq 3} h^2/(4\pi^3)\cos(\alpha_i x)\cos(\beta_j y)\cos(\gamma_k z)S^{-1}(\alpha_i,\beta_j,\gamma_k)\pi^3 l^{-3} -$$

$$- h^2/(4\pi^3)\iiint_{V_{ijk}} \frac{\cos(\alpha x)\cos(\beta y)\cos(\gamma z)}{S(\alpha,\beta,\gamma)}\,d\alpha d\beta d\gamma|$$

$$\leq S_1(l) + \sum_{i^2+j^2+k^2\geq 3} \{1/(4\pi)(|x|+|y|+|z|)(r_{ijk}-3^{1/2}\pi/(2l))^{-2} +$$

$$+ (3^{1/2}/4)(r_{ijk}-3^{1/2}\pi/(2l))^{-3}\}\pi^4/(l^4) \leq$$

$$\leq S_1(l) + \sum_{i^2+j^2+k^2\geq 3} 1/(4\pi)[(|x|+|y|+|z|)2\pi^4/(r_{ijk}^2 l^4) +$$

$$+ 2\pi^4\, 3^{1/2}/(4l^4 r_{ijk}^3)] \leq$$

$$\leq S_1(l) + 1/(2\pi)\ [|x|+|y|+|z|]C_2 + 3^{1/2}/(1-3^{1/2}C)\frac{l\pi^2}{hl^2} +$$

$$+ 3^{1/2}\ 4\ \frac{\pi}{l}[C_1+C_2(h)+C_3\ln(l)] \xrightarrow{l\to\infty} 0.$$

By making use of Proposition 6.1.7 and Proposition 6.1.9, it is now easy to see that $S_1(l)$ converges to zero if l tends to infinity, because the integral in (6.14) possesses a weak singularity in the point $(0,0,0)^T$. #

Representation (6.12) allows to deduce a dependence on h.

6.1.11 Corollary

Let $h>0$ be fixed and t an arbitrary positive number. Then we have

$$\frac{1}{t}\, E_{h/t}(x,y,z) = E_h(tx,ty,tz). \qquad (6.15)$$

Proof

This can be readily seen by a corresponding substitution in (6.12).

#

6.1.12 Remark

Corollary 6.1.11 is important for numerical computations because all calculations are to be carried out for only one meshwidth.

By the aid of the representation of E_h as a FOURIER transform of the function f_h, it is not difficult to investigate the behaviour of E_h, as $h \longrightarrow 0$. Now $E_h(s)$, $s \in R_h^3$, are to be extended in a natural manner by means of the inner product (6.3) onto R^3.

6.1.13 Theorem

Let $s=(x,y,z)^T \in R^3$, $h>0$ and $G \subset R^3$ be an arbitrary bounded domain. Then the limits for $h \longrightarrow 0$

$$E_h(s) \longrightarrow \frac{1}{4\pi|s|} \quad \text{in } L_2(G), \tag{6.16}$$

$$E_h(s) - \frac{1}{4\pi|s|} \longrightarrow 0 \quad \text{in } L_2(R^3) \tag{6.17}$$

are valid.

Proof

The proof is straightforward and therefore left to the reader.

#

6.1.14 Remark

Relation (6.17) shows that the asymptotic behaviour of $\frac{1}{4\pi|s|}$ for $|s| \longrightarrow \infty$ is well reflected by the behaviour of the function E_h since $E \notin L_{2,R}(R^3)$ but for $h \longrightarrow 0$ $E_h - E \longrightarrow 0$ in $L_{2,R}(R^3)$.

It is much more interesting to ask for suitable domains of the space R^3, in which a pointwise and even uniform convergence may be proved. Let $f(s)=|s|^{-2}$. To obtain the necessary estimates, we have to consider the second derivatives of $f_h - f$ in a distributional sense.

6.1.15 Theorem

Let $|s|>0.4\,\pi^{-2} \sup\{|\{\Delta(f_h-f)\}|_{L_{1,R}} ; h>0\}$, where by $\{\Delta g\}$ the classical part of the distribution Δg (see [V1a]) is denoted. Then $E_h(s) \leq \frac{1}{2|s|}$

Proof

Using the same notation as in [V1a], we get for

$g \in C_R^1(\bar{G}) \cap C_R(\text{co}\ \bar{G})$ in a bounded domain G with piecewise smooth boundary

$$\Delta g = \{\Delta g\} + [\frac{\partial g}{\partial n}] + \frac{\partial}{\partial n}([g]_\Gamma \delta_\Gamma), \qquad (6.18)$$

where $[\frac{\partial g}{\partial n}]$ and $[g]$ denotes the the jump by passage through Γ. By applying Δ in the described sense to $f_h - f$ we obtain the following results for the various parts:

1. $\{\Delta(f_h-f)\} \longrightarrow 0$ in $L_1(R^3)$.

Indeed, we have

$$\{\Delta(f_h-f)\}=(-h^4(\cos(\alpha h)+\cos(\beta h)+\cos(\gamma h))S^{-2}(\alpha,\beta,\gamma)/8+$$
$$+h^4(\sin^2(\alpha h)+\sin^2(\beta h)+\sin^2(\gamma h))S^{-3}(\alpha,\beta,\gamma)/8)\chi_{\mathbb{T}/h}-2r^{-4},$$

where $S(\alpha,\beta,\gamma)$ is defined in the proof of Theorem 6.1.10. Obviously, we have

$$\int_{R^3} |\{\Delta(f_h-f)\}|dG = \int_Q |\{\Delta(f_h-f)\}|dG + \int_{R^3\setminus Q} |\{\Delta(f_h-f)\}|dG .$$

As $|\{\Delta(f_h-f)\}| \le O(r^{-4})$ for $r \longrightarrow \infty$, there exists a sufficiently large square $Q=Q(\varepsilon)$ such that the integral over $R^3\setminus Q$ becomes smaller than $\varepsilon/2$ independently of h. To estimate the integral over Q, it is necessary to investigate the function $|\{\Delta(f_h-f)\}|$ for $r \longrightarrow 0$. We obtain $|\{\Delta(f_h-f)\}| \le O(h^2 r^{-2})$ if r tends to zero and hence

$$\int_Q |\{\Delta(f_h-f)\}|dG < \varepsilon/2 \quad \text{for} \quad h<h_\varepsilon .$$

2. Let Γ_h be the boundary of $Q\,\mathbb{T}/h$. It immediately follows $[\frac{\partial}{\partial n}(f_h-f)]_{\Gamma_h}=0$ and the term $\frac{\partial}{\partial n}([f_h-f]_{\Gamma_h}\delta_{\Gamma_h}) = \frac{\partial}{\partial n}([f_h]_{\Gamma_h}\delta_{\Gamma_h})$ remains to be considered. By means of the FOURIER transform we get

$$\hat{F}[\frac{\partial}{\partial n}([f_h]_{\Gamma_h}\delta_{\Gamma_h})] = \sum_{i=1}^{3} \hat{F}[D_i([f_h]_{\Gamma_{h_i}}\delta_{\Gamma_{h_i}})] = \sum_{i=1}^{3}(is_i)\hat{F}[[f_h]_{\Gamma_{h_i}}\delta_{\Gamma_{h_i}}]$$

and with $\varphi \in L_{2,R}(R^3)$

$$|\langle \hat{F}[[f_h]_{\Gamma_h}\delta_{\Gamma_h}],\varphi\rangle| \le (h^2/4)\int_{\Gamma_{h_i}} \frac{|\hat{F}[\varphi]|}{S(\alpha,\beta,\gamma)} d\Gamma_h \le \frac{h^2}{4}\,\frac{\mathbb{T}^2}{h^2}\|\varphi\|_{L_{1,R}} =$$
$$= \frac{\mathbb{T}^2}{4}\|\varphi\|_{L_{1,R}},$$

whence $|\hat{F}[[f_h]_{\Gamma_h}\delta_{\Gamma_h}]| \le \mathbb{T}^2/4$ follows .

Applying 1. and 2., we get the estimate

$$|\hat{F}[\triangle(f_h-f)]|(s)\le\|\{\triangle(f_h-f)\}\|_{L_{1,R}} + 1.5\pi^2|s|$$

and so

$$|s|^2|\hat{F}[f_h-f]|\le\|\{\triangle(f_h-f)\}\|_{L_{1,R}} + 1.5\pi^2|s|.$$

In conclusion we gather

$$|E_h(s) - \frac{1}{4\pi|s|}|\le(2\pi)^{-3}|s|^{-2}\|\{\triangle(f_h-f)\}\|_{L_{1,R}}+ \frac{3}{16\pi}|s|^{-1}. \quad \#$$

Using Theorem 6.1.14, we succeed in estimating E_h uniformly in the spaces $L_{p,h,H}$ and describing its behaviour in a certain neighbourhood of the origin.

Now we define the norm of $L_{p,h,H}(G)$ as follows:

$$\|f\|_{p,h,H} = (\sum_{s\in G_h}|f(s)|^p h^3)^{1/p}. \tag{6.19}$$

A supposition of the consideration of the limit $\lim_{h\to 0} E_h$ is a uniform estimate of E_h in $L_{p,h,H}(G)$.

6.1.16 Theorem

Let $G\subset R^3$ be an arbitrary bounded domain and $1<p<3$. Then we have for $h>0$

$$\|E_h\|_{p,h,H} \le C(p,G). \tag{6.20}$$

Proof

Let $h>0$ be fixed and $l=\mathrm{diam}\ G$. We consider for $t>0$ $E_{h/t}$ and use (6.15) and formula (6.9). We obtain

$$\|E_{h/t}\|^p_{p,h/t,H}=\sum_{s\in G_{h/t}}|E_{h/t}(s)|^p h^3 t^{-3}=\sum_{s\in G_{h/t}} t^{p-3}|E_h(ts)|^p h^3=$$

$$=\sum_{s'\in tG_h} t^{p-3}|E_h(s')|^p h^3=t^{p-3}\|E_h\|^p_{p,h,H}+\sum_{s'\in tG_h\setminus G_h} t^{p-3}|E_h(s')|^p h^3\le$$

$$\le t^{p-3}\|E_h\|^p_{p,h,H}+C^* \sum_{s'\in tG_h\setminus G_h} t^{p-3}|s'|^{-p}h^3\le$$

$$\le t^{p-3}\|E_h\|^p_{p,h,H}+t^{p-3}h^3([C_1/(3-p)]h^{p-3}(tl)^{3-p}+C_2)C \le$$

$$\le t^{p-3}(\|E_h\|^p_{p,h,H}+h^3C_2C)+C_1C\frac{1}{3-p}l^{3-p}h^p.$$

Since $p<3$ we have (6.20). #

6.1.17 Remark

It is worth noticing that the discrete fundamental solution is contained in the same L_p-spaces as $\frac{1}{4\pi|s|}$, where the behaviour near the origin and for $|s| \longrightarrow \infty$ is reflected correctly.

6.1.18 Corollary

If s satisfies the inequality

$$0<|s|<0.4\pi^{-2}\sup\{\|\Delta(f_h-f)\|_{L_{1,R}}\ ;h>0\},$$

then there exists a number $h_0>0$ with

$$|E_h(s)| \leq \frac{1}{2\pi|s|}, \quad h<h_0 .$$

Proof

It can be easily proved by using (6.15) and Theorem 6.1.14. #

6.2 Fundamental Solutions of a Discrete Generalized CAUCHY-RIEMANN Operator

For discretization of the CAUCHY-RIEMANN operator (1.6) there exist several possibilities (forward-, backward-, central differences). If we demand as in the continuous case that the generalized CAUCHY-RIEMANN operator and its L_2-adjoint operator define a factorization of the Laplacian, then the different possibilities of the discretization of the operator D are restricted. Denote by $V^{\pm}_{i,h}(x)$ the displacement of the point x about $\pm h$ in the x_i-direction. Furthermore we determine with $(D^{\pm}_{i,h}f)(x)= \pm (f(V^{\pm}_{i,h}x)-f(x))/h$ approximations of partial derivations $(D_i f)(x)$. Then let the following disrete analogues of the generalized CAUCHY-RIEMANN operator be defined

$$(D^{\pm}_h f)(x) = \sum_{i=1}^{3} e_i(D^{\pm}_{i,h}f)(x). \tag{6.21}$$

We settle that the signs "+" or "-" are to be omitted if the corresponding statements are valid in the same manner for both operators.

6.2.1 Proposition

$$D^+_h D^-_h f = D^-_h D^+_h f = -\Delta_h f \ , \tag{6.22}$$

where $f:Z\times Z\times Z \longrightarrow H$, with $Z=\{m,\ m \text{ integer}\}$.

Proof

The proof immediately follows from the definitions of the operators $D^{\pm}_h$. #

6.2.2 Proposition

The lattice functions

$$e_h^+(s) = (D_h^- E_h)(s) \quad \text{and} \quad e_h^-(s) = (D_h^+ E_h)(s), \quad s \in R_h^3,$$

are fundamental solutions of the operators D_h^+ and D_h^-.

Proof

It is readily seen by using (6.22) and Theorem 6.1.8. #

In ordér to carry out functional analytic investigations based on the functions e_h^+ and e_h^-, we deduce a closed analytical representation. For the sake of brevity we only consider $e_h^+(s)$. Maintaining the denotation of Section 6.1 we get

$$-2l^3 h^{-1}(D_h^+ E_{h,l})(s) S(\alpha_i, \beta_j, \gamma_k) = \tag{6.23}$$

$$= \sum_{i,j,k=0}^{N-1} \{ \sin\frac{(2i+1)\pi(2x+h)}{4l} \sin\frac{(2i+1)\pi h}{4l} \cos\frac{(2j+1)\pi y}{2l} \cos\frac{(2k+1)\pi z}{2l}\, e_1 +$$

$$+ \cos\frac{(2i+1)\pi x}{2l} \sin\frac{(2j+1)\pi(2y+h)}{4l} \sin\frac{(2j+1)\pi h}{4l} \cos\frac{(2k+1)\pi z}{2l}\, e_2 +$$

$$+ \cos\frac{(2i+1)\pi x}{2l} \cos\frac{(2j+1)\pi y}{2l} \sin\frac{(2k+1)\pi(2z+h)}{4l} \sin\frac{(2k+1)\pi h}{4l}\, e_3 \}.$$

Taking (6.23) as a quadrature formula we find with $h^* = \pi/l$

$$D_h^+ E_{h,l} = -h/(4\pi^3) \int_0^{\pi/h}\int_0^{\pi/h}\int_0^{\pi/h} \{ \frac{\sin(\alpha h)}{S(\alpha,\beta,\gamma)} \sin(\alpha x)\cos(\beta y)\cos(\gamma z) +$$

$$+ \frac{2\sin^2(\alpha h/2)\cos(\alpha x)\cos(\beta y)\cos(\gamma z)}{S(\alpha,\beta,\gamma)} \} d\alpha\, d\beta\, d\gamma\, e_1 +$$

$$+ \ldots + h^*/(4\pi^3) R_{h,l}(x,y,z). \tag{6.24}$$

An estimation of the error of quadrature $R_{h,l}$ in the cube V_{ijk} yields for $i,j,k \neq 0$

$$2R_{h,l} \leq (r_{ijk} - 0.5\, 3^{1/2} h^*)^{-2} [\pi^2 + 1.5\pi^4] +$$

$$+ (r_{ijk} - 0.5 \cdot 3^{1/2} h^*)^{-1} [\pi^2(|x|+|y|+|z|) + \pi^2 h^*/2 + (3\pi^4/4) h^*] +$$

$$+ \pi^2 h^* (|x|+|y|+|z|),$$

whence follows

$$|D_h^+ E_{h,l} - \frac{h}{32\pi^3} i\hat{F}[(\sin(\alpha h)e_1 + \sin(\beta h)e_2 + \sin(\gamma h)e_3) S^{-1}(\alpha,\beta,\gamma) \chi_{\pi/h}]$$

$$+\frac{h}{16\pi^3}\hat{F}[(\sin^2(\tfrac{\alpha h}{2})e_1+\sin^2(\tfrac{\beta h}{2})e_2+\sin^2(\tfrac{\gamma h}{2})e_3)S^{-1}(\alpha,\beta,\gamma)\chi_{\pi/h}]| \le$$

$$\le [A_1\sum_{i,j,k=1}^{n-1}(r_{ijk}-0.5\ 3^{1/2}h^*)^{-2}(h^*)^3+$$

$$+A_2(h^*)\sum_{i,j,k=1}^{n-1}(r_{ijk}-0.5\cdot 3^{1/2}h^*)^{-1}(h^*)^3+$$

$$+A_3\sum_{i,j,k=1}^{n-1}(h^*)^4+O(h^*)]h^*/(4\pi^3)=O(h^*)\ . \qquad \#$$

To obtain the latter estimates the assertion of Proposition 6.1.7 is used outside of a certain neighbourhood of the origin, namely for $m^2+j^2+k^2\ge n_0$ and the weak singularity of $S^{-1}(\alpha,\beta,\gamma)$ for the estimate in the case of points (m,j,k) which satisfy the inequality $m^2+j^2+k^2<n_0$. The symbol i stands for the imaginary unit. In this way we have proved the following representation formulas for e_h^+ and e_h^-:

6.2.3 Theorem

The discrete fundamental solutions e_h^+ and e_h^- of the operators D_h^+ and D_h^-, respectively, have the representations

$$e_h^{\pm}(s)=\frac{\pm h}{32\pi^3}i\hat{F}[(\sin(\alpha h)e_1+\sin(\beta h)e_2+\sin(\gamma h)e_3)S^{-1}(\alpha,\beta,\gamma)\chi_{\pi/h}]+$$

$$-\frac{\pm h}{16\pi^3}\hat{F}[(\sin^2(\alpha h/2)e_1+\sin^2(\beta h/2)e_2+\sin^2(\gamma h/2)e_3)S^{-1}(\alpha,\beta,\gamma)\chi_{\pi/h}],$$

$$s\in R_h^3 .$$

6.2.4 Remark

The functions $e_h^{\pm}(s)$ are real-valued in each component.
The behaviour of $e_h^{\pm}$ for $h\longrightarrow 0$ remains to be investigated.
We restrict our considerations to $e_h^+(s)$.

6.2.5 Theorem

We have

$$|e_h^+(s)| \le C|s|^{-2}\ ,\quad s\in R_h^3\setminus\{0\},$$

$$|e_h^-(s)| \le C|s|^{-2}\ ,\quad s\in R_h^3\setminus\{0\}.$$

Proof

First we consider the term

$$\hat{F}[(\sin^2(\alpha h/2)e_1+\sin^2(\beta h/2)e_2+\sin^2(\gamma h/2)e_3)S^{-1}(\alpha,\beta,\gamma)\chi_{\pi/h}]\frac{h}{16\pi^3}.$$

Because of symmetry it is sufficient to discuss the term $\hat{F}[\sin^2(\alpha h/2)S^{-1}(\alpha,\beta,\gamma)\chi_{\pi/h}]\frac{h}{16\pi^3}$. If the derivations occurred will be understood in a distributional sense we get

$$(x^4+y^4+z^4)\hat{F}[h\,\sin^2(\alpha h/2)S^{-1}(\alpha,\beta,\gamma)\chi_{\pi/h}] =$$

$$= \hat{F}[(D_\alpha^4+D_\beta^4+D_\gamma^4)\{h\,\sin^2(\alpha h/2)S^{-1}(\alpha,\beta,\gamma)\chi_{\pi/h}\}] +$$

$$+ (ix)\hat{F}[h^3\cos(\alpha,n)(S^{-1}(\alpha,\beta,\gamma)/2)\delta_{\Gamma_h}] -$$

$$- (ix)\hat{F}[h^3\cos(\alpha,n)(S^{-2}(\alpha,\beta,\gamma)/2)\delta_{\Gamma_h}] +$$

$$+ ix^3\hat{F}[h\,\cos(\alpha,n)S^{-1}(\alpha,\beta,\gamma)\delta_{\Gamma_h}] +$$

$$+ iy^3\hat{F}[h\,\sin^2(\alpha h/2)\cos(\beta,n)S^{-1}(\alpha,\beta,\gamma)\delta_{\Gamma_h}] +$$

$$+ iy\hat{F}[h^3\sin^2(\alpha h/2)\cos(\beta h)\cos(\gamma,n)(S^{-2}(\alpha,\beta,\gamma)/2)\delta_{\Gamma_h}] +$$

$$+ iz^3\hat{F}[h\,\sin^2(\alpha h/2)\cos(\gamma,n)S^{-1}(\alpha,\beta,\gamma)\delta_{\Gamma_h}] +$$

$$+ iz\hat{F}[h^3\sin^2(\alpha h/2)\cos(\gamma h)\cos(\gamma,n)(S^{-2}(\alpha,\beta,\gamma)/2)\delta_{\Gamma_h}].$$

All terms which contain the factor h^3 may be estimated as follows:

$$|\hat{F}[h^3\cos(\alpha,n)(S^{-1}(\alpha,\beta,\gamma)/2)\delta_{\Gamma_h}]| \le h^3\int_{\Gamma_h}|S(\pi/h,\beta,\gamma)|^{-1}d\Gamma \le \pi^2 h.$$

Denote

$$d(\alpha,\beta,\gamma):=96h\alpha^2r^{-8}-384h\alpha^2(\alpha^4+\beta^4+\gamma^4)r^{-10}-288h\alpha^4r^{-8}+16hr^{-4}\ ,$$

then we have

$$|\hat{F}[\{(D_\alpha^4+D_\beta^4+D_\gamma^4)(h\,\sin^2(\alpha h/2)S^{-1}(\alpha,\beta,\gamma)\chi_{\pi/h})\}]| \le$$

$$\le |\hat{F}[\{(D_\alpha^4+D_\beta^4+D_\gamma^4)(h\,\sin^2(\alpha h/2)S^{-1}(\alpha,\beta,\gamma)\chi_{\pi/h})\}-d(\alpha,\beta,\gamma)]| +$$

$$+ |\hat{F}[d(\alpha,\beta,\gamma)]| \le Ch^2 + |f_1(s)|h. \tag{6.25}$$

In this case we made use of the fact that $\hat{F}[d(\alpha,\beta,\gamma)]$ is a homogeneous function $f_1 \in C_R^\infty(R^3\setminus\{0\})$ of the degree +1 and the function $\{(D_\alpha^4+D_\beta^4+D_\gamma^4)(h\,\sin^2(\alpha h/2)S^{-1}(\alpha,\beta,\gamma)\chi_{\pi/h})\}$ includes terms which can be estimated in $L_{1,R}(R^3)$ by Ch^2 (for instance $h^5\cos(\alpha h)S^{-1}(\alpha,\beta,\gamma)\chi_{\pi/h}$, $h^5\sin^2(\alpha h)S^{-2}(\alpha,\beta,\gamma)\chi_{\pi/h},\dots$) as well as those terms which, after subtracting suitable parts of $d(\alpha,\beta,\gamma)$, become $L_{1,R}(R^3)$-functions. For example, we consider $g_h=h^5\cos(2h\alpha)S^{-2}(\alpha,\beta,\gamma)\chi_{\pi/h}-16hr^{-4}$. Substituting $\alpha' t=\alpha$, $\beta' t=\beta$, $\gamma' t=\gamma$ in the integral $\|g_h\|_{L_{1,R}}$, we obtain

$\|g_{h/t}\|_{L_{4,R}} = t^{-2}\|g_h\|_{L_{4,R}}$, whence $\|g_h\|_{L_{4,R}} \le Ch^2$ follows and (6.25) is verified.

The expressions of the form $\hat{F}[h\cos(\alpha,n)S^{-1}(\alpha,\beta,\gamma)\delta_{\Gamma_h}]$ remain to be investigated. A straightforward computation yields by setting $\xi=(\alpha,\beta,\gamma)^T$, $s=(x,y,z)^T$

$$\hat{F}[h\cos(\alpha,n)S^{-1}(\alpha,\beta,\gamma)\delta_{\Gamma}](s) = -\int_0^{\pi/h}\int_0^{\pi/h} hS^{-1}(-\pi/h,\beta,\gamma)e^{i(\xi,s)}d\Gamma + \int_0^{\pi/h}\int_0^{\pi/h} hS^{-1}(\pi/h,\beta,\gamma)e^{i(\xi,s)}d\Gamma =$$

$$= 2i\sin(\pi x/h)\int_0^{\pi/h}\int_0^{\pi/h} S^{-1}(\pi/h,\beta,\gamma)\cos(\beta y)\cos(\gamma z)d\beta d\gamma = 0 \quad \text{for}$$

$x=mh$. In a similar manner we obtain

$$\hat{F}[h\sin^2(\alpha h/2)\cos(\beta,n)S^{-1}(\alpha,\beta,\gamma)\delta_{\Gamma_h}](s) = 0 \quad \text{for} \quad y=jh,$$

$$\hat{F}[h\sin^2(\alpha h/2)\cos(\gamma,n)S^{-1}(\alpha,\beta,\gamma)\delta_{\Gamma_h}](s) = 0 \quad \text{for} \quad z=kh.$$

Note that

$$\int_0^{\pi/h}\int_0^{\pi/h} S^{-1}(\pi/h,\beta,\gamma)\cos(\beta y)\cos(\gamma z)d\beta d\gamma =: \hat{F}_2[hS^{-1}(\tfrac{\pi}{h},\beta,\gamma)\chi_{\pi/h}^{(2)}](y,z),$$

where $\hat{F}_2$ means a FOURIER transform in R^2. This interpretation allows the behaviour in the points $s \in R^3\setminus R_h^3$ (see Theorem 6.2.12) to be investigated. If we collect the found estimates obtained, we get for s R_h^3

$$|\hat{F}[h\sin^2(\alpha h/2)S^{-1}(\alpha,\beta,\gamma)\chi_{\pi/h}](s)| \le Ch^2|s|^{-4} + |f_1(s)|\|h\||s|^{-4} + C_1\|h\||s|^{-3}.$$

Now we consider the part

$$h\hat{F}[(\sin(\alpha h)e_1+\sin(\beta h)e_2+\sin(\gamma h)e_3)S^{-1}(\alpha,\beta,\gamma)\chi_{\pi/h}].$$

We again restrict our considerations to the term

$$h\hat{F}[\sin(\alpha h)S^{-1}(\alpha,\beta,\gamma)e_1\chi_{\pi/h}].$$

In this case it is sufficient to analyse the third partial derivations of $\sin(\alpha h)S^{-1}(\alpha,\beta,\gamma)\chi_{\pi/h}$ in a distributional sense. This leads to

$$-i(x^3+y^3+z^3)\hat{F}[h\sin(\alpha h)S^{-1}(\alpha,\beta,\gamma)\chi_{\pi/h}](s) =$$

$$= \hat{F}[\{(D_\alpha^3+D_\beta^3+D_\gamma^3)(h\sin(\alpha h)S^{-1}(\alpha,\beta,\gamma)\chi_{\pi/h})\}](s) =$$

$$-\ ix\hat{F}[h^2\cos(\alpha h)\cos(\alpha,n)S^{-1}(\alpha,\beta,\gamma)\delta_{\Gamma_h}](s) -$$

$$- y^2\hat{F}[h \sin(\alpha h)\cos(\beta,n)S^{-1}(\alpha,\beta,\gamma)\delta_{\Gamma_h}](s) -$$

$$- z^2\hat{F}[h \sin(\alpha h)\cos(\beta,n)S^{-1}(\alpha,\beta,\gamma)\delta_{\Gamma_h}](s) +$$

$$+ \hat{F}[-0.5h^3\sin(\alpha h)\cos(\beta h)\cos(\beta,n)S^{-2}(\alpha,\beta,\gamma)\delta_{\Gamma_h}](s) +$$

$$+ \hat{F}[-0.5h^3\sin(\alpha h)\cos(\gamma h)\cos(\gamma,n)S^{-2}(\alpha,\beta,\gamma)\delta_{\Gamma_h}](s) . \quad (6.26)$$

The first term can be estimated by inserting a well-defined correction term. We have

$$|\hat{F}[\{(D_\alpha^3+D_\beta^3+D_\gamma^3)(h \sin(\alpha h)S^{-1}(\alpha,\beta,\gamma)\chi_{\pi/h})\}](s)| \le$$

$$\le |\hat{F}[\{ \ldots \}+24r^{-4}-192\alpha^2r^{-6}+192\alpha(\alpha^3+\beta^3+\gamma^3)r^{-8}-96\alpha\beta r^{-6}-$$

$$-96\alpha\gamma r^{-6}](s)|+|\hat{F}[24r^{-4}-192\alpha^2r^{-6}+\ldots-96\alpha\gamma r^{-6}](s)|.$$

Similarly to the proof of Theorem 6.1.14, the statement

$$\|\{(D_\alpha^3+\ldots)\}+24r^{-4}-192\alpha^2r^{-6}+\ldots-96\alpha\gamma r^{-6}\|_{L_1(R\ )} \le Ch^2 \quad (6.27)$$

may be proved.

The second term in (6.26) is a homogeneous function $g_1 \in C_R^\infty(R^3\setminus\{0\})$ of the degree +1 [Hö]. Furthermore we have

$$|\hat{F}[0.5h^3\sin(\alpha h)\cos(\beta h)\cos(\beta,n)S^{-2}(\alpha,\beta,\gamma)\delta_\Gamma](s)| \le \pi^2 h,\ s\in R^3.$$

In the same manner the last term of (6.26) can be bounded. The expressions which contain h or h^2 will be considered as mentioned above. For instance, we obtain

$$\hat{F}[h \sin(\alpha h)\cos(\beta,n)S^{-1}(\alpha,\beta,\gamma)\delta_{\Gamma_h}](s) =$$

$$= 2i\sin(\pi y/h)\int_{-\pi/h}^{\pi/h}\int_{-\pi/h}^{\pi/h} h \sin(\alpha h)S^{-1}(\alpha,\pi/h,\gamma)\sin(\alpha x)\cos(\gamma z)d\beta d\gamma = 0 \quad \text{for } y=jh.$$

Therefore all these terms vanish on R_h^3.

Now we collect all estimates and obtain the following result

$$|e_h^+(s)| \le Ch^2|s|^{-4}+Ch|s|^{-3}+C|s|^{-2}. \quad (6.28)$$

e_h^- is to be deduced in a similar way as e_h^+ and e_h^- only differ in the sign of some terms. #

<u>6.2.6 Remark</u>

The estimate of Theorem 6.2.5 is uniform with respect to h.

<u>6.2.7 Corollary</u>

Let $h>0$, $t>0$. Then we have for all $s \in R_h^3\setminus\{0\}$

$$t^{-2}e^{+}_{h/t}(x/t,y/t,z/t) = e^{+}_{h}(x,y,z) \quad .$$

Proof

A substitution of the variables in Theorem 6.2.3 yields the assertion. #

6.2.8 Theorem

Let $G\subset R^3$ be a bounded domain. In the discrete $L_{p,h,H}$-space we have the uniform estimate

$$\|e_h\|_{p,h,H} \leq K_{p,G} \;,\quad h\leq h_o,\quad p<3/2 \;.$$

Proof

Setting $\|e^{\pm}_{h}\|^{p}_{p,h,H} = M$, by using Corollary 6.2.7 we obtain

$$\|e^{\pm}_{h/t}\|^{p}_{p,h/t,G} = \sum_{s\in G_{h/t}} |e^{\pm}_{h/t}(s)|^{p}h^{3}t^{-3} = \sum_{s\in G_{h/t}} |t^{2}e^{\pm}_{h}(ts)|^{p}h^{3}t^{-3} =$$

$$= \sum_{r\in tG_h} |e^{\pm}_{h}(r)|^{p}h^{3}t^{2p-3} = t^{2p-3}M + \sum_{r\in tG_h\setminus G_h} t^{2p-3}|e_h(r)|^{p}h^{3} \leq$$

$$\leq t^{2p-3}M + \sum_{r_{ijk}\in tG_h\setminus G_h} t^{2p-3}r_{ijk}^{-2p}h^{3}C \leq$$

$$\leq t^{2p-3}M + Ct^{2p-3}\{C_1(3-2p)^{-1}(\mathrm{diam}\; tG)^{3-2p}+C_2\} =$$

$$= t^{2p-3}(M+C_3h^3) + C_4(3-2p)^{-1}h^{2p}\leq K_{p,G} \quad \text{for} \quad t\longrightarrow\infty \,.$$

To deduce the last inequality, we used Theorem 6.2.5 and Proposition 6.1.7. #

6.2.9 Corollary

The estimate

$$|e_h(0)| \leq Ch^{-2}$$

is true.

Proof

This result immediately follows from Corollary 6.2.7. #

6.2.10 Corollary

Let $e^{*}(s) = \begin{cases} e(s) \;, & s\neq 0 \\ 0 \;, & s=0 \end{cases}$. Then we have the inequality

$$\|e_h - e^{*}\|_{p,h,H} \leq N_{p,G} \;,\quad h<h_0,\quad p<3/2 \;.$$

Proof

e^{*} fulfils the relations $|e^{*}(s)|\leq|s|^{-2}$ and $|e^{*}(ts)|=t^{-2}|e^{*}(s)|$. For that reason, the proof of Theorem 6.2.8 may be generalized for this case. #

6.2.11 Theorem

The estimate

$$\|e_h - e^*\|_{p,h,H} \le Ch^{-2+3/p}, \quad p<3/2, \qquad (6.29)$$

is valid.

Proof

Repeating the considerations which led to the statement of Theorem 6.2.5 for the function $e_h - e^*$, similarly to inequality (6.28) we obtain

$$|e_h(s) - e^*(s)| \le Ch^2|s|^{-4} + Ch|s|^{-3}, \quad s \in R_h^3\setminus\{\emptyset\}.$$

Utilizing Corollary 6.2.9, we get

$$\|e_h - e^*\|_{p,h,H}^p = \sum_{s\in G_h} |e_h(s) - e^*(s)|^p h^3 \le$$

$$\le \sum_{s\in G_h\setminus\{\emptyset\}} C^p(h^2|s|^{-4} + h|s|^{-3})^p h^3 + C^p h^{-2p+3} \le$$

$$\le \sum_{s\in G_h\setminus\{\emptyset\}} C^* h^{3+p}|s|^{-3p} + C^p h^{3-2p} \le$$

$$\le h^{3+\alpha} \sum_{s\in G_h\setminus\{\emptyset\}} C^*(h^{p-\alpha}|s|^{\alpha-p})|s|^{-2p-\alpha} + C^p h^{3-2p} \le$$

$$\le C^* h^{3+\alpha} \sum_{s\in G_h\setminus\{\emptyset\}} |s|^{-2p-\alpha} + C^p h^{3-2p} \le$$

$$\le C^* h^{3+\alpha}[C_2 + h^{2p-3}C_1(3-2\alpha-p)^{-1}(\mathrm{diam}\ G)^{3-2p-\alpha}] + C^p h^{3-2p} \le$$

$$\le C^p h^{3-2p} + C^* C_2 h^{3+\alpha} + Dh^{2p+\alpha}(3-2\alpha-p)^{-1}. \qquad \#$$

The last open problem in this connection is the question for the behaviour of e_h outside of the lattice R_h^3. To deduce statements considering the approximation of the discrete analogues of the operators T_G and F_Γ it is necessary to find a satisfactory answer to this question. Furthermore, by the help of these statements we can also obtain assertions for the stability as to the calculation of e_h.

6.2.12 Theorem

Let $p \in (\frac{6}{5}, \frac{3}{2})$. Then it holds for all $h<h_\emptyset$

$$\|e_h - e\|_{L_{p,H}} \le Ch^{(3-2p)/p}.$$

Proof

First of all we consider the second derivatives of those functions which define e_h by the aid of the FOURIER transform. We have

$$D_\alpha^2[h\sin^2(\alpha h/2)S^{-1}(\alpha,\beta,\gamma)] = 0.5h^3[\cos(\alpha h)S^{-1}(\alpha,\beta,\gamma) -$$

$$- \sin^2(\alpha h)S^{-2}(\alpha,\beta,\gamma) -$$

$$- \cos(\alpha h)\sin^2(\alpha h/2)S^{-2}(\alpha,\beta,\gamma)+\sin^2(\alpha h)\sin^2(\alpha h/2)S^{-3}(\alpha,\beta,\gamma)],$$

$$D_\beta^2[h\sin^2(\alpha h/2)S^{-1}(\alpha,\beta,\gamma)] =$$

$$= -0.5h^3\sin^2(\alpha h/2)[\cos(\beta h)S^{-2}(\alpha,\beta,\gamma)-\sin^2(\beta h)S^{-3}(\alpha,\beta,\gamma)].$$

The expressions for $D_\gamma^2[\ldots]$ has an analogous form to $D_\beta^2[\ldots]$. Therefore it holds

$$-|s|^2\hat{F}[h\sin^2(\alpha h/2)S^{-1}(\alpha,\beta,\gamma)](s) =$$

$$= \hat{F}[\{\Delta(h\sin^2(\alpha h/2)S^{-1}(\alpha,\beta,\gamma)\}](s) -$$

$$- ix\hat{F}[h\cos(\alpha,n)S^{-1}(\alpha,\beta,\gamma)\delta_{\Gamma_h}](s) -$$

$$- iy\hat{F}[h\sin^2(\alpha h/2)\cos(\beta,n)S^{-1}(\alpha,\beta,\gamma)\delta_{\Gamma_h}](s) -$$

$$- iz\hat{F}[h\sin^2(\alpha h/2)\cos(\gamma,n)S^{-1}(\alpha,\beta,\gamma)\delta_{\Gamma_h}](s). \qquad (6.30)$$

A straightforward computation shows

$$\|\{\Delta(h\sin^2(\alpha h/2)S^{-1}(\alpha,\beta,\gamma))\}\|_{L_{p,H}} \le Ch^{(3p-3)/p}, \quad 1<p<3/2,$$

whence

$$\|\hat{F}[\{\Delta(\ldots)\}]\|_{L_{q,H}(R^3)} \le C_1h^{(3p-3)/p}, \quad 3<q<\infty, \frac{1}{p}+\frac{1}{q}=1.$$

For an arbitrary bounded domain $G\subset R^3$ now follows

$$\||s|^{-2}\hat{F}[\{\Delta(\ldots)\}]\|_{L_{p,H}} \le C_2h^{(3p-3)/p} \quad 1<p<3/2$$

Next we have to estimate these parts of (6.30) which contain h. In detail we only consider the term $\hat{F}[\frac{h\cos(\alpha,n)}{S(\alpha,\beta,\gamma)}\delta_{\Gamma_h}]$. We have already deduced (proof of Theorem 6.2.5) that

$$\hat{F}[\frac{h\cos(\alpha,n)}{S(\alpha,\beta,\gamma)}\delta_{\Gamma_h}](s) = 2i\sin(\pi x/h)\hat{F}_2[f_h](y,z),$$

where $f_h=hS^{-1}(\frac{\pi}{h},\beta,\gamma)\chi^{(2)}_{\pi/h}$. Now we consider the term $\hat{F}_2[D_\beta(f_h)]$ and get with $\alpha=\frac{\pi}{h}$

$$-(iy)\hat{F}_2[f_h](y,z) = \hat{F}_2[\{-0.5h^2\sin(\alpha h)S^{-2}(\alpha,\beta,\gamma)\chi^{(2)}_{\pi/h}\}](y,z) +$$

$$+ \hat{F}_2[[f_h]_{\Gamma_h}\delta_{\Gamma_h}]. \qquad (6.31)$$

The right-hand side of (6.31) belongs for every h to L_q, $q\le 2$. Indeed, the argument of the first item obviously belongs to L_p, while the second term may be decomposed to a product (similarly as above made for $\hat{F}[h\cos(\alpha,n)S^{-1}(\alpha,\beta,\gamma)\chi_{\pi/h}]$) where one factor is a bounded trigonometric function, and the second factor may be interpreted as a FOURIER transform in the space R^1. The investigation of this FOURIER transform and the use of HOELDER's inequality yields the formulated L_q-result. An analogous consideration can be carried out for $\hat{F}_2[D_\gamma(f_h)]$. In conclusion we obtain

$$\hat{F}_2[f_h] \in L_q(G) \quad \text{for all } G \text{ with } \emptyset \notin \bar{G} \text{ and } q \le 2.$$

Now all preliminaries to estimate

$$|s|^{-1}\hat{F}[h\cos(\alpha,n)S^{-1}(\alpha,\beta,\gamma)\delta_{\Gamma_h}](s) \quad \text{in} \quad \mathrm{co}\,\bar{G}$$

have been accomplished. We assume G to be a star-shaped domain with respect to a certain ball $K_r(\emptyset)$. It holds

$$\| \, |s|^{-1}\hat{F}[h\cos(\alpha,n)S^{-1}(\alpha,\beta,\gamma)\delta_{\Gamma_h}]\|^p_{L_{p,H}(\mathrm{co}\,G)} \le$$

$$\le \| \, |s|^{-p}\|_{L_{p',H}(\mathrm{co}\,G)} \, \| \, |\hat{F}_2[f_h]|^p\|_{L_{q',H}(\mathrm{co}\,G)}$$

if $pp'>3$ and $pq'>2$. These inequalities lead to $p>6/5$. We briefly write for $\sin(\pi x/h)|s|^{-1}\hat{F}_2[f_h]=H_h(s)$. It is easy to show that $H_{h/t}(s)=t^2H_h(ts)$. Furthermore it follows

$$\|H_{h/t}\|^p_{L_{p,H}(G)} = t^{2p-3}\|H_h\|^p_{L_{p,H}(tG)} = t^{2p-3}\|H_h\|^p_{L_{p,H}(G)}$$

$$+ t^{2p-3}\|H_h\|^p_{L_{p,H}(tG\setminus G)}.$$

If $p<3/2$, then it holds

$$\|H_h\|^p_{L_{p,H}(G)} \le \pi^p h^{-p}(\mathrm{diam}\,G) \, \|\hat{F}_2[f_h]\|^p_{L_{2,H}(R)} |G|^{(2-p)/2}$$

and also

$$\|H_h\|^p_{L_{p,H}(G)} \le C(h_\emptyset)h_\emptyset^{3-2p}, \quad h<h_\emptyset, \quad 6/5<p<3/2.$$

Thence we have for all $p \in (\frac{6}{5},\frac{3}{2})$ and $h<h_\emptyset$

$$\|\hat{F}[h\sin^2(\alpha h/2)S^{-1}(\alpha,\beta,\gamma)\delta_{\Gamma_h}]\|_{L_{p,H}(G)} \le C(h_\emptyset)h^{(3-2p)/p} .$$

Therefore there remains the consideration of

$$\hat{F}[\frac{h\sin(\alpha h)}{S(\alpha,\beta,\gamma)}\chi_{\pi/h}].$$

For this purpose we investigate the term $\{\Delta(\frac{h\sin(\alpha h)}{S(\alpha,\beta,\gamma)})\}$. There it holds

$$\{\Delta(\frac{h\sin(\alpha h)}{S(\alpha,\beta,\gamma)})\}+8\alpha r^{-4} = \{\Delta(\frac{h\sin(\alpha h)}{S(\alpha,\beta,\gamma)} +4\alpha r^{-2})\} =$$

$$= -h^3\sin(\alpha h)/S-0.75h^3\sin(2\alpha h)/S^2+24\alpha r^{-4}+0.5h^3\sin^3(\alpha h)/S^3-$$
$$-32\alpha^3 r^{-6}-0.5h^3\sin(\alpha h)\cos(\beta h)/S^2+8\alpha r^{-4}+$$
$$+0.5h^3\sin(\alpha h)\sin^2(\beta h)/S^3-32\alpha\beta^2 r^{-6}-0.5h^3\sin(\alpha h)\cos(\beta h)/S^2+$$
$$+8\ r^{-4}+0.5h^3\sin(\alpha h)\sin^2(\gamma h)/S^3-32\alpha\gamma^2 r^{-6}.$$

As done above, these differences may be estimated in $L_{p,R}(R^3)$ by $Ch^{(3p-3)/p}$ for $p<3/2$. Therefore we have

$$\|\hat{F}[\{\Delta(\frac{h\sin(\alpha h)}{S(\alpha,\beta,\gamma)}\chi_{T/h} + 4\alpha r^{-2}\|_{L_{q,H}(R)}\leq C_q h^{(3p-3)/p}\ ,\quad q\geq 2.$$

Furthermore the consideration of the expression

$$|s|^2\hat{F}[\frac{h\sin(\alpha h)}{S(\alpha,\beta,\gamma)}\chi_{T/h}+4\alpha r^{-2}](s)=\hat{F}[\{\Delta(\frac{h\sin(\alpha h)}{S(\alpha,\beta,\gamma)}\chi_{T/h}+4\alpha r^{-2})\}](s)$$
$$+\ \hat{F}[h^2\cos(\alpha,n)S^{-1}(\alpha,\beta,\gamma)\delta_{\Gamma_h}](s)\ -$$
$$-\ iy\hat{F}[h\ \sin(\alpha h)\cos(\beta,n)S^{-1}(\alpha,\beta,\gamma)\delta_{\Gamma_h}](s)\ -$$
$$-\ iz\hat{F}[h\ \sin(\alpha h)\cos(\beta,n)S^{-1}(\alpha,\beta,\gamma)\delta_{\Gamma_h}](s)$$

turns out in the same way as for the term

$\hat{F}[h\ \sin^2(\alpha h/2)S^{-1}(\alpha,\beta,\gamma)\chi_{T/h}]$. #

6.2.13 Remark

The demand for the star-shapedness can be omitted, because it is possible to enclose any bounded domain in a star-shaped domain with respect to zero.

6.2.14 Remark

As in the case of the discrete Laplacian, the fundamental solution of the discrete CAUCHY-RIEMANN operator (including the properties of convergence) belongs to the discrete and continuous L_p-spaces, in which the fundamental solution e(s) is also contained.

6.3 Elements of a Discrete Quaternionic Function Theory

On the basis of the results of the foregoing section we are able to define discrete analogues of the operator T_G in such a way that these operators are right-inverse to D_h^+, D_h^-, respectively. The main difficulty to introduce a discrete analogue to the operator calculus which we treated in the previous chapters consists in finding out a suitable definition of a discrete analogue to the operator F. Only the existence of such an operator makes it possible to obtain a discrete generalized CAUCHY integral formula which is needed for the foundation of a discrete quaternionic function theory. We succeed in introducing corresponding operators F_h^+ and F_h^- such that a discrete BOREL-POMPEIU formula is valid. With this formula we are given a decisive aid to deal with the properties of discrete H-regular functions. Note that the already cited papers [Fe], [Du], [DD], [Z], [Hay] and [DL] do not contain a discrete formula of the BOREL-POMPEIU type. The concept of the discrete fundamental solutions seems to play an essential role for the construction of a BOREL-POMPEIU formula.

Additionally we introduce some notations

$\partial G_h = \{x \in G_h : \mathrm{dist}(x, \mathrm{co}\, G_h) \leq 3^{1/2} h\}$,

$\partial G_{h,l} = \{x \in \partial G_h : \exists i \in \{1,2,3\}, \text{ with } V_{i,h}^- x \notin G_h\}$,

$\partial G_{h,r} = \{x \in \partial G_h : \exists i \in \{1,2,3\}, \text{ with } V_{i,h}^+ x \notin G_h\}$,

$\partial G_{h,l,i} = \{x \in \partial G_h : V_{i,h}^- x \notin G_h\}$, $i=1,2,3$,

$\partial G_{h,r,i} = \{x \in \partial G_h : V_{i,h}^+ x \notin G_h\}$, $i=1,2,3$,

$\partial G_{h,l,i,j} = \partial G_{h,l,i} \cap \partial G_{h,l,j}$, $i,j \in \{1,2,3\}$,

$\partial G_{h,l,i,j,k} = \partial G_{h,l,i} \cap \partial G_{h,l,j} \cap \partial G_{h,l,k}$, $i,j,k \in \{1,2,3\}$.

We introduce the discrete analogue to the operator T_G as follows:

$$(T_h^+ f)(x) := \sum_{y \in \mathring{G}_h \cup \partial G_{h,l}} e_h^+(x-y) f(y) h^3 - \sum_{\substack{k,j=1 \\ j>k}}^{3} \sum_{y \in \partial G_{h,l,j,k}} e_h^+(x-y) f(y) h^3 + \sum_{\substack{y \in \partial G_{h,l,i,j,k} \\ i \neq j \neq k}} e_h^+(x-y) f(y) h^3,$$

$$(T_h^- f)(x) := \sum_{y \in \mathring{G}_h \cup \partial G_{h,r}} e_h^-(x-y)f(y)h^3 - \sum_{\substack{k,j=1 \\ j>k}}^{3} \sum_{y \in \partial G_{h,r,j,k}} e_h^-(x-y)f(y)h^3 + $$
$$+ \sum_{y \in \partial G_{h,r,i,j,k}} e_h^-(x-y)f(y)h^3 .$$

6.3.1 Remark

In the theory of finite differences, the idea of the boundary point is occasionally introduced somewhat differently, namely $x \in \partial G_h$ iff there exists an index i with $V_{i,h}^+ x \notin G_h$ $(V_{i,h}^- x \notin G_h)$. Because we also have to stress internal corners or rather internal edges, this approach is not possible for our purposes, otherwise we are not able to construct a discrete BOREL-POMPEIU formula.

For the sake of brevity we sometimes write for

$$[\sum_{\partial G_{h,l}} - \sum_{j>k} \sum_{\partial G_{h,l,i,k}}] e_h^+(x-y)f(y)h^3$$

$$\sum_{y \in \Gamma} e_h^+(x-y)f(y)h^3 \quad .$$

We carry out further investigations only for the operators which are marked by "+". Operators signified by "-" can be dealt within the same manner.

6.3.2 Theorem

We have the identity

$$(D_h^+ T_h^+ f)(x) = f(x) , \quad x \in \text{int } G_h . \tag{6.32}$$

Proof

Denoting the action of D_h^+ with respect to x by $D_{h,x}^+$, it is clear that

$$(D_h^+ T_h^+ f)(x) = \sum_{y \in \Gamma} D_{h,x}^+ e_h^+(x-y)f(y)h^3 = f(x)$$

holds. #

Using the structure of BOREL-POMPEIU's formula, we introduce the following discrete analogue F_h^+ of the operator F_Γ:

$$(F_h^+ f)(x) = f(x) - (T_h^+ D_h^+ f)(x). \tag{6.33}$$

We intend to verify that the operator F_h^+ depends only on the boundary values of the function f. The desired algebraic properties are easy to prove by the help of (6.33). Since the determination of the domain of the operator F_h^+ is decisive for all further considerations, we shall carry out the very extensive calculations in detail.

6.3.3 Theorem

Let $n_i(y)$ be the components of the unit vector of the outward-pointing normal on Γ at the point y,

$$n(y)=\sum_{i=1}^{3} n_i(y)e_i,$$

$$R_{ij} = V_{i,h}^{+}\partial G_{h,1,j}\setminus\partial G_{h,1,j},\quad S_{ij} = \partial G_{h,1,j}\setminus V_{i,h}^{+}\partial G_{h,1,j},$$

$$R_{ijk}=V_{i,h}^{+}\partial G_{h,1,j,k}\setminus\partial G_{h,1,j,k}, \text{ and}$$

$$S_{ijk}=\partial G_{h,1,j,k}\setminus V_{i,h}^{+}\partial G_{h,1,j,k}.$$

The operator F_h^+ allows the following representation:

$$\begin{aligned}(F_h^+f)(x) = &- \sum_{i=1}^{3} \sum_{y\in\partial G_{h,r,i}\cup\partial G_{h,1,i}} e_h^+(x-V_{i,h}^{-}y)\ n\ f(y)h^2 +\\ &+ \sum_{i=1}^{3} \sum_{\substack{j=1\\ j\neq i}}^{3} \sum_{y\in S_{ij}} e_h^+(x-V_{i,h}^{-}y)e_i f(y)h^2+\\ &- \sum_{i=1}^{3} \sum_{\substack{j,k=1,\, j>k\\ S_{ijk}}}^{3} e_h^+(x-V_{i,h}^{-}y)e_i f(y)h^2 +\\ &+ \sum_{i=1}^{3} \sum_{\substack{\partial G_{h,1,m,j,k}\\ m\neq j\neq k}} e_h^+(x-y)e_i f(y)h^2 ,\end{aligned} \tag{6.34}$$

$$x\in \operatorname{int} G_h.$$

Proof

$$(F_h^+f)(x) = f(x) - (T_h^+D_h^+f)(x) =$$

$$=f(x)-\Big(\sum_{\mathring{G}_h\cup\partial G_{h,1}} -\sum_{j>k}\sum_{\partial G_{h,1,j,k}}+\sum_{\partial G_{h,1,m,j,k}}\Big)\sum_{i=1}^{3} e_h^+(x-y)e_i[f(V_{i,h}^{+}y)-f(y)]h^2$$

$$=f(x)-\sum_{i=1}^{3}\Big(\sum_{V_{i,h}^{+}(\mathring{G}_h\cup\partial G_{h,1})} - \sum_{j>k} \sum_{V_{i,h}\partial G_{h,1,j,k}} + \sum_{V_{i,h}^{+}\partial G_{h,1,m,j,k}} \Big)e_h^+(x-V_{i,h}^{-}y)e_i f(y)h^2 +$$

$$+\sum_{i=1}^{3}\Big(\sum_{(\mathring{G}_h\cup\partial G_{h,1})} - \sum_{j>k}\sum_{\partial G_{h,1,j,k}}+\sum_{\partial G_{h,1,m,j,k}}\Big)e_h^+(x-y)e_i f(y)h^2$$

$$\doteq f(x)-$$

$$-\sum_{i=1}^{3}\Big(\sum_{V_{i,h}^{+}(\mathring{G}_h\cup\partial G_{h,1,i})}+\sum_{j\neq i}\sum_{V_{i,h}^{+}\partial G_{h,1,j}}-\sum_{j>k}\sum_{V_{i,h}^{+}\partial G_{h,1,j,k}}+\sum_{V_{i,h}^{+}\partial G_{h,1,m,j,k}}\Big)e_h^{+}(x-V_{i,h}^{-}y)e_i f(y)h^2$$

$$+\sum_{i=1}^{3}\Big(\sum_{\mathring{G}_h\cup\partial G_{h,1}}-\sum_{j>k}\sum_{\partial G_{h,1,j,k}}+\sum_{\partial G_{h,1,m,j,k}}\Big)e_h^{+}(x-y)e_i f(y)h^2$$

$$=-\sum_{i=1}^{3}\Big(\sum_{\partial G_{h,1,j}}+\sum_{j\neq i}\sum_{V_{i,h}^{+}\partial G_{h,1,j}}-\sum_{j>k}\sum_{V_{i,h}^{+}\partial G_{h,1,m,j,k}}\Big)e_h^{+}(x-V_{i,h}^{-}y)e_i f(y)h^2$$

$$+\sum_{i=1}^{3}\Big(\sum_{\mathring{G}_h\cup\partial G_{h,1}}-\sum_{j>k}\sum_{\partial G_{h,1,j,k}}+\sum_{\partial G_{h,1,m,j,k}}\Big)e_h^{+}(x-y)e_i f(y)h^2. \tag{6.35}$$

To obtain the latter formula we use

$$\sum_{i=1}^{3}\sum_{y\in\mathring{G}_h}[e_h^{+}(x-V_{i,h}^{-}y)-e_h^{+}(x-y)]e_i f(y)h^2=$$

$$=\sum_{y\in\mathring{G}_h}[(e_h^{+}(x-y))\,D_{h,x}^{+}]f(y)h^3 = f(x) \quad x\in \operatorname{int} G_h,$$

because it is obviously that $e_h^{+} = D_h^{-}E_h = (E_h D_h^{-})$. With the relation $V_{i,h}^{+}\partial G_{h,1,j}=(\partial G_{h,1,j}\setminus S_{ij})\cup R_{ij}$ formula (6.35) may be further simplified. We find

$$\sum_{\substack{j=1\\ j\neq i}}^{3}\sum_{V_{i,h}^{+}\partial G_{h,1,j}} e_h^{+}(x-V_{i,h}^{-}y)e_i f(y)h^2-\sum_{i=1}^{3}\sum_{\partial G_{h,1}}e_h^{+}(x-y)e_i f(y)h^2 =$$

$$=\sum_{\substack{j=1\\ j\neq i}}^{3}\sum_{\partial G_{h,1,j}}e_h^{+}(x-V_{i,h}^{-}y)e_i f(y)h^2-\sum_{i,j=1}^{3}\sum_{\partial G_{h,1,j}} e_h^{+}(x-y)e_i f(y)h^2 -$$

$$-\sum_{\substack{j=1\\ j\neq i}}^{3}\sum_{S_{ij}}e_h^{+}(x-V_{i,h}^{-}y)e_i f(y)h^2+\sum_{\substack{j=1\\ j\neq i}}^{3}\sum_{R_{ij}}e_h^{+}(x-V_{i,h}^{-}y)e_i f(y)h^2 =$$

$$=-\sum_{i=1}^{3}\sum_{\partial G_{h,1,i}}e_h^{+}(x-V_{i,h}^{-}y)e_i f(y)h^2+\sum_{j=1}^{3}\sum_{\partial G_{h,1,j}}[e_h^{+}(x-y)\,D_{h,x}^{+}]f(y)h^3 -$$

$$-\sum_{\substack{j=1\\ j\neq i}}^{3}\sum_{S_{ij}} e_h^+(x-V_{i,h}^- y)e_i f(y)h^2+\sum_{\substack{j=1\\ j\neq i}}^{3}\sum_{R_{ij}} e_h^+(x-V_{i,h}^- y)e_i f(y)h^2 =$$

$$=\sum_{\partial G_{h,l}} e_h^+(x-V_{i,h}^- y)n(y)f(y)h^2-\sum_{\substack{j=1\\ j\neq i}}^{3}\Big(\sum_{S_{ij}}-\sum_{R_{ij}}\Big)e_h^+(x-V_{i,h}^- y)e_i f(y)h^2 .$$

Using (6.35) leads to

$$(F_h^+f)(x)=\sum_{i=1}^{3}\Big(-\sum_{\partial G_{h,l}}-\sum_{\partial G_{h,r}}\Big)e_h^+(x-V_{i,h}^- y)n(y)f(y)h^2+$$

$$+\sum_{i=1}^{3}\Big(\sum_{j>k}\sum_{V_{i,h}^+\partial G_{h,l,j,k}}-\sum_{V_{i,h}^+\partial G_{h,l,m,j,k}}+\sum_{j\neq i}\sum_{S_{ij}}-\sum_{j\neq i}\sum_{R_{ij}}\Big)e_h^+(x-V_{i,h}^- y)e_i f(y)h^2+$$

$$+\sum_{i=1}^{3}\Big(-\sum_{j>k}\sum_{\partial G_{h,l,j,k}}+\sum_{\substack{\partial G_{h,l,m,j,k}\\ m\neq j\neq k}}\Big)e_h^+(x-y)e_i f(y)h^2 .$$

Utilizing $V_{i,h}^+\partial G_{h,l,j,k}=(\partial G_{h,l,j,k}\setminus S_{ijk})\cup R_{ijk}$ and taking into consideration that

$$\sum_{i=1}^{3}\Big[\sum_{V_{i,h}^+\partial G_{h,l,m,j,k}}+\sum_{\substack{j=1\\ j\neq i}}^{3}\sum_{R_{ij}}-\sum_{\substack{i,k=1\\ j>k}}^{3}\sum_{R_{ijk}}\Big]e_h^+(x-V_{i,h}^- y)e_i f(y)h^2=0$$

we obtain

$$(F_h^+f)(x)=\sum_{i=1}^{3}\Big(-\sum_{\partial G_{h,l,i}}-\sum_{\partial G_{h,r,i}}\Big)e_h^+(x-V_{i,h}^- y)n(y)f(y)h^2+$$

$$+\sum_{i=1}^{3}\Big\{\sum_{j>k}\sum_{\partial G_{h,l,j,k}}-\sum_{j>k}\sum_{S_{ijk}}+\sum_{\substack{j=1\\ j\neq i}}^{3}\sum_{S_{ij}}\Big\}\, e_h^+(x-V_{i,h}^- y)e_i f(y)h^2+$$

$$+\sum_{i=1}^{3}\Big(-\sum_{j>k}\sum_{\partial G_{h,l,j,k}}+\sum_{\partial G_{h,l,m,j,k}}\Big)e_h^+(x-y)e_i f(y)h^2 =$$

$$=-\sum_{i=1}^{3}\Big(\sum_{\partial G_{h,l,i}}+\sum_{\partial G_{h,r,i}}\Big)e_h^+(x-V_{i,h}^- y)n(y)f(y)h^2+$$

$$+\sum_{i=1}^{3}\Big(-\sum_{j>k}\sum_{S_{ijk}}+\sum_{j=1}^{3}\sum_{S_{ij}}\Big)e_h^+(x-V_{i,h}^- y)e_i f(y)h^2+$$

$$+\sum_{i=1}^{3}\sum_{\substack{\partial G_{h,l,m,j,k} \\ m \neq j \neq k}} e_h^+(x-y)e_i f(y)h^2 \ . \qquad \#$$

6.3.4 Remark

Of course, it is possible to define F_h^+ by formula (6.34). Then we have to prove the discrete BOREL-POMPEIU formula.

6.3.5 Definition

An H-valued function f defined on G_h is called H-discrete-plus-regular (H-discrete-minus-regular) if

$f \in \ker D_h^+(\text{int } G_h)$ $(f \in \ker D_h^-(\text{int } G_h))$.

6.3.6 Remark

As in the case of continuous H-regular functions, we only consider left-regular functions. Therefore we shall omit the "left".

In the following we consider in detail the operators we have introduced.

6.3.7 Corollary

We have

(i) $F_h^+ f \in \ker D_h^+(\text{int } G_h)$, (6.36)

(ii) $((F_h^+)^2 f)(x) = (F_h^+ f)(x)$, $x \in \text{int } G_h$, (6.37)

(iii) $(F_h^+ f)(x) = f(x)$, $f \in \ker D_h^+(\text{int } G_h)$, $x \in \text{int } G_h$. (6.38)

Proof

The proof may be given by the help of (6.32) and (6.33). #

6.3.8 Corollary

It holds

(i) $(T_h^+ D_h^+ f)(x) = -(F_h^+ f)(x)$ for all $x \in \text{co } G_h$, (6.39)

(ii) $T_h^+ f \in \ker D_h^+(\text{co } G_h)$, (6.40)

(iii) $(F_h^+ f)(x) = 0$ for all $x \in \text{co } G_h$ and for all $f \in \ker D_h^+(\text{int } G_h)$. (6.41)

Proof

Relation (6.40) comes by simply operating with D_h^+ on $T_h^+ f$ and by using the properties of e_h^+. Equation (6.39) is to be verified by repetition of the proof of Theorem 6.3.3 for

$x \in co\, G_h$. Finally (6.41) follows from (6.39). #

In addition to the discrete BOREL-POMPEIU formula, it is of great importance for the development of a solution theory for discrete boundary value problems that the orthogonal complement of $\ker D_h^+(\mathrm{int}\, G_h)$ in $L_{2,h,H}(G_h)$ can be described.

6.3.9 Theorem

The following orthogonal decomposition of $L_{2,h,H}(G_h)$ holds:

$$L_{2,h,H}(G_h) = \ker D_h^+(\mathrm{int}\, G_h) \oplus D_h^-(\mathring{W}^1_{2,H}(G_h)). \tag{6.42}$$

Proof

First we show that

$$[D_h^-(\mathring{W}^{1,-}_{2,h}(G_h))]^\perp \subset \ker D_h^+(\mathrm{int}\, G_h).$$

Assume that

$$\sum_{x \in G_h} \overline{f(x)} g(x) h^3 = 0 \quad \text{for all} \quad g \in D_h^-(\mathring{W}^{1,-}_{2,h}(G_h)),$$

whence

$$\sum_{x \in G_h} \overline{f(x)} (D_h^- s)(x) h^3 = 0 \quad \text{for all} \quad s \in \mathring{W}^{1,-}_{2,h}(G_h).$$

This leads to

$$\sum_{i=1}^{3} \sum_{x \in G_h} \overline{f(x)} e_i (s(x) - s(V^-_{i,h} x)) h^2 =$$

$$= \sum_{i=1}^{3} \sum_{x \in G_h} \overline{f(x)} e_i s(x) h^2 - \sum_{i=1}^{3} \sum_{V^-_{i,h} G_h} \overline{f(V^+_{i,h} x)} e_i s(x) h^2 =$$

$$= -\sum_{x \in \mathrm{int} G_h} \sum_{i=1}^{3} [\overline{f(x)} - \overline{f(V^+_{i,h} x)}] \overline{e_i} s(x) h^2 =$$

$$= \sum_{x \in \mathrm{int} G_h} \overline{(D_h^+ f)(x)} s(x) h^3 = 0 \quad \Longrightarrow \quad (D_h^+ f)(x) = 0$$

and so $(D_h^+ f)(x) = 0$ for all $x \in \mathrm{int}\, G_h$.

On the other hand, we have for every $f \in \ker D_h^+(\mathrm{int}\, G_h)$

$$\langle f, g \rangle = \langle f, D_h^- s \rangle = \sum_{x \in \mathrm{int}\, G_h} \overline{(D_h^+ f)(x)} s(x) h^3 = \langle D_h^+ f, s \rangle = 0,$$

which means

$$\ker D_h^+(\mathrm{int}\, G_h) \subset [D_h^-(\mathring{W}^{1,-}_{2,h}(G_h))]^\perp .$$

Together the two inclusions yield

$$\ker D_h^+(\mathrm{int}\, G_h) = [D_h^-(\mathring{W}_{2,h}^{1,-}(G_h))]^{\perp}.$$ #

6.3.10 Remark

Theorem 6.3.9 is formulated for a fixed h. In this case the orthogonal complement may also be given in the form

$$(\ker D_h^+)^{\perp} = D_h^-(\{f : \mathrm{tr}\, f = 0\}),$$

because the set $\{f : \mathrm{tr}\, f = 0\}$ belongs to all discrete $W_{p,h,H}^k$-spaces. The special form (6.42) is in accordance with our further investigations and is chosen similarly to the continuous case.

The orthoprojections onto $\ker D_h^+(\mathrm{int}\, G_h)$, $D_h^-(\mathring{W}_{2,h}^{1,-}(G_h))$ will be denoted by $\mathbf{P}_h^+$, $\mathbf{Q}_h^+$, respectively. Obviously, there because $D_h^+\mathbf{P}_h^+=0$ we have

$$D_h^+\mathbf{Q}_h^+ f = D_h^+ f \quad \text{for all } f. \tag{6.43}$$

These orthoprojections allow statements to be formulated for the solution of the discrete boundary value problem

$$-\Delta_h u = f \quad \text{in } \mathrm{int}\, G_h,$$
$$u = g \quad \text{on } \partial G_h.$$

Note that up to here only concrete boundary value problems of the form

$$-\Delta u = K_h \quad \text{in } \mathrm{int}\, Q_1,$$
$$u = 0 \quad \text{on } \partial Q_1$$

have been considered (see Proposition 6.1.1) which are solved there explicitly. The aim of our further considerations is to gain statements of the existence and uniqueness of the solution of DIRICHLET's problem for the discrete LAPLACE equation in bounded lattices.

6.3.11 Theorem

The boundary value problem

$$-\Delta_h u = f \quad \text{in } \mathrm{int}\, G_h$$
$$u = 0 \quad \text{on } \partial G_h$$

has a solution for every f.

Proof

The $\mathbf{H}$-valued discrete function $u=T_h^-\mathbf{Q}_h^+T_h^+f$ satisfies the equation

$$-\Delta_h u = D_h^- D_h^+ u = f.$$

From $\operatorname{im} \mathbf{Q}_h^+ = D_h^-(\mathring{W}^{1,-}_{2,h}(G_h))$ follows $u=0$ on ∂G_h. Indeed, let $v \in D_h^-(\mathring{W}^{1,-}_{2,h}(G_h))$, then with a suitable $w \in \mathring{W}^{1,-}_{2,h}(G_h)$ we obtain

$$T_h^- v = T_h^- D_h^- w = w - F_h^- w = w$$

and

$$\operatorname{tr} T_h^- v = \operatorname{tr} w = 0.$$

#

6.3.12 Theorem

The boundary value problem

$$\begin{aligned} -\Delta_h u &= 0 \quad \text{in} \quad \operatorname{int} G_h, \\ u &= g \quad \text{on} \quad \partial G_h \end{aligned} \tag{6.44}$$

is solvable for any g.

Proof

Let w be any extension of g into $\operatorname{int} G_h$. We consider

$$u = w + T_h^- \mathbf{Q}_h^+ T_h^+ \Delta_h w. \tag{6.45}$$

There holds

$$-\Delta_h u = -\Delta_h w + D_h^+ D_h^- T_h^- \mathbf{Q}_h^+ T_h^+ \Delta_h w = 0,$$

$$\operatorname{tr} u = \operatorname{tr} w = g .$$

#

6.3.13 Corollary

Every solution of the boundary value problem (6.44) has the following representation:

$$u = f_1 + T_h^- f_2$$

with $f_1 \in \ker D_h^-(\operatorname{int} G_h)$, $f_2 \in \ker D_h^+(\operatorname{int} G_h)$.

Proof

We know from (6.45) that $u = w + T_h^- \mathbf{Q}_h^+ T_h^+ \Delta_h w$. This implies

$$u = w - T_h^- \mathbf{Q}_h^+ T_h^+ D_h^+ (D_h^- w) = w - T_h^- \mathbf{Q}_h^+ (D_h^- w - F_h^+ D_h^- w),$$

whence since $\mathbf{Q}_h^+ F_h^+ = 0$

$$u = w - T_h^- \mathbf{Q}_h^+ D_h^- w .$$

We further get

$$u = w - T_h^- D_h^- w + T_h^- \mathbf{P}_h^+ D_h^- w = F_h^- w + T_h^- \mathbf{P}_h^+ D_h^- w$$

and finally

$f_1=F_h^-w,\ f_2=P_h^+D_h^-w$. #

6.3.14 Remark

The concept of an orthogonal decomposition of $L_{2,h,H}(G_h^-)$ which we have described leads in a simple way to representations of the solution similar to those we could obtain by the help of the generalized VEKUA theory in the continuous case.

6.3.15 Theorem

The boundary value problem

$$-\Delta_h u = f \quad \text{in} \quad \text{int}\ G_h, \qquad u = g \quad \text{on} \quad \partial G_h \tag{6.46}$$

has for any f and g a unique solution.

Proof

Theorem 6.3.11 and Theorem 6.3.12 at once yield the existence of a solution of the problem (6.46). Assume there are two different solutions of (6.46). This leads to the problem

$-\Delta_h u=0,\quad \text{tr}\ u=0$.

For the solution of this boundary value problem, $u=T_h^-D_h^-u$ holds, using BOREL-POMPEIU's formula. Furthermore, because of $D_h^-u \in \ker D_h^+ \cap D_h^-(\mathring{W}^{1,-}_{2,h}(G_h))$ we find $Q_h^+D_h^-u=P_h^+D_h^-u$.

So we finally conclude $D_h^-u=0$, and therefore $u=T_h^-(D_h^-u)=0$. This is a contradiction. #

6.3.16 Theorem

The discrete space $L_{2,h,H}(\partial G_h)$ permits the following direct decompositions:

$$L_{2,h,H}(\partial G_h) = \text{im tr}\ F_h^- \oplus \text{im tr}\ T_h^- , \tag{6.47}$$

$$L_{2,h,H}(\partial G_h) = \text{im tr}\ F_h^+ \oplus \text{im tr}\ T_h^+ . \tag{6.48}$$

Proof

It is sufficient to show (6.47). Let $g \in L_{2,h,H}(\partial G_h)$ be an arbitrary H-valued function. The solution of the boundary value problem (6.46) possesses the representation (Corollary 6.3.13)

$u=f_1+T_h^-f_2,\ f_1 \in \ker D_h^-,\ f_2 \in \ker D_h^+$,

Consequently, we have

$g=\text{tr}\ F_h^-f_1+\text{tr}\ T_h^-f_2$.

The uniqueness of this representation remains to be verified. Let $g=w_1+w_2$ with $w_1=\text{tr } F_h^- v_1$, $w_2=\text{tr } T_h^- v_2$ be another representation, then for the difference with the first one is valid

$$\emptyset = \text{tr } F_h^-(f_1-v_1) + \text{tr } T_h^-(f_2-v_2). \tag{6.49}$$

BOREL-POMPEIU's formula at once yields for $T_h^- f$

$$F_h^- \text{tr } T_h^- f = 0.$$

By application of F_h^- in (6.49) relations (6.49) and (6.37) lead to

$$\text{tr } F_h^-(f_1-v_1)=\emptyset .$$

Therefore $w_1=\text{tr } F_h^- f_1$ and in the same way $w_2=\text{tr } T_h^- f_2$. #

6.3.17 Remark

The direct decompositions (6.47) and (6.48) are analogous to the continuous case. The validity of such a result could not be expected at first, because for proving the continuous case PLEMELJ-SOKHOTZKIJ's formulas were used, which are not available in the discrete theory.

Starting with the result of Theorem 6.3.15, we investigate the invertibility of the operator $\text{tr } T_h^- F_h^+$.

6.3.18 Theorem

The operator

$$\text{tr } T_h^- F_h^+ : \text{im tr } F_h^+ \longrightarrow \text{im tr } T_h^-$$

is an isomorphism.

Proof

Starting from $\text{tr}T_h^- F_h^+ v=\emptyset$, it follows that $T_h^- F_h^+ v=\emptyset$ in int G_h. Using Theorem 6.3.15, we obtain $F_h^+ v=\emptyset$, which implies (Theorem 6.3.2) $v=\emptyset$ for $v \in \text{im tr } F_h^+$. Therefore $\text{tr } T_h^- F_h^+$ is a 1-1-mapping.

Now let $w \in \text{im tr } T_h^-$, consequently $w=\text{tr } T_h^- v$. The boundary value problem

$$-\Delta_h u=\emptyset \quad \text{in} \quad \text{int } G_h,$$

$$u=w \quad \text{on} \quad \partial G_h$$

has a solution of the form (Theorem 6.3.12)

$$u=f_1+T_h^- f_2, \quad f_1 \in \ker D_h^-, \ f_2 \in \ker D_h^+,$$

whence

$tr\ u=w=tr\ F_h^- f_1 + tr\ T_h^- F_h^+ f_2$.

Therefore

$tr\ T_h^- v - tr\ T_h^- F_h^+ f_2 = tr\ F_h^- f_1 = 0$

and finally

$w = tr\ T_h^- v = tr\ T_h^- F_h^+ f_2 \in im\ tr\ T_h^- F_h^+$. #

6.3.19 Theorem

The orthoprojections P_h^+ and Q_h^+ allow the representations

$$P_h^+ = F_h^+ (tr\ T_h^- F_h^+)^{-1} tr\ T_h^- : L_{2,h,H} \xrightarrow{onto} L_{2,h,H} \cap \ker D_h^+(\mathrm{int}\ G_h), \tag{6.50}$$

$$Q_h^+ = (I - P_h^+) : L_{2,h,H} \xrightarrow{onto} L_{2,h,H} \cap D_h^-(\mathring{W}^{1,-}_{2,h}(G_h)). \tag{6.51}$$

Proof

Obviously, $P_h^+ f \in \ker D_h^+(\mathrm{int}\ G_h)$. Let $f \in \ker D_h^+(\mathrm{int}\ G_h)$. Then we have $f = F_h^+ f$ and so

$P_h^+ f = F_h^+ (tr\ T_h^- F_h^+)^{-1} tr\ T_h^- F_h^+ f = F_h^+ f = f$.

We have shown that P_h^+ is an operator which is acting onto $\ker D_h^+(\mathrm{int}\ G_h)$. An easy calculation yields

$(P_h^+)^2 = P_h^+,\ (Q_h^+)^2 = Q_h^+,\ P_h^+ Q_h^+ = Q_h^+ P_h^+$.

Considering $T_h^- Q_h^+ f$ it follows

$tr\ T_h^- Q_h^+ f = tr\ T_h^- f - (tr\ T_h^- F_h^+)(tr\ T_h^- F_h^+)^{-1} tr\ T_h^- f = 0$.

That means $Q_h^+ f \in D_h^-(\mathring{W}^{1,-}_{2,h}(G_h))$. On the other hand, let $g = D_h^- u$ with $u \in \mathring{W}^{1,-}_{2,h}(G_h)$, thus we obtain

$Q_h^+ g = D_h^- u - F_h^+ (tr\ T_h^- F_h^+)^{-1} tr\ T_h^- D_h^- u = D_h^- u - F_h^+ (tr\ T_h^- F_h^+)^{-1} tr\ u = D_h^- u = g$.

Therefore the assertion is proved. #

Finally we intend to describe some properties of H-discrete-regular functions defined on exterior domains as well as to discuss statements which are concerned with questions of extension.

6.3.20 Theorem

Let $f : co\ (G_h \cup \partial G_h) \longrightarrow H$ be a lattice function with the property $|f(s)| \longrightarrow 0$ for $|s| \longrightarrow \infty$. Furthermore we assume the existence of $T^+_{h,co\ G_h} D^+_h f$. Then we have

$$f(x) = (F^+_{h,co\ G_h} f)(x) - (T^+_{h,co\ G_h} D^+_h f)(x) , \quad x \in co\ G_h . \tag{6.52}$$

Proof

Consider $G_L=(co\ G_h \cup \partial G_h) \cap Q_L$, $Q_L=[-L,L]^3$.

In $int\ Q_L$ we have

$$f(x)=(F^+_{h,G_L} f)(x)+(T^+_{h,G_L} D^+_h f)(x)$$

with

$$(F^+_{h,G_L} f)(x)=\sum_{i=1}^{3} \sum\nolimits_1 e^+_h(x-V^-_{i,h}y)e_i f(y)h^2+$$

$$+\sum_{i=1}^{3} \sum\nolimits_2 e^+_h(x-y)e_i f(y)h^2+(F^+_{h,co\ G} f)(x) ,$$

where $\sum_1$, $\sum_2$ denote the summation over the appropriately chosen subsets of ∂Q_L with respect to y. The exact structure of these sums is not necessary here.

Let x be an arbitrary point. Then for sufficiently large L we have

$$\left|\sum_{i=1}^{3} \sum\nolimits_1 e^+_h(x-V_{i,h}y)e_i f(y)h^2+\sum_{i=1}^{3} \sum\nolimits_2 e^+_h(x-y)e_i f(y)h^2\right| \le$$

$$\le 18L^2h^{-2}C|x-y|^{-2}\sup\{|f(y)|, y\in\partial Q_L\}h^2 \le$$

$$\le C_1 \sup\{|f(y)|, y\in\partial Q_L\} \longrightarrow 0 \quad \text{for } L \longrightarrow \infty .$$

#

6.3.21 Remark

$F^+_{h,co\ G_h}$ is defined similarly to the case of bounded domains. We have to take into consideration that the subsets of boundary points used must be considered as boundary points of the set $co\ G_h \cup \partial G_h$.

6.3.22 Corollary

For $x \in co\ G_h$ and $f \in ker\ D^+_h(co\ G_h)$ we have

$$f(x)=(F^+_{h,co\ G_h} f)(x). \tag{6.53}$$

6.3.23 Remark

Note that by using the subsets $S_{ij}(G_h)$, $S_{ijk}(G_h)$ and

$G_{1,h,m,j,k}$ (analogous constructions for co G_h) the operators F^+_{h,G_h} and $F^+_{h,co\,G_h}$ are not given on the same subsets of G_h in general. For that reason the operators F^+_{h,G_h} and $F^+_{h,co\,G_h}$ do differ not only the sign as it is in the continuous case. A simple connection between them has not been found up to now.

6.3.24 Corollary

Let $f\in \ker D^+_h(co\,G_h)$ with $|f(x)|\longrightarrow 0$ for $|x|\longrightarrow \infty$. Then we have the estimate

$$|f(x)| \leq C|x|^{-2} \quad \text{for} \quad |x|>R. \tag{6.54}$$

Proof

Obviously, when $x\in co\,G_h$

$$f(x)=(F^+_{h,co\,G_h}f)(x)\ ,\ x\in co\,G_h,$$

whence follows for x with $|x|>R$

$$|f(x)|\leq C\sum_{y\in\partial G_h}|x-y|^{-2}h^2\leq C_1\sup\{|x-y|^{-2}, y\in G_h\}\leq C_2|x|^{-2}.$$ #

With this result we finish our considerations of the H-discrete-regular functions. The reader can continue further investigations into the subject of elliptic equations on exterior domains by the help of the methods presented in [B] also, in the case of discrete functions. Considering $D_{h,i}e^+_h$ for $|x|\longrightarrow\infty$, a discrete variant of LIOUVILLE's theorem can be found.

6.4 Main Properties of Discrete Operators

In the preceding section the definitions of the discrete analogues of the operators T_G, D and F_Γ as well as the proof of simple algebraic properties were in the centre of our considerations. This already led to statements of existence and uniqueness of the solution of DIRICHLET's problem. Statements relative to correctness and a-priori estimates of the solutions require the answer of the question in which way we have to choose a pair of spaces in which the discrete operators are uniformly bounded. In order to obtain assertions about the relations between the solutions of discrete and continuous boundary value problems, it is necessary to investigate the behaviour of the discrete operators for $h\longrightarrow 0$. Both questions are dealt within the present section. First we shall deal with considerations of the continuity and estimates of the norm for fixed $h>0$.

6.4.1 Theorem

For $1<p<3$ and $q<\frac{3p}{3-p}$ the operator

$$T_h^+ : L_{p,h,H}(G_h) \longrightarrow L_{q,h,H}(G_h)$$

is continuous.

Proof

Assume $p<q$. An application of HOELDER's inequality (for sums) to a product consisting of three factors yields

$$|T_h^+ f(x)| \le \Big(\sum_{y\in G_h}\{|f(y)|^{p/q}|x-y|^{-3/q+3/(2q)+3/(2p')-1}\}^q h^3\Big)^{1/q} *$$

$$* \Big(\sum_{y\in G_h}\{|f(y)|^{p(1/p-1/q)}\}^{pq/(q-p)} h^3\Big)^{1/p-1/q} *$$

$$* \Big(\sum_{y\in G_h}\{|x-y|^{-3/p'+3/(2q)+3/(2p')-1}\}^{p'} h^3\Big)^{1/p'} \le$$

$$\le \|f\|_{p,h,H}^{1-p/q}\big(C(\operatorname{diam} G)^{(3p'/(2q))+(3/2-p')}(3p'/(2q)-3/2-p')^{-1}\big)^{1/p'} *$$

$$* \Big(\sum_{y\in G_h}\{|f(y)|^p|x-y|^{-3/2+3q/(2p')-q}h^3\}\Big)^{1/q} =$$

$$= \|f\|_{p,h,H}^{1-p/q} K(p',q,G)^{1/p'} *$$

$$* \Big(\sum_{y\in G_h}|f(y)|^p|x-y|^{-3/2+3q/(2p')-q}h^3\Big)^{1/q}.$$

Thence

$$\|T_h^+ f\|_{q,h,H}^q \le \|f\|_{p,h,H}^{q-p} K(p',q,G)^{q/p'} \sum_{y\in G_h}|f(y)|^p h^3 K(q,p',G),$$

and so

$$\|T_h^+ f\|_{q,h,H} \le K(p',q,G)^{1/p'} K(q,p',G)^{1/q}\|f\|_{p,h,H}. \qquad (6.55)$$

Necessary conditions for the validity of the above estimates are

$$\frac{3}{2}+\frac{3q}{2p'}-q>0 \quad\text{and}\quad \frac{3p'}{(2q)}+\frac{3}{2}-p'>0.$$

Both inequalities are fulfilled for $q<\frac{3p}{3-p}$. p' is defined by $\frac{1}{p}+\frac{1}{p'}=1$. The function $K(r,s,G)$ is given by

$$K(r,s,G)=C(\operatorname{diam} G)^{3r/(2s)+3/(2-r)}(3r/(2s)+3/2-r)^{-1}$$

and arises from Proposition 6.1.7. The principle of this proof is similar to the continuous case. #

6.4.2 Corollary

$$T_h^+ \in L(L_{p,h,H};L_{2,h,H}) \quad \text{for} \quad p>6/5,$$

$$T_h^+ \in L(L_{2,h,H};L_{q,h,H}) \quad \text{for} \quad q<6.$$

Proof

The proof can be obtained by a special choice of p and q and the application of Theorem 6.4.1. #

6.4.3 Theorem

$$T_h^+ \in L(L_{p,h,H};C_{h,H}) \quad \text{for} \quad p>3.$$

Proof

$$|T_h^+f(x)| = |\sum_{y\in G_h} e_h^+(x-y)f(y)h^3| \le$$

$$\le (\sum_{y\in G_h} |e_h^+(x-y)|^q h^3)^{1/q} (\sum_{y\in G_h} |f(y)|^p h^3)^{1/p} \le$$

$$\le \|f\|_{p,h,H} \|e_h^+\|_{q,h,H}.$$

The last inequality follows by using Theorem 6.2.8 and HOELDER's inequality. #

Next we restrict the domain of T_h^+ to $D_h^-(W_{2,h}^{1,-}(G_h))$ and verify the following embedding theorem:

6.4.4 Theorem

Let $u \in \dot{W}_{2,h,H}^{1,-}(G_h)$ and $q<6$. Then it holds

$$\|u\|_{q,h,H} \le C\|u\|_{W_{2,h,H}^1}.$$

Proof

It is clear that $D_h^- u \in L_{2,h,H}$, whence follows

$$\|u\|_{q,h,H} \le \|T_h^- D_h^- u\|_{q,h,H} \le K(2,q,G)^{1/2} K(q,2,G)^{1/q} \|D_h^- u\|_{2,h,H} \le$$

$$\le C\|u\|_{W_{2,h,H}^1}.$$

In this case we made use of the discrete BOREL-POMPEIU's formula and Theorem 6.4.1. #

In the discrete case it is also possible to deduce a relation between the smallest eigenvalue $\lambda_{1,h}(G_h)$ of $-\Delta_h$ and the norm of T_h^+. A simple calculation yields

$$\langle -\Delta_h u, v\rangle = \langle u, -\Delta_h v\rangle \quad \text{for } u,v \in \overset{\circ}{W}{}^{1,-}_{2,h}(G_h).$$

Besides we can obtain

$$\langle -\Delta_h u, u\rangle = \langle D_h^+ D_h^- u, u\rangle = \langle D_h^- u, D_h^- u\rangle \geq 0 \quad \text{for } u \in \overset{\circ}{W}{}^{1,-}_{2,h}(G_h).$$

Because we have for $u \in \overset{\circ}{W}{}^{1,-}_{2,h}(G_h) \cap \ker D_h^-(\operatorname{int} G_h)$ by the help of the discrete BOREL-POMPEIU's formula $u=0$, the positive definiteness of $-\Delta_h$ is shown. Furthermore we get

$$\sup\{\|T_h^- f\|_{2,h,H}/\|f\|_{2,h,H}; f \in \operatorname{im} Q_h^+\} =$$

$$= \sup\{\|g\|_{2,h,H}/\|D_h^- g\|_{2,h,H}; g \in W^{1,-}_{2,h}(G)\} = (\lambda_{1,h}(G_h))^{-1/2}. \quad \#$$

6.4.5 Theorem

(i) $\|T_h^- f\|_{2,h,H} \leq (\lambda_{1,h}(G_h))^{-1/2} \|f\|_{2,h,H}, \quad f \in \operatorname{im} Q_h^+,$ (6.56)

(ii) $\|T_h^- f\|_{2,1,h,H} \leq (1+1/\lambda_{1,h}(G_h))^{1/2} \|f\|_{2,h,H}, \quad f \in \operatorname{im} Q_h^+.$ (6.57)

Proof

Inequality (6.56) has already been verified. Estimate (6.57) may be proved similarly to the continuous case (see Theorem 4.1.14). #

6.4.6 Remark

One can verify that estimates (6.56) and (6.57) cannot be improved.

By the aid of (6.55) it is possible to find out lower bounds for $\lambda_{1,h}(G_h)$ (independently of h). These problems may be solved with methods which are applied to the continuous case. We can prove that $\lambda_{1,h}(G_h) \longrightarrow \lambda_1(G)$ for $h \longrightarrow 0$. Obviously holds $\|Q_h^+\|_{[L_{2,h,H}; L_{2,h,H}]} = 1$. Therefore we can obtain all the estimates which are needed for investigating elliptic boundary value problems even in the discrete case (in corresponding spaces) for fixed h, and it enables us to formulate the results of Chapter IV in a discrete language. Note that the norms of T_h^+ and T_h^- as well as the embedding constants of Theorem 6.4.4 may be uniformly estimated with respect to h. By the aid of a lower estimate of $\lambda_{1,h}$ the assertion of Theorem 6.4.5 becomes independent of h. To generate a relation between the discrete and the continuous boundary value problems, we have to consider the behaviour of the operators

T_h^+, T_h^-, Q_h^+, F_h^+, F_h^-, D_h^- and D_h^+ for $h \longrightarrow 0$. In particular, statements about the corresponding properties in $L_{p,H}(G)$ outside of the lattice G_h are also needed.

6.4.7 Theorem

Let $f \in \mathfrak{R}_H \cap L_{\infty,H}(G)$, then

$$\| T_h^+ f - T_G f \|_{C_{h,H}(G)} \longrightarrow 0 .$$

If $f \in C_H^{0,\beta}(\bar{G})$, $0 < \beta \leq 1$, then we have

$$\| T_h^+ f - T_G f \|_{C_{h,H}(G)} \leq C(G, \| f \|_{q,h,H}, \| f \|_{q,H}) h^{-2+3/p} |\ln(h)| +$$

$$+ K_{p,G} |h|^\beta \| f \|_{C_H^{0,\beta}(G)}$$

for $p < 3/2$, $1/p + 1/q = 1$.

Proof

Let $W(y)$ be a cube with side h and midpoint y. We obtain the following estimates:

$$|(T_h^+ f)(x) - (Tf)(x)| \leq \Big| \sum_{y \neq x} e_h^+(x-y) f(y) h^3 - \sum_{x \notin W(y)} \int_{W(y)} e(x-z) f(z) dG_z \Big| +$$

$$+ \Big| \int_{W(x)} e(x-y) f(y) dG_y \Big| \leq$$

$$\leq \sum_{y \neq x} | e_h^+(x-y) - e(x-y) | \, |f(y)| h^3 +$$

$$+ \sum_{y \neq x} \Big| e(x-y) f(y) h^3 - \int_{W(y)} e(x-z) f(z) dG_z \Big| + \int_{W(x)} |x-y|^{-2} |f(y)| dG_y \leq$$

$$\leq \Big(\sum_{y \neq x} | e_h^+(x-y) - e(x-y) |^p h^3 \Big)^{1/p} \| f \|_{q,h,H} + h^{3/p-2} \| f \|_{q,W(x)} +$$

$$+ \sum_{y \neq x} \Big| e(x-y) f(y) h^3 - \int_{W(y)} e(x-z) f(z) dG_z \Big| .$$

Furthermore we decompose the last sum

$$\sum_{y \neq x} \Big| e(x-y) f(y) h^3 - \int_{W(y)} e(x-z) f(z) dG_z \Big| \leq$$

$$\leq \underbrace{\sum_{y \neq x} \Big| \int_{W(y)} (e(x-y)[f(y) - f(z)] dG_z \Big|}_{= S_1} + \underbrace{\sum_{y \neq x} \int_{W(y)} | e(x-y) - e(x-z) | \, |f(z)| dG_z}_{= S_2} .$$

By the help of a TAYLOR expansion of the function $e(x-z)$ we obtain the following estimate of S_2:

$$\sum_{y\neq x}\int_{W(y)}|e(x-y)-e(x-z)||f(z)|dG_z\leq$$

$$\leq\sum_{y\neq x}\int_{W(y)}\sum_{i=1}^{3}|D_{i,z}e(x-z)|_{z=\theta(y)}||z_i-y_i||f(z)|dG_z\leq$$

$$\leq 3C^*\sum_{y\neq x}\left(\int_{W(y)}h^p2^{-p}(|x-y|-3^{1/2}0.5h)^{-3p}dG_z\right)^{1/p}\|f\|_{q,W(y)}\leq$$

$$\leq h^{-2+3/p}((1.5)C')\sum_{y\neq x}h^3(|x-y|-3^{1/2}(0.5)h)^{-3}\|f\|_{q,W(y)}\leq$$

$$\leq h^{-2+3/p}(1.5)C\ [C\sum_{y\neq x}h^3|x-y|^{-3}\ \|f\|_{q,W(y)}]\leq$$

$$\leq h^{-2+3/p}C^*\|f\|_{q,G}[C_1+C_2(\text{diam } G)-C_3|\ln(h)|]. \tag{6.58}$$

In this proof of (6.58) we use Proposition 6.1.7 and the inequality

$$(|x-y|-3^{1/2}(0.5)h)^{-1}\leq C|x-y|^{-1}\quad \text{for } x,y\in R_h^3,\ x\neq y,$$

with $C>(1-3^{1/2}\cdot 0.5)^{-1}$.

Now follows the estimation of the term S_1. We have

$$S_1(x)\leq\sum_{y\neq x}\left(\int_{W(y)}|e(x-y)|^pdG_z\right)^{1/p}\left(\int_{W(y)}|f(y)-f(z)|^qdG_z\right)^{1/q}\leq$$

$$\leq\sum_{y\neq x}(|e(x-y)|^ph^3)^{1/p}\left(\int_{W(y)}|f(y)-f(z)|^qdG_z\right)^{1/q}\leq$$

$$\leq\|\hat{e}\|_{p,h,H}\left(\sum_{y\neq x}\int_{W(y)}|f(y)-f(z)|^qdG_z\right)^{1/q}\leq$$

$$\leq\|\hat{e}\|_{p,h,H}\left(\sum_{y\neq x}|\sup\{|f(z)|;z\in W(y)\}-\inf\{|f(z)|;z\in W(y)\}|h^3\right)^{1/q}*$$

$$*(2V)^{1-1/q}\longrightarrow 0\quad\text{for}\quad h\longrightarrow 0\ ,$$

if $f\in\mathcal{R}_H$(RIEMANN integrable) , $|f|\leq V$ and $p<3/2$.

Summing all estimates and using Theorem 6.2.8 and Theorem 6.2.11 we get the inequality

$$|(T_h^+f)(x)-(Tf)(x)| \le C(G,\|f\|_{q,h,H},\|f\|_{q,G})h^{-2+3/p}|\ln(h)| +$$

$$+ K_{p,G}(\sum_{y\in G_h}\int_{W(y)}|f(y)-f(z)|^q dG_z)^{1/q},$$

where $x \in G_h$.

Since the right-hand side of the last inequality does not depend on x, we may find uniform estimates in G_h. #

6.4.8 Theorem

Let $f \in \mathcal{R}_H \cap L_{\infty,H}(G)$. For $p \in (6/5,3/2)$

$$\|T_h^+f-Tf\|_{p,H} \longrightarrow 0$$

is valid, and for $f \in C_H^{0,\beta}(\bar{G})$, $p \in (6/5,3/2)$, $0<\beta\le 1$ we have

$$\|T_h^+f-Tf\|_{p,H} \le C\|f\|_{\infty,G}h^{-2+3/p} +$$

$$+ C_1(G,\|f\|_{q',h,H},\|f\|_{q',G})h^{-2+3/p'}|\ln(h)| +$$

$$+ K_{p',G}\|f\|_{C_H^{0,\beta}(G)}h^{\beta}.$$

Proof

By using the natural extension of e_h^+ from R_h^3 into R^3 it is possible to define $(T_h^+f)(x)$ for $x \notin G_h$. To get $L_{p,H}$-estimates for T_h^+, we can repeat the same considerations as in the foregoing Theorem 6.4.7. As $x \in R_h^3$ we do not have to add up all y with $y\neq x$, but all y with $|y-x|\ge h$. The remaining terms as well as the estimation of the sums $S_1(x)$ and $S_2(x)$ can be done similarly to the above accomplished calculations. This leads to

$$|(T_h^+f)(x)-(Tf)(x)| \le \sum_{y;|y-x|\ge h}|e_h^+(x-y)-e(x-y)||f(y)|h^3 +$$

$$+ Ch^{-2+3/p'}|\ln h| +$$

$$+ \|\hat{e}\|_{p',h,H}(\sum_{y\in G_h}\int_{W(y)}|f(y)-f(z)|^{q'}dG_z)^{1/q'}.$$

The assertion follows from Theorem 6.2.12. #

6.4.9 Remark

Note that the speed of convergence cannot exceed $h^{1/2}$, where $h^{1/2}$ will not be reached. This is caused by the condition $p>6/5$. Of course, it is possible to prove the convergence in

$L_{p,H}(G)$ for $1\le p\le 6/5$, but it does not lead to an increased speed of convergence. The true reason for this is not known.

To prepare a statement of convergence for Q_h^+, we consider the operator F_h^+. For the sake of simplicity we only deal with the case $F_h^+ f$ for $f \in C_H^{1,\beta}(\bar{G})$ in order to prove convergence with a certain speed.

6.4.10 Theorem

Let $f \in C_H^{1,\beta}(\bar{G}) \cap (\ker D)(G)$. Then we have the estimate

$$\| f - F_h^+ f \|_{2,h,H} \le C h^{\beta},$$

where the constant C does not depend on h.

Proof

BOREL-POMPEIU's formula yields

$$f(x) - (F_h^+ f)(x) = (T_h^+ D_h^+ f)(x) = (T_h^+ (D_h^+ f - Df))(x).$$

Besides we have

$$|(D_h^+ f)(x) - (Df)(x)| \le C \| f \|_{C_H^{1,\beta}} h^{\beta},$$

and therefore

$$\| f - F_h^+ f \|_{2,h,H} \le \| T_h^+ \|_{[C_H, L_{2,h,H}]} \| D_h^+ f - Df \|_{2,h,H} \le C \| f \|_{C_H^{1,\beta}} h^{\beta}. \qquad \#$$

6.4.11 Theorem

Let $f \in C_H^{1,\beta}(\bar{G})$. Then

$$\| Q_h^+ f - Qf \|_{2,h,H} \le C h^{\beta}, \qquad 0 < \beta \le 1 .$$

Proof

It is readily seen that

$$Q_h^+ f - Qf = Q_h^+ Pf + Q_h^+ Qf - QQf.$$

On the other hand

$$Q_h^+ Pf = Q_h^+ FPf = Q_h^+ FPf - Q_h^+ F_h^+ Pf = Q_h^+ (F - F_h^+) Pf.$$

It follows

$$\| Q_h^+ Pf \|_{L_{2,h,H}} \le C h^{\beta}.$$

Using $Qf = Dg$ with $g \in \overset{\circ}{W}{}_2^1(G) \cap C_H^{2,\beta}(\bar{G})$ we thus obtain considering $D_h^- g \in \operatorname{im} Q_h^+$

$$Q_h^+Dg-Dg=Q_h^+(Dg-D_h^-g)+Q_h^+D_h^-g-Dg=Q_h^+(Dg-D_h^-g)+(D_h^-g-Dg),$$

whence

$$\| Q_h^+Qf-QQf \|_{2,h,H} \le C_2 h^{\beta} .$$ #

We can prove a somewhat weaker result in the case $f \in C_H^{0,\beta}(\bar{G})$.

6.4.12 Theorem

$$\| Q_h^+f-Qf \|_{2,h,H} \longrightarrow 0 \quad \text{for} \quad h \longrightarrow 0 \quad \text{if} \quad f \in C_H^{0,\beta}(\bar{G}).$$

Proof

Let $f \in C_H^{0,\beta}(\bar{G})$ and $\varepsilon > 0$. Then holds

$$\| Q_h^+f-Qf \|_{2,h,H} \le \| Q_h^+f-Q_h^+f^* \|_{2,h,H} + \| Q_h^+f^*-Qf^* \|_{2,h,H} + $$

$$+ \| Qf^*-Qf \|_{2,h,H} ,$$

where $f^* \in C_H^{1,\beta}(\bar{G})$ with $\| f^*-f \|_{C_H^{0,\beta}} < \varepsilon/4$. Furthermore we obtain

$$\| Q_h^+(f-f^*) \|_{2,h,H} \le \| f-f^* \|_{2,h,H} < \varepsilon/4 .$$

Because of the validity of Theorem 6.4.11 we have for all $h>0$

$$\| Q_h^+f^*-Qf^* \|_{2,h,H} \le C(f^*)h^{\beta},$$

and so

$$\| Q(f^*-f) \|_{2,h,H} \le 2 \| Q(f^*-f) \|_{2,G} \le \varepsilon/2 .$$

The last estimate holds for $h<h_1$ on account of the RIEMANN-integrability of the function $Q(f^*-f)$. With $C(f^*)h^{\beta} < \varepsilon/4$ for $h<h_2$ we obtain

$$\| Q_h^+f-Qf \|_{2,h,H} < \varepsilon \quad \text{for} \quad h<\min\{h_1,h_2\}.$$ #

6.4.13 Remark

The consideration of the operator Q_h^+ emphasizes the efficiency of this approach. In particular, the part $(\mathrm{tr}\, T_h^-F_h^+)^{-1}$ of the projection Q_h^+ is hard to deal with using other inadequate approximations of T_G and F_Γ, because the conditions of the invertibility in most of the cases are not fulfilled uniformly.

6.5 Numerical Solution of Boundary Value Problems of NAVIER-STOKES Equations

In this chapter we intend to demonstrate the numerical solution of elliptical boundary value problems by the help of the discrete theory we have developed. In this connection we shall deal with boundary value problems of STOKES' equations and then consider an iterative method of solving NAVIER-STOKES equations.

6.5.1 Proposition

Let $f,g \in \overset{\circ}{W}{}^{1,-}_{2,h,H}(G_h)$. Then

$$\langle D_h^+ f,g\rangle = \langle f,D_h^- g\rangle .$$

Proof

We get from the definition of the scalar product

$$\langle D_h^+ f,g\rangle=\sum_{\mathrm{int}\,G_h}\sum_{i=1}^{3} \overline{e_i(f(V_{i,h}^+x)-f(x))}g(x)h^2=$$

$$=-\sum_{i=1}^{3}\sum_{V_{i,h}^+\overset{\circ}{G}_h} \overline{f(x)e_i}g(V_{i,h}^-x)h^2+ \sum_{i=1}^{3} \overline{f(x)e_i}g(x)h^2=$$

$$=\sum_{i=1}^{3}\sum_{\mathrm{int}\,G_h}\overline{f(x)e_i}(g(x)-g(V_{i,h}^-x))h^2=\langle f,D_h^- g\rangle \ . \qquad \#$$

6.5.2 Remark

The functions f and g may be extended by zero into the domain $\mathrm{co}\ G_h$. Therefore the definition of D_h^+f and D_h^-f does not cause any difficulties in points on ∂G_h.

6.5.3 Proposition

Let $f,g \in \overset{\circ}{W}{}^{1,-}_{2,h,H}(G_h)$. Then

$$\mathrm{Re}\langle D_h^+ \,\mathrm{Re}\, f,g\rangle = \mathrm{Re}\langle f,\mathrm{Re}\, D_h^- g\rangle \ . \tag{6.59}$$

Proof

Proposition 6.5.1 immediately yields

$$\langle D_h^+ \,\mathrm{Re}\, f,g\rangle=\langle \mathrm{Re}\, f,\, D_h^- g\rangle,$$

and therefore

$$\mathrm{Re}\langle D_h^+ \,\mathrm{Re}\, f,g\rangle=\mathrm{Re}\langle \mathrm{Re}\, f,D_h^- g\rangle=\langle \mathrm{Re}\, f,\mathrm{Re}\, D_h^- g\rangle=\mathrm{Re}\langle f,\mathrm{Re}\, D_h^- g\rangle. \qquad \#$$

In the following we shall abbreviate $\mathrm{Re}\langle\ ,\ \rangle$ to $[\ ,\ ]$. Now we define

$$\mathrm{grad}_h^+ f:=(D_{1,h}^+,D_{2,h}^+,D_{3,h}^+)^T,$$

$$\operatorname{div}_h^- v := \sum_{i=1}^{3} D_{i,h}^- v_i \ .$$

Identity (6.59) may be written in the form

$$[\operatorname{grad}_h^+ u, v] = [u, -\operatorname{div}_h^- v] \ ,$$

where $u, v \in W_{2,h,H}^{1,-}(G_h)$, $u : G \longrightarrow R^1$, $v : G \longrightarrow R^3$. Thence

$$(\operatorname{grad}_h^+)^* = -\operatorname{div}_h^- \ ,$$

and consequently we obtain

$$\ker \operatorname{div}_h^- = (\operatorname{im} \operatorname{grad}_h^+)^\perp \ ,$$

where the orthogonality is to be understood with respect to the scalar product (6.59).

6.5.4 Proposition

Let $u \in \mathring{W}_{2,h,H}^{1,-}(G_h) \cap \ker \operatorname{div}_h^-$, $p \in L_{2,h,R}(G_h)$ with $\operatorname{Im} p = 0$. Then

$$[D_h^- u, Q_h^+ p] = 0 \ .$$

Proof

Because of $D_h^- u \in \operatorname{im} Q_h^+$, $\operatorname{Re} D_h^- u = 0$ and $\operatorname{Im} p = 0$ we get

$$[D_h^- u, Q_h^+ p] = [Q_h^+ D_h^- u, p] = [D_h^- u, p] = 0.$$ #

6.5.5 Theorem

For each $f \in L_{2,h,H}$ there exist H-valued functions $u \in W_{2,h}^{1,-}(G_h) \cap \ker \operatorname{div}_h^-$ and $p \in L_{2,h,R}$ with $\operatorname{Im} p = 0$ such that

$$\frac{\varrho}{\eta} Q_h^+ T_h^+ f = D_h^- u + \frac{1}{\eta} Q_h^+ p \ . \tag{6.60}$$

Proof

Obviously $D_h^- u \in \operatorname{im} Q_h^+$, $Q_h^+ p \in \operatorname{im} Q_h^+$. With respect to the validity of Proposition 6.5.4 it is necessary to show that the relations

$$[D_h^- u, Q_h^+ T_h^+ f] = 0 \quad \text{for} \quad u \in \mathring{W}_{2,h,H}^{1,-}(G_h) \cap \ker \operatorname{div}_h^- \ ,$$

$$[Q_h^+ p, Q_h^+ T_h^+ f] = 0 \quad \text{for} \quad p \in L_{2,h,R} \quad \text{with} \quad \operatorname{Im} p = 0$$

imply $f = 0$.

First we obtain

$[D_h^- u, Q_h^+ T_h^+ f] = [D_h^- u, T_h^+ f] = [u, f] = 0$ and therefore $f = \operatorname{grad}_h^+ q$ with $\operatorname{Im} q = 0$. On the other hand, we have $\| Q_h^+ q \|_{2,h,H} = 0$, whence $\operatorname{grad}_h^+ q = D_h^+ Q_h^+ q = 0$ and $f = 0$. #

If the operator T_h^- is applied to representation (6.60) then there follows, by making use of the discrete BOREL-POMPEIU formula the existence of a decomposition of the function $T_h^- Q_h^+ T_h^+ f$ into the sum
into the sum

$$u + \frac{1}{\eta} T_h^- Q_h^+ p = \frac{\varrho}{\eta} T_h^- Q_h^+ T_h^+ f$$

with suitably chosen discrete functions

$u \in \mathring{W}^{1,-}_{2,h,H}(G_h) \cap \ker \operatorname{div}_h^-$, $p \in L_{2,h,R}$, $\ker \operatorname{Im} p = 0$.

Summarizing we may formulate for the solution of the discrete STOKES' problem.

6.5.6 Theorem

The discrete boundary value problem

$$-\Delta_h u + \frac{1}{\eta} \operatorname{grad}_h^+ p = \frac{\varrho}{\eta} f \quad \text{in int } G_h, \tag{6.61}$$

$$\operatorname{div}_h^- u = 0 \quad \text{in int } G_h, \tag{6.62}$$

$$u = 0 \quad \text{on } \partial G_h \tag{6.63}$$

has for every $f \in L_{2,h,H}$ a solution (u,p), where u and p are uniquely defined (p up to a real constant !).

Proof

The existence has already been shown. Formula (6.60) and Proposition 6.5.4 yield

$$\| Q_h^+ T_h^+ f \|^2_{2,h,H} = \| D_h^- u \|^2_{2,h,H} + \frac{1}{\eta^2} \| Q_h^+ p \|^2_{2,h,H},$$

whence

$$\| D_h^- u \|_{2,h,H} + \frac{1}{\eta} \| Q_h^+ p \|_{2,h,H} \le 2^{1/2} \| Q_h^+ T_h^+ f \|_{2,h,H}. \tag{6.64}$$

This a-priori estimate leads us to the uniqueness of u. Assuming the existence of two solutions (u,p_1) and (u,p_2), thus it immediately implies $p_1 - p_2 \in \ker Q_h^+$, therefore $p_1 - p_2 \in \ker D_h^+(\text{int } G_h)$ and $p_1 - p_2 = \text{const.} \in \mathbb{R}^1$. #

6.5.7 Corollary

There is valid the a-priori estimate

$$(\lambda_{1,h}/(1+\lambda_{1,h}))^{1/2} \| u \|_{2,1,h,H} + \frac{1}{\eta} \| Q_h^+ p \|_{2,h,H} \le 2^{1/2} \| T_h^+ f \|_{2,h,H}.$$

Proof

It may be proved by using Theorem 6.4.4 and Theorem 6.4.5. #

6.5.8 Remark

The treatment of the discrete STOKES problem points out a wide correspondence with the continuous case. This relates to both the method and the concrete formulation of the results. The only speciality of the discrete theory is the change-over from D_h^+ to D_h^- in the partial summation (partial integration in the continuous case).

6.5.9 Remark

The discrete boundary value problem (6.61)-(6.62)-(6.63) may be interpreted by a scheme of finite differences. That means our discrete function theory can be regarded as a new approach to the construction and analytical investigation of finite difference methods.

In a simple way a-priori estimates, for instance (6.64) allow the investigation of stability problems.

Now we shall deal with DIRICHLET's problem for NAVIER-STOKES equations. Considering the latter formulated results for the STOKES problem, the non-linear term $M^*(u)=\frac{\varrho}{\eta}(u,\mathrm{grad})u$ remains to be discretized in a proper way.

Defining

$$M_h^{*,-}(u):=\frac{\varrho}{\eta}(u,\mathrm{grad}_h^-)u \quad \text{and} \quad M_h^-(u):=M_h^{*,-}-\frac{\varrho}{\eta}f,$$

we have

$$\left\|\frac{\eta}{\varrho}M_h^{*,-}(u)\right\|_{p,h,H}^p \le \left\|\sum_{i,j=1}^{3} u_i D_{i,h}^- u_j e_j\right\|_{p,h,H}^p \le \sum_{i,j=1}^{3}\|u_i D_{i,h}^- u_j\|_{p,h,H}=$$

$$=\sum_{i,j=1}^{3}\sum_{y\in G_h}|u_i(y)D_{i,h}^-u_j(y)|^p h^3 \le \sum_{i,j=1}^{3}\|u_i\|_{q,h,H}^p\|D_{i,h}^-u_j\|_{2,h,H}^p \le$$

$$\le \sum_{i,j=1}^{3} C^p\|u_i\|_{2,1,h,H}^p\|u_j\|_{2,1,h,H}^p \le 9C^p\|u\|_{2,1,h,H}^{2p}$$

with $q<6$, $p=\frac{2q}{2+q}$. To obtain this estimate, we used Theorem 6.4.4 and HOELDER's inequality for sums. Consider the boundary value problem

$$-\Delta_h u+\frac{1}{\eta}\mathrm{grad}_h^+ p+\frac{\varrho}{\eta}(u,\mathrm{grad}_h^-)u=\frac{\varrho}{\eta}f \quad \text{in } \operatorname{int} G_h, \tag{6.65}$$

$$\mathrm{div}_h^- u=0 \quad \text{in } \operatorname{int} G_h, \tag{6.66}$$

$$u=0 \quad \text{on } \partial G_h. \tag{6.67}$$

Applying the discrete generalized VEKUA's theory we obtain the following problem:

$$u = -T_h^- Q_h^+ T_h^+ M_h^-(u) - \frac{1}{\eta} T_h^- Q_h^+ p \qquad \text{in } G_h, \tag{6.68}$$

$$\operatorname{Re} Q_h^+ T_h^+ M_h^-(u) = \frac{1}{\eta} \operatorname{Re} Q_h^+ p \qquad \text{in } G_h. \tag{6.69}$$

<u>6.5.10 Proposition</u>

The problems (6.65)-(6.66)-(6.67) and (6.68)-(6.69) are equivalent.

<u>Proof</u>

Let (u,p) be a solution of (6.68)-(6.69), then the equations (6.65) and (6.66) immediately follow, by applying $-\Delta_h$ and $-\operatorname{Re} D_h^-$ to (6.68) and using (6.69). Obviously equation (6.67) is fulfilled.

Conversely, let (u,p) be a solution of the discrete boundary value problem (6.65)-(6.66)-(6.67). Applying in turn T_h^+, the discrete BOREL-POMPEIU formula, Q_h^+, $D_h^- u \in \operatorname{im} Q_h^+$, T_h^- and finally BOREL-POMPEIU's formula again onto the equation (6.65) we get (6.68). (6.66) directly leads to (6.69).

#

<u>6.5.11 Remark</u>

In contrast to the continuous case, we can dispense with the weak formulation of the finite difference problem in the latter proof. Besides it is not necessary to combine the considerations of regularity and solvability. For that reason the proof of Proposition 6.5.10 gets such a short form.

Consider the following iteration procedure:

$$u_n = -T_h^- Q_h^+ T_h^+ M_h^-(u_{n-1}) - \frac{1}{\eta} T_h^- Q_h^+ p_n, \tag{6.70}$$

$$\operatorname{Re} Q_h^+ T_h^+ M_h^-(u_{n-1}) = \frac{1}{\eta} \operatorname{Re} Q_h^+ p_n, \tag{6.71}$$

$$u_0 \in \overset{\circ}{W}{}^{1,-}_{2,h,H}(G_h) \cap \ker \operatorname{div}_h^-, \quad n=1,2,3,\ldots.$$

It is known from Theorem 6.5.6 that the STOKES problems (6.70)-(6.71) have a solution in each case. Therefore the iteration procedure may be carried out. Since the norms of T_h^-, T_h^+, Q_h^+ and M_h^- can be estimated in a similar manner as in the continuous case, the whole proof of convergence of iteration is to be carried out analogously.

<u>6.5.12 Theorem</u>

System (6.70)-(6.71) has a unique solution (u,p), where

$u \in \overset{\circ}{W}{}^{1,-}_{2,h,H}(G_h) \cap \ker \operatorname{div}_h^-$, $p \in L_{2,h,R}$ (p is uniquely defined up to a real constant) if $\frac{\varrho}{\eta}\|f\|_{p,h,H} \leq (16K_h^2 C_{1,h})^{-1}$. (6.72)

For every function $u_o \in \mathring{W}^{1,-}_{2,h,H}(G_h) \cap \ker \operatorname{div}_h^-$ with

$$\|u_o\|_{2,1,h,H} \leq R_h \tag{6.73}$$

the procedure (6.70)-(6.71) converges in $\mathring{W}^{1,-}_{2,h}(G_h) \times L_{2,h,H}$ to the solution of the problem (6.68)-(6.69).

Proof

For the proof we refer to the continuous case. #

6.5.13 Remark

There hold the following relations for the constants used:

$$K_h = \|T_h^-\|_{[L_{2,h,H} \cap \operatorname{im} Q_h^+, \mathring{W}^{1,-}_{2,h,H}]} \|T_h^+\|_{[L_{p,h,H}, L_{2,h,H}]},$$

$$C_{1,h} = 8^{1/p} C_h \frac{\varrho}{\eta},$$

$$W_h = \{(4K_h C_{1,h})^{-2} - \varrho\|f\|_{p,h,H}(\eta C_{1,h})^{-1}\}^{1/2},$$

$$R_h = (4K_h C_{1,h})^{-1}.$$

An analysis of the proof of Theorem 6.4.1 shows that the norms $\|T_h^-\|_{[L_{p,h,H}, L_{2,h,H}]}$ and $\|T_h^-\|_{[L_{2,h,H}, L_{q,h,H}]}$ (similarly $\|T_h^+\|$) can be uniformly estimated with respect to h. In this way the embedding constant $C_{1,h}$ is also to be bounded uniformly. Using the monotony property of the eigenvalues $\lambda_{1,h}(G_h)$ in case of the embedding in a larger domain (for instance, in a described cube), we finally can find a uniform estimate for K_h and later also for R_h.

If f is RIEMANN-integrable, then $\|f\|_{L_{p,h,H}(G)}$ converges such that W_h is uniformly bounded. This enables us to formulate Theorem 6.5.12 in such a way that the right-hand sides contain only terms which do not depend on h.

6.5.14 Theorem

Suppose that f is RIEMANN-integrable,

$$\frac{\varrho}{\eta}\|f\|_{p,H} < (16K^2C_1)^{-1},$$

$$\|u_o\|_{2,1,H} < (4KC_1)^{-1}. \tag{6.74}$$

Then the system (6.68)-(6.69) has for $h<h_o$ a unique solution

$u \in \mathring{W}^{1,-}_{2,h}(G_h) \cap \ker \operatorname{div}_h^-$, $p \in L_{2,h,R}$

(p unique up to a real constant). For every function

$u_o \in \overset{\circ}{W}{}^{1,-}_{2,h}(G_h) \cap \ker \operatorname{div}^-_h$ with (6.74) the iteration procedure (6.70)-(6.71) converges in $\overset{\circ}{W}{}^{1,-}_{2,h}(G_h) \times L_{2,h,H}$ to the solution of (6.68)-(6.69).

Proof

It may be proved that $\lambda_{1,h}(G_h)$ converges to $\lambda_1(G)$ if h tends to zero. Using this property and the RIEMANN-integrability of $|x|^p$, $p>-3$ in bounded domains of R^3, we have $K_h \longrightarrow K$, $C_{1,h} \longrightarrow C_1$, $W_h \longrightarrow W$, $R_h \longrightarrow R$ and our assertion follows.

#

6.5.15 Remark

Comparing the suppositions of Theorem 4.6.8 and Theorem 6.5.14 we state that there is only a little change of the domain of convergence and the necessary solvability conditions in case of the transition to the discrete problem.

6.5.16 Corollary

(i) There holds

$$\|u\|_{2,1,h,H} \le (4K_h C_{1,h})^{-1} - W_h.$$

(ii) Let $L_h = (4K_h C_{1,h})^{-1} - 4K_h C_{1,h} W$. Then we have

$$\|u_n - u\|_{2,1,h,H} \le L_h^n \|u_o - u\|_{2,1,h,H}.$$

If $u_o = 0$, then is valid

$$\|u_n - u\|_{2,1,h,H} \le L_h^n \{(4K_h C_{1,h})^{-1} - W_h\}.$$

6.5.17 Remark

Corollary 6.5.16 relates to the situation of Theorem 6.5.12. Starting from Theorem 6.5.14 we can find estimates which do not depend on h.

By it we can finish our considerations of the discrete boundary value problem for NAVIER-STOKES equations.

For every $h>0$ the question of existence and uniqueness was clarified, an a-priori estimate of the solution could be given, and the speed of the convergence was obtained. The fixed-point principle ensures the stability of the iteration procedure introduced. For the discrete STOKES problems which are to be solved in each step of the iteration method we could prove the unique solvability. For all constants which occurred explicit bounds could be found. Now we shall turn to the numerical realization of the proposed method.

6.5.18 Theorem

Set $v_h = T_h^- Q_h^+ T_h^+ M_h^-(u) + u + \frac{1}{\eta} T_h^- Q_h^+ p$.

If $f \in L_{p,H}(G)$ then

$$v_h \longrightarrow 0 \quad \text{for} \quad h \longrightarrow 0.$$

If $f \in C_H^{0,\beta}(\bar{G})$, $0 < \beta < 1$, then

$$\| v_h \|_{W^4_{2,h,H}} (G) \leq C_1 h^\beta + C_2 h \, |\ln(h)| .$$

Proof

Denote the exact solution of (6.70)-(6.71) for fixed h by u_h. We have to consider the behaviour of u_h for $h \longrightarrow 0$. In the first step we estimate the approximation error by applying the discrete operators onto the solution (u,p) of the problem (6.68)-(6.69).

$$\| Q_h^+ T_h^+ M_h^-(u) - QTM(u) + D_h^- u - Du + \frac{1}{\eta} Q_h^+ p - \frac{1}{\eta} Qp \|_{2,h,H} \leq$$

$$\leq \| Q_h^+ T_h^+ (M_h^{*,-}(u) - M^*(u) \|_{2,h,H} + \| Q_h^+ (T_h^+ - T) M^*(u) \|_{2,h,H} +$$

$$+ \| (Q_h^+ - Q) TM^*(u) \|_{2,h,H} + \| D_h^- u - Du \|_{2,h,H} + \frac{1}{\eta} \| Q_h^+ p - Qp \|_{2,h,H} +$$

$$+ \frac{\varrho}{\eta} \| Q_h^+ T_h^+ f - Q_h^+ Tf \|_{2,h,H} + \frac{\varrho}{\eta} \| (Q_h^+ - Q) Tf \|_{2,h,H} .$$

Assume $f \in L_{p,H}, p > 3$, we obtain by making use of Theorem 4.6.11 $u \in W^2_{p,H}$, whence follows by the aid of embedding theorems $M^*(u) \in C_H^{0,\beta}(\bar{G})$, $\beta < 1 - \frac{3}{p}$. In detail, we find the following estimates for the above mentioned items. It holds because of $u \in C_H^{1,\beta}(\bar{G})$

$$\| D_h^- u - Du \|_{2,h,H} \leq Ch^\beta \quad , \quad \beta < 1 - \frac{3}{p} ,$$

and

$$\| Q_h^+ T_h^+ (M_h^{*,-}(u) - M^*(u)) \|_{2,h,H} \leq C_1 \| M^{*,-}_h(u) - M^*(u) \|_{2,h,H}$$

$$\leq C_2 h^\beta \quad , \beta < 1 - \frac{3}{p} .$$

Furthermore we get by using Theorem 6.4.8 and Theorem 6.4.11

$$\| Q_h^+ (T_h^+ - T) M^*(u) \|_{2,h,H} \leq C_3 h^\beta + C_4 h^{2-3/p} | \ln(h) | \quad , \quad \beta < 1 - \frac{3}{p} .$$

Owing to $TM^*(u) \in C_H^{1,\beta}(\bar{G})$ and Theorem 6.4.11, we find

$$\|(Q_h^+-Q)TM^*(u)\|_{2,h,H} \le C_5 h^\beta \quad , \quad \beta<1-\frac{3}{p} .$$

If $f \in L_{p,H}$, $p>3$, then $Tf \in C_H^{0,\beta}(\bar{G})$. In this case Theorem 6.4.12 yields

$$\|(Q_h^+-Q)Tf\|_{2,h,H} \longrightarrow 0 \quad \text{for} \quad h \longrightarrow 0 .$$

In connection with Theorem 6.4.8 and $\|Q_h^+\|=1$ we get

$$\|Q_h^+T_h^+f-Q_h^+Tf\|_{2,h,H} \longrightarrow 0 \quad \text{for} \quad h \longrightarrow 0.$$

Setting, instead of $f \in L_{p,H}(G)$, $f \in C_H^\beta(\bar{G})$ we obtain the estimates

$$\|(Q_h^+-Q)Tf\|_{2,h,H} \le Ch^\beta,$$

$$\|Q_h^+T_h^+f-Q_h^+Tf\|_{2,h,H} \le C_1 h\, |\ln(h)| + C_2 h^\beta.$$

By the help of Theorem 6.4.12 and the statement of regularity Theorem 4.6.11 we finally get

$$\|Q_h^+p-Qp\|_{2,h,H} \longrightarrow 0 \quad \text{for} \quad h \longrightarrow 0 .$$

Assume $p \in C_R^{1,\beta}(\bar{G})$, then it follows by using Theorem 6.4.11

$$\|Q_h^+p-Qp\|_{2,h,H} \le Ch^\beta .$$

Summing all these estimates we come to

$$Q_h^+T_h^+M_h^-(u) + D_h^-u + \frac{1}{\eta}Q_h^+p = QTM(u) + Du + \frac{1}{\eta}Qp + w_h ,$$

where

$$\|w_h\|_{2,h,H} \longrightarrow 0 \quad \text{for} \quad h \longrightarrow 0 \quad \text{and} \quad f \in L_{p,H} \quad , \ p>3,$$

and

$$\|w_h\|_{2,h,H} \le C_1 h^\beta + C_2 h\, |\ln(h)| \quad \text{for} \quad f \in C_H^{0,\beta}(\bar{G}),\ 0<\beta<1.$$

Bearing in mind that (u,p) is a solution of the problem (4.69)-(4.70) we find

$$Q_h^+T_h^+M_h^-(u) + D_h^-u + \frac{1}{\eta}Q_h^+p = w_h .$$

Obviously $w_h \in \operatorname{im} Q_h^+$. Therefore we apply T_h^- to w_h and obtain an estimate of the approximation error. #

Our next aim is to find an error estimate in the space $\mathring{W}^{1,-}_{2,h,H}(G_h)$. We prove the following result:

6.5.19 Theorem

Let $f \in \mathcal{R} \cap L_\infty$. Then there is valid

$$\|u - u_h\|_{W^1_{2,h,H}(G_h)} \longrightarrow 0 \quad \text{for} \quad h \longrightarrow 0.$$

By the assumption $f \in C_H^{0,\beta}(\bar{G})$, $0<\beta<1$, we even have

$$\|u - u_h\|_{W^1_{2,h,H}(G_h)} \le C(h^\beta + h\,|\ln(h)|), \; h<h_o .$$

Proof

Theorem 6.5.18 and (6.74) yield

$$D_h^-(u-u_h) + \frac{1}{\eta}Q_h^+(p-p_h) = Q_h^+T_h^+(M_h^-(u)-M_h^-(u_h)) + w_h . \tag{6.75}$$

Applying D_h^+ to (6.74) we find $\operatorname{Re} D_h^+ w=0$. Furthermore we have

$$w_h = Q_h^+T_h^+D_h^+w_h = Q_h^+T_h^+g_h .$$

Denote $M_h^-(u) - M_h^-(u_h) + g_h$ with f_h, then (6.75) can be given in the form

$$D_h^-(u-u_h) + \frac{1}{\eta}Q_h^+(p-p_h) = Q_h^+T_h^+f_h.$$

Therefore $(u-u_h, p-p_h)$ are solutions of a discrete STOKES problem. Proposition 6.5.4 and Theorem 6.5.5 yield the orthogonality of $D_h^-(u-u_h)$ and $\frac{1}{\eta}Q_h^+(p-p_h)$.

Scalar multiplication of this equation with $Q_h^+(p-p_h)$ leads to

$$\frac{1}{\eta}\|Q_h^+(p-p_h)\|^2_{2,h,H} = -[D_h^-(u-u_h), Q_h^+(p-p_h)] +$$

$$+ [Q_h^+T_h^+(M_h^-(u)-M_h^-(u_h)), Q_h^+(p-p_h)] + [w_h, Q_h^+(p-p_h)] \le$$

$$\le \{\|Q_h^+T_h^+(M_h^-(u)-M_h^-(u_h))\|_{2,h,H} + \|w_h\|_{2,h,H}\}\|Q_h^+(p-p_h)\|_{2,h,H},$$

and therefore

$$\frac{1}{\eta}\|Q_h^+(p-p_h)\|_{2,h,H} \le \|Q_h^+T_h^+(M_h^-(u)-M_h^-(u_h))\|_{2,h,H} + \|w_h\|_{2,h,H} .$$

The identity

$$u-u_h = -TQTM(u) - \frac{1}{\eta}TQp + T_h^-Q_h^+T_h^+M_h^-(u_h) + \frac{1}{\eta}T_h^-Q_h^+p_h$$

yields

$$\|u-u_h\|_{2,1,h,H} \le \|TQTM(u)-T_h^-Q_h^+T_h^+M_h^-(u) + \frac{1}{\eta}(TQp-T_h^-Q_h^+p)\|_{2,1,h,H} +$$

$$+ \|T_h^-Q_h^+T_h^+(M_h^-(u)-M_h^-(u_h))\|_{2,1,h,H} + \frac{1}{\eta}\|T_h^-Q_h^+(p-p_h)\|_{2,1,h,H} \le$$

$$\le \|v_h\|_{2,1,h,H} + \|T_h^-Q_h^+T_h^+(M_h^-(u)-M_h^-(u_h))\|_{2,1,h,H} +$$

$$+ \frac{1}{\eta}(1+\lambda_{1,h}^{-1})^{1/2}\| Q_h^+(p-p_h)\|_{2,h,H} \le$$

$$\le \| v_h\|_{2,1,h,H} + 2(1+\lambda_{1,h}^{-1})^{1/2}\| Q_h^+T_h^+(M_h^-(u)-M_h^-(u_h))\|_{2,h,H} +$$

$$+ (1+\lambda_{1,h}^{-1})^{1/2}\| w_h\|_{2,h,H} \le$$

$$\le 2K_h\| w_h\|_{2,h,H} +$$

$$+ 2K_hC_{1,h}\| u-u_h\|_{2,1,h,H}(\| u\|_{2,1,h,H}+\| u_h\|_{2,1,h,H}).$$

With the condition $\frac{6}{5}<p<\frac{3}{2}$ immediately follows

$$\| u-u_h\|_{2,1,h,H} \le$$

$$\le 2K_h\| w_h\|_{2,h,H}\{1-2K_hC_{1,h}(\| u\|_{2,1,h,H}+\| u_h\|_{2,1,h,H})\}^{-1}.$$

Supposing (6.72) and (6.73) we have

$$\| u_h\|_{2,1,h,H} \le (4K_hC_{1,h})^{-1} - W_h , \tag{6.76}$$

and for u Corollary 4.6.9 yields

$$\| u\|_{2,1,G} \le (4KC_1)^{-1} - W . \tag{6.77}$$

These two inequalities, the possibility of a uniform estimate of K_h and $C_{1,h}$ and the RIEMANN-integrability of u and D_iu $(i=1,2,3)$ ensure that for sufficiently small h we have uniformly with respect to h

$$1-2K_hC_{1,h}(\| u\|_{2,1,h,H}+\| u_h\|_{2,1,h,H}) > c > 0 .$$

Now we can describe the convergence $u_h \longrightarrow u$ in dependence on the properties of w_h which we have already considered. #

6.6 Concluding Remarks

6.6.1 Remark

For every fixed $h>0$ the discrete and continuous boundary value problems may be considered in the same way.

6.6.2 Remark

Discrete problems may be considered as finite difference methods. That means that the presented procedure can be understood as a unique principle of the construction and the analytic investigation of finite difference schemes. The considerations of a fixed h yield the solvability and stability of the corresponding finite difference problems.

6.6.3 Remark

Unlike the line of action in [Sam] and [SLM], the concrete structure of the matrices of the arising linear and nonlinear systems of equations is not used for proving solvability and stability. This is very good for the treatment of nonlinear problems (see [Ko]).

6.6.4 Remark

Since there are no requirements for the matrices of finite difference schemes, it is possible to investigate scalar and vector equations in the same manner. For linear problems it is known that the validity of approximation and the stability of a finite difference method lead to convergence. The discrete representations of the solutions permit us to carry out the proof of convergence of nonlinear problems, using the same concept.

6.6.5 Remark

The constants which occur in the a-priori estimate and in the proof of convergence may be explicitly expressed by $|G|$, diam G and the norm of the right side of the equation.

6.6.6 Remark

A decisive difference between our theory and the theory of finite difference methods consists in the fact that by formulating representation formulas the solutions of discrete boundary value problems are given. We only have to calculate DIRICHLET's problem of the LAPLACE equation if we do not have a matrix representation of the orthoprojection Q_h^+.

6.6.7 Remark

The calculation of e_h^+ and e_h^- is a necessary assumption for numerical application of our results. Numerical tests in this direction [Ei] produced satisfactory results.

6.6.8 Remark

This theory enables us to prove assertions for continuous boundary value problems by means of the consideration of discrete boundary value problems and by the investigation of the limit procedure $h \longrightarrow 0$. This line of action is often used. The reader can find it in [LU], for instance.

6.6.9 Remark

The creation of a discrete variant of the classical potential theory appears possible since no obstacles were seen for discretizing the boundary integral methods by the aid of our concept.

6.6.10 Remark

In addition to numerical significance, the procedure presented here pointed out essential methodological advantages, especially through the uniformity of the arguments.

APPENDIX

1. First Steps in Hypercomplex Structures

Two discoveries in the middle of the last century were decisive for the development of a hypercomplex analysis. The Irishman W.R.HAMILTON found out, by loss of the commutativity, a generalization of the field of complex numbers, the so-called skew-field of quaternions, while the German mathematician H.GRASSMANN introduced the algebraic ideas of the inner and outer product of multivectors in a paper in 1844. W.R.HAMILTON was also the first to use quaternions for the definition of the ∇-operator $\nabla = \partial_1 e_1 + \partial_2 e_2 + \partial_3 e_3$ which acts on the vector functions $u=u_1e_1+u_2e_2+u_3e_3$. In this connection we note that J.C.MAXWELL gave the definition of the rotation and divergence of a vector function, such that we get

$$\nabla u = -\text{div } u + \text{rot } u$$

which means a separation of the scalar and the vector part of u. W.K.CLIFFORD succeeded in joining these frameworks in a unique algebra. He created the basis of usual vector calculus in the three-dimensional space, which has been completed by J.W.GIBBS and O.HEAVISIDE. It is known that such algebras were advantageously applied to problems in quantum physics only in the thirties, worked out by the famous physicists PAULI and DIRAC.

2. Universal CLIFFORD Algebras $C(V_{n,s})$

The so-called universal CLIFFORD algebra $C(V_{n,s})$ is defined in the following way:

Let $n \in \mathbb{N}$, $n \geq 1$ and let $e_1,\dots,e_n$ be an orthonormal basis of the n-dimensional real linear space $V_{n,s}$. Then for $C(V_{n,s})$ over $V_{n,s}$ (with respect to the quadratic form $[a \cdot a]=\sum_{i=1}^{s} a_i^2 - \sum_{i=s+1}^{n} a_i^2$) is given a basis by $\{e_A : A \subseteq \{1,\dots,n\}\}$ where $e_A=e_{\alpha_1}\dots e_{\alpha_r}$ when $A=\{\alpha_1,\dots,\alpha_r\}$ and $1 \leq \alpha_1 < \alpha_2 < \dots < \alpha_r \leq n$. It is clear that $e_{\{i\}}=e_i$ $(i=1,\dots,n)$. The product in $C(V_{n,s})$ is defined by the relations

$$e_i^{\,2} = e_\emptyset\,, \quad i=1,\dots,s,$$

$$e_i^{\,2} = -e_\emptyset\,, \quad i=s+1,\dots,n,$$

$$e_i e_j + e_j e_i = \emptyset\,, \qquad i \neq j,$$

and $e_\emptyset = e_\phi$ ($e_\emptyset$ denotes the identity of $C(V_{n,s})$). Notice that each element of $C(V_{n,s})$ can be written by

$$a = \sum_A a_A e_A \qquad a_A \in R\,.$$

Define an involution in $C(V_{n,s})$ by

$$\bar{a} = \sum_A a_A \bar{e}_A$$

with

$$\bar{e}_A = \bar{e}_{\alpha_r} \dots \bar{e}_{\alpha_1}\,,\ \bar{e}_i = -e_i \ (i=1,\dots,n).$$

Finally a norm is given by

$$|a|_\emptyset^2 = Re(a\,\bar{a}) = \sum_A |a_A|_A^2.$$

Notice that for each $a \in C(V_{n,s})$ $a_\emptyset = Re\ a$. The reader can find information about CLIFFORD algebras in [BDS], [Lou1], [Lou2], [Lou3], [Por], [Cru] and [Che].

3. Complexified CLIFFORD Algebras

To study special physical questions, it is necessary to work in certain complex algebras. In [Ry1] a complex CLIFFORD algebra is introduced very beautifully. We follow this explanation. Starting from $A_n = C(V_{n,\emptyset})$ we get a complex CLIFFORD algebra by taking the real symmetric tensor product $A_n \otimes_R C$, which is denoted by $A_n(C)$. The algebra $A_n(C)$ is associative and its complex dimension is 2^n. Each element Z can be written as

$$Z = z_\emptyset e_\emptyset + \dots + z_n e_n + \dots + z_{j_1 \dots j_r}\ e_{j_1} \dots\ e_{j_r} + \dots + $$
$$+ z_{1,\dots,n} e_1 \dots e_n\,,$$

where $z_\emptyset, \dots, z_n, z_{j_1 \dots j_r}, z_{1 \dots n} \in C$, and therefore

$$z_{j_1 \dots j_r} = x_{j_1 \dots j_r} + i\ y_{j_1 \dots j_r}$$

with $x_{j_1 \dots j_r}\,, y_{j \dots j_r} \in R$.

The norm of Z is given by

$$|Z| = (x_\emptyset^2 + \dots + x_{1\dots n}^2)^{1/2} + (y_\emptyset^2 + \dots + y_{1\dots n}^2)^{1/2} .$$

The complex vector subspace spanned by $\{e_i\}_{i=\emptyset}^n$ will be denoted by C^{n+1}, and an arbitrary vector

$$z_\emptyset e_\emptyset + z_1 e_1 + \dots + z_n e_n$$

is denoted by z. The mapping $e_{j_1} \dots e_{j_r} \longrightarrow (-1)^r e_{j_r} \dots e_{j_1}$ describes a linear automorphism on $A_n(C)$. The images of Z will be denoted by $\bar{Z}$. It is clear that

$$(e_\emptyset + ie_1)(\overline{e_\emptyset + ie_1}) = \emptyset .$$

Therefore $A_n(C)$ is not a division algebra.
The set $\{z+z_\emptyset : z\bar{z}=\emptyset\}$ is called complex null cone or singularity cone and is denoted by CN_z . The complex-quaternion algebra $A_2(C)$ (cf. [Sou1]), now and then denoted by CH, is of exceptional interest. K.IMAEDA proved in 1976 (cf. [Ima]) that this algebra is isomorphic to the PAULI algebra, and the vector space $A_2(C)$ may be identified with the complex MINKOWSKI space CM. This author named the complex quaternions biquaternions. H.G.HAEFELI proved in [Hae] that among all CLIFFORD algebras the algebras of complex numbers, quaternions and biquaternions are the only ones with real norms and real traces.
Now in [BDS] the concept of the even subalgebra of a CLIFFORD algebra is illustrated in a very nice matter. We start with the so-called DIRAC-algebra $C(V_{4,1})$. We have the relations

$$\gamma_i \gamma_j + \gamma_j \gamma_i = \emptyset, \qquad i \neq j,$$

$$\gamma_1^2 = e_\emptyset,$$

$$\gamma_2^2 = \gamma_3^2 = \gamma_4^2 = -e_\emptyset,$$

where $\{\gamma_1, \gamma_2, \gamma_3, \gamma_4\}$ form a basis of $V_{4,1}$. It is clear that $C(V_{4,1})$ is of the dimension 2^4. In physics it is usual to describe this basis by the aid of the following complex matrices:

$$\gamma_1 = \begin{pmatrix} 1 & \emptyset & \emptyset & \emptyset \\ \emptyset & 1 & \emptyset & \emptyset \\ \emptyset & \emptyset & -1 & \emptyset \\ \emptyset & \emptyset & \emptyset & -1 \end{pmatrix}, \quad \gamma_2 = \begin{pmatrix} \emptyset & \emptyset & \emptyset & 1 \\ \emptyset & \emptyset & 1 & \emptyset \\ \emptyset & -1 & \emptyset & \emptyset \\ -1 & \emptyset & \emptyset & \emptyset \end{pmatrix}, \gamma_3 = \begin{pmatrix} \emptyset & \emptyset & \emptyset & -i \\ \emptyset & \emptyset & i & \emptyset \\ \emptyset & i & \emptyset & \emptyset \\ -i & \emptyset & \emptyset & \emptyset \end{pmatrix}, \gamma_4 = \begin{pmatrix} \emptyset & \emptyset & 1 & \emptyset \\ \emptyset & \emptyset & \emptyset & -1 \\ -1 & \emptyset & \emptyset & \emptyset \\ \emptyset & 1 & \emptyset & \emptyset \end{pmatrix},$$

where i denotes the usual imaginary unit.
Choosing the subalgebra of $C(V_{4,1})$, which is generated by all products with an even number of factors of the elements γ_i, i=1,2,3,4, we get an isomorphic algebra to the 2^3-dimensional PAULI-algebra where the product is governed by the relations

$$\sigma_i \sigma_j + \sigma_j \sigma_i = 0 \quad , \; i \neq j, \; i,j=1,2,3,$$

$$\sigma_1^2 = \sigma_2^2 = \sigma_3^2 = 1,$$

$$\sigma_\emptyset = e_\emptyset \; .$$

The new notation is defined by

$$\sigma_1 = \gamma_2 \gamma_1 \; , \; \sigma_2 = \gamma_3 \gamma_1 \; , \; \sigma_3 = \gamma_4 \gamma_1 \; .$$

Obviously, there is the universal CLIFFORD algebra $C(V_{3,3})$. In physics, the elements σ_i, i=1,2,3, are written by the so-called PAULI matrices

$$\sigma_1^* = \begin{bmatrix} 0 & 1 \\ 1 & 0 \end{bmatrix}, \qquad \sigma_2^* = \begin{bmatrix} 0 & -i \\ i & 0 \end{bmatrix}, \qquad \sigma_3^* = \begin{bmatrix} 1 & 0 \\ 0 & -1 \end{bmatrix},$$

which may be identified with the 4X4-matrices found above. Finally, repeating the application of this procedure to the CLIFFORD algebra $C(V_{3,3})$ we get the algebra of the real quaternions $H=C(V_{2,0})$ with the corresponding multiplication rules (1.5). The quaternionic basis may be obtained by

$$e_\emptyset, \; e_1 = \sigma_1 \sigma_2 \; , \; e_2 = \sigma_3 \sigma_1 \; , \; e_3 = \sigma_2 \sigma_3 .$$

Our considerations about CLIFFORD algebras yield that the 8 elements $e_\emptyset, \sigma_1, \sigma_2, \sigma_3, e_1, e_2, e_3, \sigma_1\sigma_2\sigma_3$ form a basis in $C(V_{3,3})$. By identifying σ_i with σ_i^* for i=1,2,3 and $e_\emptyset$ with $\begin{bmatrix} 1 & 0 \\ 0 & 1 \end{bmatrix}$ we shall observe, following P.LOUNESTO in [Lou2], the special matrices

$$\begin{bmatrix} \psi_1 & 0 \\ \psi_2 & 0 \end{bmatrix},$$

$\psi_i \in C$, which are quite the same as the spinor vectors $(\psi_1, \psi_2)^T$ in quantum physics. These matrices form a left ideal V of $C(V_{3,3})$. Considered as a real linear space, V has the basis $\{f_\emptyset, f_1, f_2, f_3\}$, where

$$f_\emptyset = \frac{1}{2}(e_\emptyset + \sigma_3),\ f_1 = \frac{1}{2}(e_3+\sigma_3),\ f_2 = \frac{1}{2}(e_2-\sigma_1), \text{ and}$$

$$f_3 = \frac{1}{2}(\sigma_1 + \sigma_1\sigma_2\sigma_3),$$

which correspond with the matrices

$$f_\emptyset = \begin{pmatrix} 1 & 0 \\ 0 & 0 \end{pmatrix},\ f_1 = \begin{pmatrix} 0 & 0 \\ i & 0 \end{pmatrix},\ f_2 = \begin{pmatrix} 0 & 0 \\ -1 & 0 \end{pmatrix},\ f_3 = \begin{pmatrix} i & 0 \\ 0 & 0 \end{pmatrix}.$$

For applications in physics and quantum mechanics we refer to the very beautiful books [HS], [Hes2] and [BT].

4. CAUCHY's Integral Formulas

K.HABETHA proved in [Hab1] that in order to have an analogue to the classical CAUCHY integral formula it is necessary to restrict the considerations to the algebra of complex numbers, the algebra of quaternions or CLIFFORD algebra. In [BDS] the following general BOREL-POMPEIU formula in a CLIFFORD algebra $A_n = C(V_{n,\emptyset})$ is given:
Let $m \leq n$, $M \subset G$ (G an open subset of R^{m+1}) be an m+1-dimensional compact differentiable and oriented manifold-with-boundary $\partial M = \Gamma$. Then if $u \in C^1_A(G)$

$$\frac{1}{|S_1|}\int_\Gamma \frac{\bar{y}-\bar{x}}{|y-x|^{m+1}}\alpha(y)\,d\Gamma\, u(y) - \frac{1}{|S_1|}\int_M \frac{\bar{y}-\bar{x}}{|y-x|^{m+1}}(Du)(y)\,dy = \begin{cases} u(x), & x \in \mathring{M} \\ \emptyset, & x \in G\setminus M \end{cases},$$

where $|S_1| = 2\pi^{\frac{m+1}{2}} \frac{1}{\Gamma(\frac{m+1}{2})}$ is the area of the n+1-dimensional unit sphere S_1, $\alpha(y) = \sum_{i=1}^{m} e_i \alpha_i(y)$ denotes the outer normal at the point y and $D = \sum_{i=1}^{m} e_i D_i$ stands for the generalized CAUCHY-RIEMANN operator.
When $Du = \emptyset$ (u is called monogenic), then we have a hypercomplex generalization of CAUCHY's integral formula.

Firstly, hypercomplex generalizations of CAUCHY's integral formula were given by A.W.BIZADSE [Bi 1] (1955), V.IFTIMIE [I 1] (1965), D.HESTENES [Hes1] (1968), and R.DELANGHE in a series of papers [Del1], [Del2], [Del3], [Del4] (1970-1972).

J.RYAN deduces in [Ry1] a generalized integral formula within a complexified CLIFFORD algebra, especially for biquaternions. Now we follow his considerations. Let

$$E(z) = \frac{z_\emptyset e_\emptyset - z_1 e_1 - \dots - z_n e_n}{(z_\emptyset^2 + \dots + z_n^2)^{(n+1)/2}}$$

be a function which is defined on $C\backslash N_\emptyset$ and $n \equiv 1 \mod 2$. Suppose M is a manifold with the properties:

1. M is an $n+1$-dimensional manifold lying in C^{n+1}.
2. $N_{z'} \cap M = \{z'\}$.
3. For each $z \in M$ the tangent space TM_z is spanned by the $n+1$ vectors $\{\tau_{j,z} e_j\}_{j=\emptyset}^n$, $\tau_{j,z} \in C\backslash\{\emptyset\}$.

Further suppose $u: C^{n+1} \longrightarrow A_n(C)$ is a $C^1_{A_n}$-function. Then we have the formula

$$u(z) = \frac{1}{|S_1|} \int_{\partial M} E(z'-z)(Dz')u(z') - \frac{1}{|S_1|} \int_M E(z'-z)d(Dz')u(z'),$$

where $Dz' = \sum_{i=0}^{n} (-1)^i e_i d\hat{z}'_i$ with

$$d\hat{z}_j = dz'_\emptyset \wedge \dots \wedge dz'_{j-1} \wedge dz'_{j+1} \wedge \dots \wedge dz'_n .$$

One of the most general results in this direction is given by K.HABETHA [Hab2]. His investigations are carried out in a certain algebra A which is constructed as follows:

Let V, W be real finite-dimensional vector spaces furnished with the usual EUCLIDean topology, G is a domain in V. Assume the existence of the mentioned algebra with $A \supset V, W$ and the distributive (not necessarily associative) multiplication. We have

$$\lambda(u\ v) = (\lambda u)v = u(\lambda v)$$

for $\lambda \in R, u, v \in A$. The multiplication in A is defined by the rule

$$e_i e_j = \sum_k \varepsilon_{ijk} e_k ,$$

where ε_{ijk} are real constants, $\{e_i : i \in I\}$ is a basis in V, $\{e_j : j \in J\}$ is a basis in W and $I,J \subset K$. Furthermore let $I=\{0,1,\ldots,n\}$ and $\dim W=m$. Now we can formulate the following result:

Let $u,v : G \longrightarrow W$ be continuously differentiable, w a closed differentiable differential form in G with values in A and of degree p, let M_{p+1} be a differentiable manifold of dimension $p+1$ in G with sufficiently smooth boundary, then

$$\int_{\partial M_{p+1}} v\ w\ u = \int_{M_{p+1}} (dv \wedge w\ u) + \int_{M_{p+1}} (-1)^p v\ w \wedge du).$$

K.HABETHA referred to papers of EDENHOFER [Ede] and JOHN [J] who obtained for different parameters p CAUCHY formulas with information on u at a point z_o in M_{p+1} under a special choice of v.

In the function theory of several complex variables there are also integral formulas of CAUCHY's type. It can be readily seen that for a holomorphic function $u(z)$ in the polydisc $G=G_1 \times \ldots \times G_n$, which belongs to $C(\bar{G})$, we have

$$u(z) = \frac{1}{(2\pi i)^n} \int_\Gamma \frac{u(t)}{t-z} dt ,$$

where

$$\Gamma = \partial G_1 \times \ldots \times \partial G_n , \quad \frac{dt}{t-z} = \frac{dt_1 \wedge \ldots \wedge dt_n}{(t_1 - z_1) \ \ldots \ (t_n - z_n)} .$$

Now let $G \subset C_n$ be a domain with a smooth boundary Γ.

Further let $u \in C^1(\bar{G})$, then holds a generalization of BOREL-POMPEIU's formula

$$u(z) = \frac{(n-1)!}{(2\pi i)^n} \left[\int_\Gamma \frac{u(t)\sigma(t-z)}{|t-z|^{2n}} - \int_G \sum_{\nu=1}^{n} (\bar{t}_\nu - \bar{z}_\nu) \frac{\partial u}{\partial \bar{t}_\nu} \frac{d\bar{t} \wedge dt}{|t-z|^{2n}} \right],$$

where $\sigma(z) = \sum_\nu (-1)^{\nu-1} \bar{z}_\nu \, d\bar{z}_1 \wedge \ldots \underset{\nu}{\wedge} \ldots \wedge d\bar{z}_n \wedge dz_1 \wedge \ldots \wedge dz_n$.

In most of the papers this relation is called CAUCHY-GREEN's formula. If u fulfils the CAUCHY-RIEMANN equations, then the well-known integral formula of MARTINELLI-BOCHNER arises. Note that the kernel function $(t-z)|t-z|^{-2n}$ has not the property of holomorphy. For special domains the above given formula allows further generalizations such as formula of LERAY-GREEN or formula of WEILL, cf. [Scha]. V.SOUCEK describes in [Sou2] the basic correspondence between the complexified FUETER equation and the spin-$\frac{1}{2}$-massless field equation. There a connection is found between boundary value type integral formulation on EUCLIDean space-time and initial value type integral formulation on MINKOWSKI space by the aid of the complexified CAUCHY's integral formula. A connection between the integral formula given by R.PENROSE [Pen] and the complexified CAUCHY integral formula by the aid of explicit formulas is also shown. By translating the CH-CAUCHY integral formula into a spinor version the door is opened to various applications of problems in mathematical physics. For this reason

$$z = [z_\emptyset, z_1, z_2, z_3] \in CM$$

are identified with

$$z^1 = z_\emptyset + ie_1z_1 + ie_2z_2 + ie_3z_3 \in CH$$

and finally with

$$z^2 = \begin{bmatrix} z_\emptyset+z_3 & z_1-iz_2 \\ z_1+iz_2 & z_\emptyset-z_3 \end{bmatrix} \in C(2).$$

CAUCHY's type formulas are also described in [Sou3] and [Sou4] for the spin-$\frac{n}{2}$-case. We also refer to the very nice paper [BS].

5. Hypercomplex Differentiability

In 1848 in the paper [Mej] A.S.MEJHLIHZON obtained the following result:
Let G be a connected open subset of H, let $u: H \longrightarrow H$ be a quaternionic-differentiable (on the left) function in each

point $x \in G$ which means that there exists

$$\frac{du}{dx} = \lim_{h \longrightarrow 0} h^{-1}\{u(x+h)-u(x)\},$$

then $u(x)$ allows the representation

$$u(x) = a + xb,$$

where a and b are constant quaternions. It is clear that the class of quaternionic-differentiable functions arising is too small for further investigations. H.MALONEK found a way out of this delicate situation. In his Habilitationsschrift he defined a hypercomplex derivative as a linear mapping in a certain sense. It holds in greater detail:

Let u be a continuous mapping of a neighbourhood of a point $\vec{z} \in H^m=\{\vec{z}=(z_1,\ldots,z_m),\quad z_k=x_k e_0 - x_0 e_k,\quad k=1,\ldots,m\}$ in the CLIFFORD algebra $\mathcal{A}$. The function $u(x)$ is called hypercomplex left-differentiable in x if there exists a left-$\mathcal{A}$-linear mapping $l \in \mathcal{L}(H^m,\mathcal{A})$ such that

$$\lim_{\Delta\vec{z}\to 0} \frac{|f(z+\triangle z)-f(z)-l(\triangle z)|}{|\triangle z|}.$$

l is called left-derivative and denoted by u_l'. Furnished with this idea of differentiation we can prove the following theorem:

A function $u(\vec{z})$ is hypercomplex left-differentiable in $\vec{z}$ iff $u(z)$ is left monogenic at this point.

A similar result holds for right-derivatives. In this way the equivalence of RIEMANN's approach and CAUCHY's approach is verified. Note that in the function theory of several complex variables for the validity of HARTOG's theorem this equivalence is guaranteed. We also mention the paper [Som3] in which the definition of left(right)-monogenic functions is given in terms of differentiability by the aid of

$$df = \sum_{i=1}^{m} dz_i a_i \ .$$

Finally we refer to [Hab2] who makes interesting remarks about generalizations of the complex differentiability to the case of higher dimensions. Results of some other authors are cited there.

6. Differential Forms

In [Hae] we can find that the differential of any H-valued function $f(\bar{x}) \in C^1_H$ may be written as

$$df = a_1 d\bar{x}\, e_1 + a_2 d\bar{x}\, e_2 + a_3 d\bar{x}\, e_3 + a_\emptyset d\bar{x}\, e_\emptyset,$$

where $d\bar{x} = dx_\emptyset e_\emptyset - dx_1 e_1 - dx_2 e_2 - dx_3 e_3$ and a_i, $i=\emptyset,1,2,3$ certain H-valued functions. If $a_\emptyset=\emptyset$, then $f(x)$ is right-regular (in FUETER's sense cf. App. 14). Besides, if one has $a_1\bar{a}_2 a_3 = a_3\bar{a}_2 a_1$, then the inverse function $x(f)$ is left-regular. If $a_1\bar{a}_2 a_3 \neq a_3\bar{a}_2 a_1$, then the quaternion $q=a_1\bar{a}_2 a_3 - a_3\bar{a}_2 a_1$ is orthogonal to the quaternions a_1, a_2 and a_3 in the sense of the scalar product $\langle a,b\rangle = \mathrm{Re}\,\bar{a}b$ and $\frac{1}{2}|q|$ means the volume of the 3-dimensional parallelepiped whose edges are a_1, a_2 and a_3 (cf. [Sud]). Furthermore it is always possible to transform every differential of right-regular function f by left-multiplication with the hypercomplex function $k=(\bar{a}_1 a_2 \bar{a}_3 - \bar{a}_3 a_2 \bar{a}_1)^{-1}$ into a differential of a function which is regular on both sides. Identifying the tangent space at each point of H with H itself we can consider the differential as the H-valued 1-form

$$df = \frac{\partial f}{\partial x_0} dx_\emptyset + \frac{\partial f}{\partial x_1} dx_1 + \frac{\partial f}{\partial x_2} dx_2 + \frac{\partial f}{\partial x_3} dx_3 .$$

On the other hand, any H-valued 1-form may be regarded as the R-linear map $\phi: H \longrightarrow H$ by

$$\phi\Big(\sum_{i=\emptyset}^{3} x_i e_i\Big) = \sum_{i=\emptyset}^{3} a_i x_i .$$

An H-valued r-form may be considered as a map from H to the space of alternating R-multilinear maps from $H\times\ldots\times H$ into H. The exterior product of the r-form ϕ and the s-form ψ can be defined by

$$(\phi \wedge \psi)(a_1, \dots, a_{r+s}) = \frac{1}{r!\,s!}\sum_{p} \mathrm{sgn}\, p\, \phi(a_{p_1} \dots a_{p_r})\psi(a_{p_{r+1}} \dots a_{p_{r+s}}),$$

where p denotes the permutations of r+s objects. For

information we refer to [Sud].
In [Som3] F.SOMMEN gets the following representation of the classical exterior differential operator $d = \sum_{j=\emptyset}^{n} dx_j \frac{\partial}{\partial x_j}$:

$$d = D\, dx_{\emptyset} + \partial,$$

where $D=\sum_{j=\emptyset}^{n} e_j \frac{\partial}{\partial x_j}$ is the generalized CAUCHY-RIEMANN operator and ∂ denotes the hypercomplex exterior differential operator

$$\partial = \sum_{j=1}^{n} dz_j \frac{\partial}{\partial x_j} .$$

Another decomposition of de RHAM's operator d is proposed in [Sou5]. We have

$$du = \mathfrak{D} u + \vec{\mathfrak{D}} u,$$

where $\mathfrak{D} u = dx\, \partial u$, $\vec{\mathfrak{D}} u = dx_1 \partial^1 u + dx_2 \partial^2 u + dx_3 \partial^3 u$,

$$\partial u = \frac{1}{4}[D_{\emptyset} - e_1 D_1 - e_2 D_2 - e_3 D_3], \quad \partial^i = -e_i\, \partial u e_i .$$

Solutions of the equation $\mathfrak{D} u = \emptyset$ correspond to the antiregular functions in FUETER's sense. After the identification of C_2 with H we get to interpretations of $\mathfrak{D}$ and $\vec{\mathfrak{D}}$ with the DIRAC operator or the twistor operator, respectively, in the theory presented by ATIYAH, HITCHIN and SINGER (cf. [AHS]).
A detailed discussion of these forms is also given in [Mal].
By means of A_n-valued 1-forms $dz_j = dx_j - e_j dx_{\emptyset}$, $j=1,\dots,n$, in [Som3] the A_n-valued k-forms

$$F_l^{(k)} = \sum_{j_l} dz_{j_1} \wedge \dots \wedge dz_{j_r} F_{j_1 \dots j_r}(x)$$

are introduced in open subsets of R^{n+1}. When the coefficients $F_{j_1 \dots j_r}(x)$ all satisfy the generalized CAUCHY-RIEMANN equations $DF_{j_1 \dots j_r}(x) = \emptyset$, then $F_l^{(k)}$ is called left-monogenic k-form in G. F.SOMMEN uses the calculus of monogenic differential forms to lay the foundation of a homology theory.
Let $A_n \odot \Lambda(R^{n+1})$ denote the symmetric tensor product of the

algebra A_n with the algebra $\Lambda(R^{n+1})$. Introducing the wedge product "$\wedge$" within $A_n \otimes \Lambda(R^{n+1})$ by the formula

$$\left(\sum_{i=1}^{2^{n+1}} a_i \odot b_i\right) \wedge \left(\sum_{j=1}^{2^{n+1}} c_j \odot d_j\right) =: \sum_{i,j=1}^{2^{n+1}} a_i c_j \odot b_i \wedge d_j,$$

where $a_i, c_j \in A_n$ and $b_i, d_j \in \Lambda(R^{n+1})$ the real vector space $A_n \odot \Lambda(R^{n+1})$ becomes an associative algebra. Notice that the alternating algebra $\Lambda(R^{n+1})$ is canonically isomorphic to the space A_n. Hence it is possible to define the star map on A_n with

$$*(e_{i_1} \dots e_{i_j}) e_{i_1} \dots e_{i_j} = e_1 \dots e_n$$

for each basic element in A_n. Write $dx = \sum_{i=1}^{n} e_i \odot dx_i$ and

$Dx = \sum_{i=0}^{n} (-1)^i e_i \odot \hat{dx_i}$, where

$\hat{dx_i} = dx_0 \wedge \dots \wedge dx_{i-1} \wedge dx_{i+1} \wedge \dots \wedge dx_n$.

For a pointwise differentiable function $f: G \dashrightarrow A_n$ $n>1$, to be left regular it is necessary and sufficient that

$$\frac{1}{(n-1)!} d\{(-* \odot id)(dx \wedge \dots \wedge dx) f\} = Dx \frac{\partial f}{\partial x_0}.$$

This result is contained in [Ry2]. In [Sud] A.SUDBERY gave the quaternionic analogue to this formula.

7. Conformal Transforms

Let $u: R^n \dashrightarrow R^n$ be a continuously differentiable function $u=u(x)$. Such a function is called <u>conformal in the sense of GAUSS</u> if there exists a continuous function $\lambda: G \longrightarrow R^+$ with

$$(D_i u)^2 = \lambda(x)$$

and

$$D_i u \, D_j u = 0, \quad i,j=1,\dots,n; \; i \neq j .$$

These equations permit the same geometric interpretation as in the case of R^2. Conformal mappings preserve the angles and change the scale uniformly in regard to the direction

(cf. [Bi 2],p.232). J.LIOUVILLE (1850), S.LIE (1872) proved the following theorem (cf.[Schou],p.312):
For $n>2$ every conformal mapping of $\mathbf{R}^n$ on itself is always a similarity transform with respect to a point or a combination of an inversion at a point and a similarity transform with respect to that point.
Let $\mathbf{H}^*$ be the one-point compactification of $\mathbf{H}$. Then the general conformal transform in $\mathbf{H}^*$ is of the form

$$\gamma(x) = (ax + b)(cx + d)^{-1}$$

with $a^{-1}b \neq c^{-1}d$. In [Sud] we can find the statement:
Given $u:\mathbf{H} \longrightarrow \mathbf{H}$ and a conformal mapping γ of the form described above. Let

$$[M(\gamma)u](x) = \frac{(cx + d)^{-1}}{|b - ac^{-1}d|\ |cx + d|^2}\ u(\gamma(x)) .$$

If u is H-regular at $\gamma(x)$, then $M(\gamma)u$ is H-regular at x. Further investigations into the representation of conformal transforms employing CLIFFORD algebras are made in [Lou1]. These results base on the minimal compactification of the real orthogonal space $\mathbf{R}^{p,q}$ by the set of conformal mappings into $S_p X S_q / \mathbf{Z}_2$ (cf.[Pen1]).
Note that in the paper [Ahl] by L.AHLFORS an interesting connection is given between MÖBIUS transformations and CLIFFORD matrices which are two by two matrices whose entries are CLIFFORD numbers. This method based on a remarkable paper by K.T.VAHLEN [Vah] and yields the bridge to the classical function theory in the complex plane. More recent papers are [LS] and [AL].

8. Generalized TAYLOR and LAURENT Expansions

It is well known that the k-th power z^k is a homogeneous polynomial with respect to x and y of degree k. Obviously it is an analytic function. The definition of corresponding polynomials in the CLIFFORD algebra A goes back to R.DELANGHE [Del1]. He introduced the so-called FUETER polynomials, which are given in the following manner:
Let $\nu = (\nu_1, \ldots, \nu_n)$ be a multiindex with
$|\nu| = \nu_1 + \ldots + \nu_n$, $\nu! = \nu_1! \ldots \nu_n!$, $x = (x_0, x_1, \ldots, x_n)$.

Let $z_j = x_j - x_\emptyset e_j$, $j=1,\dots,n$. The polynomials

$$P_\nu(x) = \frac{1}{|\nu|!}\sum_{\mu\in S_{|\nu|}} z_{\mu(1)}\cdot\ldots\cdot z_{\mu(|\nu|)}, \quad |\nu|>\emptyset,$$

$$P_\nu(x) = e_\emptyset, \quad |\nu|=\emptyset,$$

where $S_{|\nu|}$ is the symmetric permutation group of $|\nu|$ elements, are left- and right-regular functions which are homogeneous of degree k.

A further possibility to deduce FUETER polynomials arises by constructing of the CAUCHY-KOWALEWSKI extension of the real powers $x_1^{\nu_1}x_2^{\nu_2}\dots x_n^{\nu_n}$ which reads $z_1^{\nu_1}\odot z_2^{\nu_2}\odot\dots z_n^{\nu_n}$, where $z_i = x_i e_\emptyset - x_\emptyset e_i$, $i=,2,\dots,n$, and $z_i|_R = x_i$. The CAUCHY-KOWALEWSKI product is exactly defined in Appendix 9. This approach was found by F.SOMMEN (cf. [BDS]).

A new interpretation of FUETER polynomials was recently given by H.MALONEK in [Mal]. He introduced the permutative product

$$a_1 \times a_2 \times \dots \times a_n = \frac{1}{n!}\sum_{\Pi(i_1,\dots,i_n)} a_{i_1}a_{i_2}\dots a_{i_n}$$

with $a_i \in V_+$ (non-commutative ring). FUETER polynomials $P_\nu(x)$ may be written as

$$P_\nu(x) = P_\nu^*(\vec{z}) = \frac{1}{\nu!} z_1 \times z_2 \times \times\dots\times z_n = \frac{1}{\nu!}\vec{z}^{\,\nu}$$

with $\vec{z}=(z_1,\dots,z_n)$, $\nu=(\nu_1,\dots,\nu_n)$. Note that the identity

$$\vec{z}^{\,\nu} = (-1)^{|\nu|}(z\times e_1)^{\nu_1}\times\dots\times(z\times e_n)^{\nu_n}$$

is valid. Therefore $P_\nu^*(\vec{z})$ may be considered as a power of $\vec{z}$ in a certain sense.

Now, following [BDS] we obtain that each function u which is left-regular for $|x|<R$ may be represented in a ball with the radius $(\sqrt{2}-1)R$ by the TAYLOR series

$$u(x) = \sum_{k=\emptyset}^{\infty}\sum_{|\nu|=k} P_\nu(x)c_\nu ,$$

where $c_\nu = \frac{1}{\nu!}(\partial^\nu u)(\emptyset)$, $\partial^\nu = \frac{\partial^{|\nu|}}{\partial x_1^{\nu_1}\dots\partial x_n^{\nu_n}}$.

To ensure convergence this series is to be considered as a

multiple power series of the real variables $x_\emptyset$, $x_1, \dots, x_n$. Using spherical polynomials F.SOMMEN got in [Som5] an expansion which converges absolutely and uniformly in the ball $B_R(\emptyset)$. In this context it should be mentioned that V.IFTIMIE in [I2] also obtained TAYLOR series in a slightly different way. If u is left-regular for $R_1 < |x| < R_2$ R.DELANGHE got in [Del3] the LAURENT series

$$u(x) = \sum_{k=\emptyset}^{\infty} \sum_{|\nu|=k} P_\nu(x) c_\nu + \sum_{k=\emptyset}^{\infty} \sum_{|\nu|=k} Q_\nu(x) d_\nu ,$$

where

$$Q_\nu(x) = \frac{1}{n-1} \partial_y^\nu \bar{D}_y C_{|\nu|,\frac{n-1}{2}}\left(\left\langle \frac{x}{|x|}, \frac{y}{|y|}\right\rangle\right) \frac{|y|^{|\nu|}}{|x|^{|\nu|+n-1}}$$

and $C_{|\nu|,(n-1)/2}(t)$ are GEGENBAUER polynomials (cf. [Mue1]). This expansion is uniformly and absolutly convergent on each compact subset of the annulus $\{x: R < |x| < R\}$ (cf. [Som5]). We also mention the very nice paper [Hab2] of K.HABETHA.

Let $H^n = \{\vec{z} : \vec{z} = (z_1, \dots, z_n),\ z_k = x_k e_\emptyset - x_k e_\emptyset ,\ k = 1, \dots, n\}$.

In [Mal] a WEIERSTRASS' approach is proved by consideration of the multiple left-power series

$$P(\vec{z}, \vec{a}) = \sum_{|\nu|=\emptyset}^{\infty} (\vec{z} - \vec{a})^\nu c_\nu \quad , \ \vec{a}, \vec{z} \in H^n,\ c_\nu \in A_n(R),$$

without application of a hypercomplex CAUCHY integral formula.

By means of the complex left- and right-regular FUETER polynomials, which may be constructed similarly to the real ones, in [Ry3] J.RYAN also established in $A_n(C)$ TAYLOR and LAURENT expansions.

Some statements in [BDS] are devoted to the isolated singularities. The terms left-meromorphic and left-entire functions are introduced. A hypercomplex version of MITTAG-LEFFLER's theorem is given.

9. CAUCHY-KOWALEWSKI-Product

Due to the non-commutativity of the CLIFFORD algebra the pointwise product of monogenic functions is not anymore monogenic (see 1.2.6). But by means of a CAUCHY-KOWALEWSKI type theorem on the extension of analytic functions in R^n to monogenic functions in R^{n+1} a product between monogenic functions is introduced in [BDS]. For instance there is the proposition that for any A_n-valued analytic function in the whole of R^n given by the power series

$$P(x) = \sum_{|\nu|=\emptyset}^{\infty} c_\nu \vec{x}^{\nu}$$

the CAUCHY-KOWALEWSKI extension is exactly

$$P(\vec{z}) = \sum_{|\nu|=\emptyset}^{\infty} c_\nu \vec{z}^{\nu} .$$

Then the C.-K.-product of those two functions

$$P(z) = \sum_{|\nu|=\emptyset}^{\infty} c_\nu \vec{z}^{\nu} \quad \text{and} \quad Q(z) = \sum_{|\lambda|=\emptyset}^{\infty} d_\lambda \vec{z}^{\lambda}$$

is defined by

$$P(\vec{z}) \odot_R Q(\vec{z}) = \sum_{|\sigma|=\emptyset}^{\infty} b_\sigma \vec{z}^{\sigma} \quad \text{where} \quad b_\sigma = \sum_{\nu+\lambda=\sigma} c_\nu d_\lambda .$$

For example we have

$$\operatorname{Exp} \vec{z} = \sum_{|\nu|=\emptyset}^{\infty} \frac{1}{\nu!} \vec{z}^{\nu} \quad \text{or} \quad \operatorname{Exp} \vec{z} = \sum_{|\nu|=0}^{\infty} \frac{1}{k!} \sum_{|\nu|=k} \binom{k}{\nu} \vec{z}^{\nu} .$$

10. Generalization of WIRTINGER's Calculus

In the complex plane there exists a simple connection between the real variables x and y and the complex variables z and $\bar{z}$ by the well-known formulas

$$x = \frac{1}{2} (\bar{z} + z) ,$$

$$y = \frac{i}{2} (\bar{z} - z) .$$

This can be transformed into higher dimensions in the following way:

$$x_\emptyset = \frac{1}{2} (\bar{z} + z),$$

$$x_k = \frac{1}{2}(\bar{z}e_k - e_k z) = \frac{1}{2}(e_k\bar{z} - ze_k)$$

with $z \in A_n(R)$. For real quaternions another representation is found in [Sud].
In matrix notation this unique connection reads as follows:

$$\begin{pmatrix} \bar{z} \\ z_1 \\ z_2 \\ \cdot \\ \cdot \\ \cdot \\ z_n \end{pmatrix} = \begin{pmatrix} e_\emptyset & -e_1 & -e_2 & \dots & -e_n \\ -e_1 & e_\emptyset & & & \\ -e_2 & \cdot & e_\emptyset & & \\ & \cdot & & & \\ & \cdot & & & \\ & \cdot & & & \\ -e_n & & & & e_\emptyset \end{pmatrix} \begin{pmatrix} x_\emptyset \\ x_1 \\ x_2 \\ \cdot \\ \cdot \\ \cdot \\ x_n \end{pmatrix},$$

and conversely for the right-linear combination

$$\begin{pmatrix} x_\emptyset \\ x_1 \\ x_2 \\ \cdot \\ \cdot \\ \cdot \\ x_n \end{pmatrix} = \frac{1}{n+1} \begin{pmatrix} e_\emptyset & e_1 & e_2 & \dots & e_n \\ e_1 & n & e_1e_2 & & e_1e_n \\ e_2 & e_2e_1 & n & & e_2e_n \\ \cdot & \cdot & \cdot & & \cdot \\ \cdot & \cdot & \cdot & & \cdot \\ \cdot & \cdot & \cdot & & \cdot \\ e_n & e_ne_1 & e_ne_2 & \dots & n \end{pmatrix} \begin{pmatrix} \bar{z} \\ z_1 \\ z_2 \\ \cdot \\ \cdot \\ \cdot \\ z_n \end{pmatrix}.$$

Note that the hypercomplex "vector" $\vec{z}$ may be written by

$$\vec{z} = (z_1, \dots, z_n) = \vec{x}\, e_\emptyset + x_\emptyset \vec{\tau}$$

with $\vec{x} = (x_1, \dots, x_n)$, $\vec{\tau} = (-e_1, \dots, -e_n)$. In a beautiful manner the difference between the two complexifications C^n and H^n is described in [Mal]. In C^n the independent variables are combined by the imaginary unit i, while in H^n an independent variable and a basis element is combined with the same independent variable in each case. These situations can be represented diagrammatically:

$\mathbb{C}^n$:

$$\begin{pmatrix} x_1 \\ x_2 \\ \cdot \\ \cdot \\ \cdot \\ x_n \end{pmatrix} + i \cdot \begin{pmatrix} y_1 \\ y_2 \\ \cdot \\ \cdot \\ \cdot \\ y_n \end{pmatrix} = \begin{pmatrix} z_1 \\ z_2 \\ \cdot \\ \cdot \\ \cdot \\ z_n \end{pmatrix},$$

H^n:

$$\begin{pmatrix} x_1 \\ x_2 \\ \cdot \\ \cdot \\ \cdot \\ x_n \end{pmatrix} + x_\emptyset \cdot \begin{pmatrix} -e_1 \\ -e_2 \\ \cdot \\ \cdot \\ \cdot \\ -e_n \end{pmatrix} = \begin{pmatrix} z_1 \\ z_2 \\ \cdot \\ \cdot \\ \cdot \\ z_n \end{pmatrix}.$$

11. Residue Theory

Suppose $u \in A_n(R)$ has an isolated left singular point a, the <u>residue of u at a</u> is defined by

$$\mathrm{Res}_a u = \frac{1}{|S_1|} \int_{S_R(a)} \alpha(y) u(y) dS_R ,$$

where α stands for the unit vector of the outer normal at the point y, and $S_R(a)$ denotes a sphere around the point a with the radius R. In [BDS] the following general result is given:

Let M be an $n+1$-dimensional compact differentiable oriented manifold-with-boundary contained in an open set G, $\partial M = \Gamma$ and u be left-regular in G, except for the isolated left singular points $a_1, \ldots, a_m$ with $a_j \notin \Gamma$, $j=1,\ldots,m$, and $a_{i_1}, \ldots, a_{i_k} \in \mathrm{int}\, M$. Then

$$\int_\Gamma \alpha(y) u(y) d\Gamma_y = |S_1| \sum_{s=1}^{k} \mathrm{Res}_{a_{i_s}} u .$$

For left poles a theorem which corresponds to the classical one is also obtained. Under special conditions similar considerations are made in complex CLIFFORD analysis, cf. [Ry3]. In [Som3] a general theory of residues for left-monogenic differential forms by generalizing the duality theory (cf.[DB3]) for monogenic functions is constructed.

12. On Winding Numbers

In [Sud] an analogue of the notion of the winding number of a curve Γ around a point in the plane, the so-called <u>wrapping number</u>, is introduced. In [Hab3] K.HABETHA defined an <u>index</u> for an n-dimensional closed and bounded manifold $M \subset \mathbb{R}^{n+1}$

with the help of CLIFFORD analysis. If we assume M sufficiently smooth, then the index calculated by

$$n(M,z) = \frac{1}{|S_1|}\int_M \frac{Re(\alpha(y)(\overline{y-z}))}{|y-z|^{n+1}}\,dM$$

is an integer in each component of the complement of M.
The definition of the <u>topological winding number</u> $c(\sum_{n-k},\sum_k)$ of $\sum_{n-k}$ around $\sum_k$ is given in 1952 by L.S. Pontrjagin in a classical manner by the aid of the so-called intersection number, where $\sum_k$ is a k-cycle in R^{n+1} and $\sum_{n-k}$ a (n-k)-cycle in $R^{n+1}\setminus\sum_k$. In [Som3] F.SOMMEN introduced the notion of the indicatrix of cycles which is defined by the formula

$$I(\textstyle\sum_{n-k},\sum_k) = \displaystyle\int_{(x,y)\in\sum_{n-k}\times\sum_k} K_k(x,y)\quad ,$$

where

$$K_k(x,y) = \frac{1}{|S_1|}\sum_{|A|=k} \operatorname{sgn} A\; dz_{N_n\setminus A}\,\frac{\bar{x}-\bar{y}}{|x-y|^{n+1}}\,d\,y_A$$

with $N_n=\{1,\ldots,n\}$, $A=\{\alpha_1,\ldots,\alpha_h\}$, $\alpha_i<\alpha_j$ for $i<j$, $dz_A = dz_{\alpha_1}\wedge\ldots\wedge dz_{\alpha_h}$, $dz_\emptyset=1$, sgn A denotes the signature of the permutation $\{\alpha_1,\ldots,\alpha_h,\beta_1,\ldots,\beta_{n-h}\}$ of N_n with $\beta_i<\beta_j$ for $i<j$, $d\,y_j=dy_j-e_jdy_\emptyset$, $dz_j=dx_j-e_jdx_\emptyset$ for $j=1,\ldots,n$, and finally $|S_1|$ is the area of the unit sphere in R^{n+1}. Note that $K_k(x,y)$ is called CAUCHY kernel of degree k. There exists a close connection between $I(\sum_{n-k},\sum_k)$ and $c(\sum_{n-k},\sum_k)$. In some cases these numbers coincide. F.SOMMEN used the indicatrix to deduce the relation

$$c(\textstyle\sum_{n-k}\sum_k) = (-1)^{k(n-k)+1}c(\sum_k,\sum_{n-k}).$$

In [BS] a definition of the index of a point z with respect to an n-dimensional cycle Γ is given. That means:
Let $\Gamma\subset C^{n+1}\setminus CN_z$, $z\in C^{n+1}$ (notation cf. [Appendix 3]). Then the index of z with respect to Γ is defined by

$$\mathrm{Ind}_\Gamma z = \frac{1}{|S_1|}\int_\Gamma \frac{z'-z}{|z'-z|^{n+1}}\, Dz'.$$

For all $z \in G = C^{n+1} \setminus \bigcup_{z\subset\Gamma} CN_z$ is proved there

(i) $\mathrm{Ind}\, z \in \mathbf{Z}$,

(ii) $\mathrm{Ind}_\Gamma z$ (Γ fixed) is locally constant on G.

13. Generalized Elementary Functions

There are various possibilities to construct analogues to elementary functions in hypercomplex analysis. By the help of the so-called left CAUCHY-KOWALEWSKI extension (cf. [BDS]) such analogues may be obtained. Using multiple TAYLOR series in [Som4] it is shown that with the complex function $f(z) = \sum_{k=\emptyset}^{\infty} c_k z^k$, $z \in C$, the series

$$F(u,z) = \sum_{k=\emptyset}^{\infty} c_k(\langle u,z\rangle - u_\emptyset z)^k \quad , (u,z) \in R^{n+1} \times C^n,$$

can be associated. For $n=1$ holds $F(u,z)=f(u\ z)$. It is easy to see that for $(1-\langle u,x\rangle)\langle u,y\rangle \neq u_\emptyset^2\langle x,y\rangle$ the function

$$P(u,z) = \sum_{k=\emptyset}^{\infty} (\langle u,z\rangle - u_\emptyset z)^k = \frac{1-\langle u,z\rangle - u_\emptyset z}{(1-\langle u,z\rangle)^2 + u_\emptyset^2 |z|^2}$$

generalizes $\frac{1}{1-uz}$, $(u,z) \in C^2$, where $e_\emptyset$ is identified with 1.

A corresponding <u>analogue to the exponential function</u> e^{uz} may be described by

$$\mathrm{Exp}(u,z) = \sum_{k=\emptyset}^{\infty} \frac{1}{k!}(\langle u,z\rangle - u_\emptyset z)^k = e^{\langle u,z\rangle}[\cos(u_\emptyset |z|) - \frac{z}{|z|}\sin(u_\emptyset |z|)].$$

In a different way an analogue of the complex logarithm in the algebra of quaternions was constructed by A.SUDBERY in [Sud]. Let $z = \sum_{k=\emptyset}^{3} x_i e_i$, then

$$L(z) = \frac{-r^2 + x_\emptyset \sum_{i=1}^{3} e_i x_i}{2r^2(r^2+x_\emptyset^2)} + \frac{\sum_{i=1}^{3} e_i x_i}{2r^3} \tan^{-1}\left(\frac{r}{x_\emptyset}\right),$$

where $r = (x_1^2+x_2^2+x_3^2)^{1/2}$, is the desired <u>generalized complex</u>

logarithm.

14. Generalized CAUCHY-RIEMANN Systems

In [Fue1] R.FUETER dealt with the so-called reduced quaternions. In this paper two different generalizations of CAUCHY-RIEMANN equations are introduced.

First let $u(x)=u_o e_o+u_1 e_1+u_2 e_2$ be given in a domain $G \subset R^3$.

The function u is called analytical if two of the non-linear differential equations

$$\text{(i)} \qquad Du_1 \cdot Du_2 = 0,$$

$$\text{(ii)} \qquad Du_0 \cdot Du_2 = 0,$$

$$\text{(iii)} \qquad Du_0 \cdot Du_1 = 0,$$

where $D=\frac{\partial}{\partial x_0}e_0+\frac{\partial}{\partial x_1}e_1+\frac{\partial}{\partial x_2}e_2$, and besides

$$\sum_{i=0}^{2}\left(\frac{\partial u_l}{\partial x_i}\right)^2 \neq 0, \quad l=0,1,2,$$

are fulfilled. If two of the following equations are fulfilled:

$$\text{(i')} \qquad \frac{\partial u}{\partial x_1}\cdot\frac{\partial u}{\partial x_2} = 0,$$

$$\text{(ii')} \qquad \frac{\partial u}{\partial x_0}\cdot\frac{\partial u}{\partial x_2} = 0,$$

$$\text{(iii')} \qquad \frac{\partial u}{\partial x_0}\cdot\frac{\partial u}{\partial x_1} = 0,$$

and we have

$$\sum_{l=0}^{2}\left(\frac{\partial u_l}{\partial x_k}\right)^2 \neq 0, \quad k=0,1,2,$$

then u is called hypo-analytical. There is duality between these two notions. If

$$u= \xi(x_0+i(x_1^2+x_2^2)^{1/2})+i\,\eta(x_0+i(x_1^2+x_2^2)^{1/2})$$

is an analytical function of the complex variable

$x_\emptyset+i(x_1^2+x_2^2)^{1/2}$ and denotes $x=x_\emptyset e_\emptyset+x_1e_1+x_2e_2$, then

$$u^*(x) = \xi^*(x) + (x_1^2+x_2^2)^{1/2}(x_1e_1+x_2e_2)\,\eta^*(x)$$

is hypo-analytical in a domain G' which arises by rotation of G about the real axis, where the values of ξ^* and ξ, respectively η^* and η coincide.

In the paper [Fue2] right- and left-regular functions are introduced where the functions $u=\sum_{i=\emptyset}^{3} u_ie_i$ which fulfil the system of differential equations

$$\begin{pmatrix} \partial_\emptyset & -\partial_1 & -\partial_2 & -\partial_3 \\ \partial_1 & \partial_\emptyset & \partial_3 & -\partial_2 \\ \partial_2 & -\partial_3 & \partial_\emptyset & \partial_1 \\ \partial_3 & \partial_2 & -\partial_1 & \partial_\emptyset \end{pmatrix} \begin{pmatrix} u_\emptyset \\ u_1 \\ u_2 \\ u_3 \end{pmatrix} = \begin{pmatrix} \emptyset \\ \emptyset \\ \emptyset \\ \emptyset \end{pmatrix}$$

are called <u>right-regular</u>, while the solutions of the system

$$\begin{pmatrix} \partial_\emptyset & -\partial_1 & -\partial_2 & -\partial_3 \\ \partial_1 & \partial_\emptyset & -\partial_3 & \partial_2 \\ \partial_2 & \partial_3 & \partial_\emptyset & -\partial_1 \\ \partial_3 & -\partial_2 & \partial_1 & \partial_\emptyset \end{pmatrix} \begin{pmatrix} u_\emptyset \\ u_1 \\ u_2 \\ u_3 \end{pmatrix} = \begin{pmatrix} \emptyset \\ \emptyset \\ \emptyset \\ \emptyset \end{pmatrix}$$

are called <u>left-regular</u>. For both classes has been developed a certain function theory.

In [Saa] E.M.SAAK introduced a system $\{e_{n,m}^{(i)}\}$ of n unitary square matrices of order m. The absolute values of their elements are equal to zero or one. Besides the condition

$$\text{(i)}\quad \det\sum_{i=\emptyset}^{n} x_ie_{n,m}^{(i)} \neq \emptyset \quad \text{for all} \quad x=(x_1,\dots,x_n)\in \mathbf{R}^n$$

and the relations

$$\text{(ii)}\quad [e_{n,m}^{(i)}]^*e_{n,m}^{(j)} + [e_{n,m}^{(j)}]^*e_{n,m}^{(i)} = \emptyset \quad , \ i\neq j,$$

shall be satisfied. The star denotes the transposed matrix. Condition (i) may be deduced by the other assumptions. A corresponding differential matrix $e_{n,m}^{(i)}(\frac{\partial}{\partial x})$ is associated to each matrix, where the entries with the value one are replaced by $\frac{\partial}{\partial x_i}$. The system of first order equations of dimension (n,m)

$$\sum_{i=1}^{n} e_{n,m}^{(i)}\left(\frac{\partial}{\partial x}\right)u = 0$$

generalizes the classical CAUCHY-RIEMANN equations. Putting n=3, m=4 and

$$e_{3,4}^{(1)}=\begin{pmatrix} 0 & 1 & 0 & 0 \\ 1 & 0 & 0 & 0 \\ 0 & 0 & 0 & -1 \\ 0 & 0 & 1 & 0 \end{pmatrix} \quad e_{3,4}^{(2)}=\begin{pmatrix} 0 & 0 & 1 & 0 \\ 0 & 0 & 0 & 1 \\ 1 & 0 & 0 & 0 \\ 0 & -1 & 0 & 0 \end{pmatrix} \quad e_{3,4}^{(3)}=\begin{pmatrix} 0 & 0 & 0 & 1 \\ 0 & 0 & -1 & 0 \\ 0 & 1 & 0 & 0 \\ 1 & 0 & 0 & 0 \end{pmatrix}$$

we get the well-known system of MOISIL-TEODORESCU (cf. [MT]). FUETER's systems can also be classified here. A certain analogue to BOREL-POMPEIU's formula is obtained.

V.S.VINOGRADOV investigated in his Habilitation (cf. [Vin]), among other things, the connection between CAUCHY's integral formula for FUETER's systems and the integral representation of BOCHNER-MARTINELLI in C^2. Furthermore he considered so-called spinor-systems of the form

$$\left(\sum_{i=1}^{n} A_i \frac{\partial}{\partial x_i}\right)u = f$$

if the matrices fulfil the relations

$$A_i^2 = E \ , \ A_iA_j + A_jA_i = 0 \ .$$

The solvability of corresponding boundary value problems is especially studied. In [Dez] A.A.DEZIN discovered first-order elliptic systems, for which boundary value problems of NOETHER's type exist. A very general class of first order systems with reference to function-theoretic properties was considered by G.HILE (cf. [Hil1],[Hil2],[HP]). In [MN] the analytic functions in the complex plane by the definition of a hyper-holomorphic function are generalized in the following way:

Let $z=z_1+z_2j$, $j^2=-1$, $z_i \in G \subset C$, let $u=u_1+u_2j$, $u_1,u_2 \in C_C^1(G)$.

A function is called hyper-holomorphic if the system of

differential equations

$$\frac{\partial u_1}{\partial \bar{z}_1} = \frac{\partial \bar{u}_2}{\partial z_2} \quad \text{and} \quad \frac{\partial u_2}{\partial \bar{z}_1} = - \frac{\partial \bar{u}_1}{\partial z_2}$$

is fulfilled. $\frac{\partial}{\partial z}$ and $\frac{\partial}{\partial \bar{z}}$ are known from classical function theory (cf. [Tu]). Recently classes of so-called p-hyperholomorphic functions have been considered. In [Mar] is given the definition of the p-hyperholomorphicity. It reads as follows

$$p \frac{\partial f_1}{\partial \bar{z}_\nu} = \frac{\partial f_2}{\partial z_{n+\nu}},$$

$$\frac{\partial f_2}{\partial \bar{z}_\nu} = -p \frac{\partial \bar{f}_1}{\partial z_{n+\nu}}, \quad 1 \leq \nu \leq n,$$

where $p \in C^1_R(G)$. The special cases $(n=2, p\equiv 1)$, $(n=1, p\equiv 1)$ were considered in [Bal] and [Riz], respectively.

A comparison of the hyperholomorphicity with the notion of the right-regularity of FUETER shows that by replacing u_2 by $\overline{u_2}$ a hyperholomorphic function will be transformed into a right-regular one. The complexification of FUETER's system for the left regularity reads

$$\frac{\partial u_1}{\partial \bar{z}_1} = \frac{\partial u_2}{\partial \bar{z}_2} \quad \text{and} \quad \frac{\partial u_1}{\partial z_2} = - \frac{\partial u_2}{\partial z_1} \qquad \text{(cf. [Sud]).}$$

E.STEIN and G.WEISS have been given another possibility to construct generalized CAUCHY-RIEMANN equations. In their paper [SW] a first-order system with constant coefficients is associated to each irreducible representation of the n-dimensional rotation group SO(n) in a natural way, which may be regarded as generalization of the classical CAUCHY-RIEMANN equations. Spinor representations, representations by means of anti-symmetric tensors and such which will be reached by spherical harmonic functions are investigated in detail. There is given an answer to the question if the function $|u|^p$ with a certain p is subharmonic when $u=(u_1,\dots,u_n)$ satisfies such a generalized CAUCHY-RIEMANN system. A lower bound for p can be given by $\frac{n-2}{n-1}$.

Above all the fundamental papers of R.DELANGHE [Del1],[Del2] and his collaborators F.BRACKX and F.SOMMEN (cf. [DB1], [DB2], [DB3] and [BDS]) have made the function theory in

CLIFFORD algebras to a well developed theory. In [Del1] left (right) monogenic functions are introduced. Now we follow the statements in [BDS].

Let $m \leq n$. An A_n-valued function $u \in C_A^1(G)$, $G \subset R^{m+1}$ is said to be <u>left (right) monogenic</u> if and only if

$$Du = \emptyset \qquad (uD = \emptyset),$$

where

$$D = \sum_{i=\emptyset}^{m} e_i D_i \;, \quad u = \sum_A e_A u_A \quad \text{and} \quad \{e_A\} \text{ form a basis in}$$

A_n. Using the components of u we have

$$Du = \sum_{i,A} e_i e_A D_i u_A \qquad (uD = \sum_{i,A} e_A e_i D_i u_A) .$$

The monogenicity condition is equivalent to a linear system of 2^n homogeneous partial differential equations of first order with constant real coefficients.

If $n=m=1$, then $A_1(R)=C$, $u=u_\emptyset e_\emptyset + u_1 e_1$ and $D=e_\emptyset \partial_{x_\emptyset} + e_1 \partial_{x_1}$ our system becomes

$$\partial_{x_\emptyset} u_\emptyset - \partial_{x_1} u_1 = \emptyset,$$
$$\partial_{x_\emptyset} u_1 + \partial_{x_1} u_\emptyset = \emptyset$$

which is the classical CAUCHY-RIEMANN system. Setting

$$u(\sum_{i=1}^{m} e_i x_i) = \sum_{i=1}^{m} e_i u_i (\sum_{i=1}^{m} e_i x_i)$$

then arises the so-called RIESZ system

$$\text{div } u = \emptyset \,,$$
$$\text{curl } u = \emptyset \,.$$

In [Ry1] J.RYAN invetigated complexified regular functions as follows:

For $G \subset C^{n+1}$ and a holomorphic function (componentwise) $u: G \longrightarrow A_n(C)$, then if

$$\sum_{i=\emptyset}^{n} e_i \frac{\partial u}{\partial z_i} = \emptyset \qquad (\sum_{i=\emptyset}^{n} \frac{\partial u}{\partial z_i} e_i = \emptyset)$$

we say that u is a complex left (right) regular function. J.RYAN proved for these functions some function-theoretic statements.

From the physicists point of view K.IMAEDA [Ima] has developed a corresponding function theory of complex quaternions (bi-quaternions).

The ring of holomorphic functions on $G \subset C^{n+1}$ with values in a complex vector space V will be denoted by $\mathcal{O}(G,V)$. $L(V_1,V_2)$ denotes the space of all C-linear mappings between the complex vector spaces V_1 and V_2. Generalization of a CAUCHY-RIEMANN operator is given in [BS] in the following way:

Let $\phi : C^{n+1} \longrightarrow L(V_1,V_2)$ a linear mapping. The operator

$$D \;:\; \mathcal{O}(G,V_1) \longrightarrow \mathcal{O}(G,V_2)$$

is defined by

$$D_\phi(u) = \sum_{i=\emptyset}^{n} \phi(e_i)\frac{\partial u}{\partial x_i}$$

where $e_\emptyset$, $e_1,\ldots,$ e_n generate the CLIFFORD algebra A_n. A special choice of the spaces V_i , $i=1,2$, and yields several relevant realizations such as some versions of the complex DIRAC operator.

In [Lou1], [LK] generalizations of CAUCHY-RIEMANN equations are presented by the help of special differential operators which are related to rotation, translation, dilatation and special conformal symmetry of differentiable functions.

Functions which satisfy these differential equations are called <u>special regular</u>.

K.HABETHA pointed out in [Hab1] a generalization of the concept of CAUCHY-RIEMANN equations for a very general situation (cf. [Appendix 4]). Let $u:G \longrightarrow W$, $G \subset V$ a domain (V, W vector spaces introduced in subsection A4, which are included in an algebra A). The function u is called <u>left-(right-) regular</u> in G if there $Du=\emptyset$ $(uD=\emptyset)$.

Left- and right-regular functions are called <u>biregular</u>. Employing the multiplication rule in A we get the following real system of first order partial differential equations:

$$Du = \sum_{i\in I}\sum_{j\in J} \varepsilon_{ijk}\frac{\partial u_j}{\partial x_i} = \emptyset, \quad k \in K ,$$

where the used notion is also described in Appendix 4.

The latter system is called <u>CAUCHY-RIEMANN system with respect to V, W and A</u>. Using the calculus of differential forms further generalization is given.

15. CAUCHY-RIEMANN Operators in Spherical Coordinates

Representation of the operator $D=\sum_{i=0}^{m} e_i D_i$ in spherical coordinates was deduced in [BDS]. Thus for each point $x \in R^{m+1}\backslash\{0\}$ we have

$$x = rw \ , \ w = \sum_{i=0}^{m} e_i w_i \ , \ w_i = x_i/r \ , \quad i=0,1,\ldots,m,$$

The operator D can be given the following spherical form:

$$D = w\frac{\partial}{\partial r} + \frac{1}{r}\frac{\partial}{\partial w}$$

with $$\frac{\partial}{\partial w} = \sum_{i=1}^{m} \frac{1}{\sin^2\theta_1 \cdot \ldots \cdot \sin^2\theta_{i-1}} \frac{\partial \omega}{\partial \theta_i}\frac{\partial}{\partial \theta_i} .$$

The angles θ_i are derived from the spherical coordinates

$$x_0 = r\cos\theta_1,$$
$$x_1 = r\sin\theta_1 \cos\theta_2,$$
$$\cdots\cdots\cdots\cdots\cdots\cdots\cdots\cdots$$
$$x_{m-1} = r\sin\theta_1 \sin\theta_2 \ldots \sin\theta_{m-1}\cos\theta_m,$$
$$x_m = r\sin\theta_1 \sin\theta_2 \ldots \sin\theta_{m-1}\sin\theta_m .$$

The operator $\Gamma = \bar{w}\frac{\partial}{\partial w}$ is called <u>spherical CAUCHY-RIEMANN operator</u>, $\tilde{\Gamma} = \overline{\frac{\partial}{\partial w}}\, w$ is its adjoint. This denotation leads to

$$D = w\left(\frac{\partial}{\partial r} + \frac{1}{r}\Gamma\right) .$$

Notice that in [Som5] it is shown that the LAPLACE-BELTRAMI operator Δ_s over the unit sphere S^{m-1} equals

$$\Delta_s = (\tilde{\Gamma} - I)\Gamma .$$

16. Pseudoanalytical Functions

In generalization of I.N.VEKUA's theory (cf.[Vek]) of the complex plane for higher dimensional spaces similar problems have been investigated. In some papers [Obo1], [Obo2] E.I.OBOLASVILI dealt with the solutions of two types of systems:

$$\operatorname{div} u + a\, u = 0,$$
$$\operatorname{rot} u + [u \times b] = 0$$

and

$$\operatorname{div} u + a\, u = 0,$$
$$\operatorname{grad} \varphi + \operatorname{rot} u + (u \times b) + \varphi a = 0 ,$$

where a, b are given vector functions, $u=(u_1,u_2,u_3)$ an unknown vector function and φ an unknown scalar function. Notice that the solutions of the first system are called generalized potential vectors. Several boundary value problems are also studied. A major contribution in this direction seems to be the papers [Gol1], [Gol2], [Gol3], and [Gol4] of B.GOLDSCHMIDT. He considered differential equations of the type

$$Dw = \sum_A c_A J_A(w) + F(x)$$

in a CLIFFORD algebra $\mathcal{A}$, where the operator on the right-hand side describes an arbitrary linear mapping in $\mathcal{A}$ (cf. [Gol1]). He also obtained the following CAUCHY integral representation with a certain kernel $K_A(x,t)$ and a system $\{w_k\}_{k=1}^N$ of linearly independent solutions of $Dw=0$, namely

$$w(x) = \sum_A \int_{\partial G} K_A(x,t) J_A(d\sigma_t w(t)) + \sum_{k=1}^N d_k w_k(x),$$

where $d_k = \langle w, w_k \rangle = \operatorname{Re} \int_G w(x)\overline{w_k(x)}\, dx$, $J_A = J_{\alpha_1} \ldots J_{\alpha_n}$

and finally $J_i(e_k) = \begin{cases} -e_i & k=i \\ e_k & k \neq i \end{cases}$.

B.GOLDSCHMIDT examined regularity properties of weak solutions of the above given system. Besides this he got a theorem about removable singularities. One of the main results is an explicit integral representation of the solutions of the considered pseudoanalytic system using FREDHOLM's theory. A detailed discussion of the kernel function is added. Generalizations of BOREL-POMPEIU's formula and jump-formulas of PLEMELJ-SOKHOTZKI's type with respect to the pseudoanalytic system are verified.

17. Polyanalytical Functions

Polyanalytical functions have been established as an important subject in complex analysis. A function u of the form

$$u(z) = \sum_{k=0}^{m-1} h_k(z)\bar{z}^{k}$$

is called m-analytical, where $h_k(z)$, $k=0,1,\dots,m-1$, are given analytical functions in a complex domain $G \subset C$. Each m-analytical function can be regarded as a solution of the equation (cf. [Ba])

$$\frac{\partial^m u}{\partial \bar{z}^m} = 0$$

with $\frac{\partial}{\partial \bar{z}} = \frac{1}{2}(\frac{\partial}{\partial x} + i\frac{\partial}{\partial y})$. The plane theory is rather well developed. Higher dimensional generalizations are contained in papers [Bra], [DB1], [DB2], [DB3], [DB4] and [BP].

An A-valued function $u \in C^k_A(G)$ is called left k-monogenic in G iff $D^k u=0$ in G, where $D=\sum_{i=0}^{m} e_i D_i$. Left-k-monogenic functions allow the following representation by a CAUCHY's type formula.

Let $u \in C^k_A(G)$, S a m+1-dimensional compact differentiable oriented manifold with boundary, $S \subset G$, if u left-k-monogenic, then

$$\int_{\partial S} \sum_{j=0}^{k-1} (-1)^j E_{j+1}(y-x)d\sigma_y D^j u(y) = \begin{cases} u(x), & x \in \overset{\circ}{S} \\ 0, & x \in \text{co } S \end{cases},$$

where

$$d\sigma = \sum_{i=0}^{m} (-1)^i e_i d\hat{x}_i, \quad d\hat{x}_i = dx_0 \wedge \dots \wedge dx_{i-1} \wedge dx_{i+1} \wedge \dots \wedge dx_m,$$

$$E_k(x) = \frac{1}{|S_1|}\frac{1}{(k-1)!}\frac{\bar{x}x_0^{k-1}}{|x|^{m+1}}.$$

This formula is an immediate consequence of the generalization of BOREL-POMPEIU's formula for functions $u \in C^k_A(G)$ which is given in [BP]. Complexification of the operator D^k

in [DB1] as well as an complexification of operators of the type $D_m D_{m-1} \ldots D_1$ is described in [BS].

18. Computational Methods in CLIFFORD Algebras

There is a CLIFFORD algebra calculator available called CLICAL [LMV]. CLICAL is a product of five years of program development by an interdisciplinary group with members having background extending from Computer Science to Mathematical Education and Theoretical Physics. CLICAL was first published in 1988 by its developers: Pertti LOUNESTO, Risto MIKKOLA and Vesa VIERROS. CLICAL is a calculator-type computer program which allows you to use your personal computer as a desktop calculator in conjunction with computations on complex numbers, vector spaces and CLIFFORD algebras. CLICAL is an interactive computer program designed for instruction in vectoralgebra and researches in CLIFFORD algebra. CLIFFORD algebra is the universal algebra unified language of geometry and physics, incorporating complex numbers, vectors, spinors and matrices. The real power of CLICAL is the unified geometrical approach provided by the CLIFFORD algebras.
The paper [BCDS] in which REDUCE 3.2 is used is worth noticing. The paper describes a simple implementation of the computational rules for CLIFFORD numbers in arbitrary dimension using the REDUCE 3.2 computer algebra package. It includes the various products and involutións and the inversion by means a complex matrix representation. A number of representative examples is contained.

19. Further Investigations in the Field of Hypercomplex Function Theory

Far-reaching results were obtained in the treatment of hypercomplex generalizations of FOURIER-, LAPLACE- and FOURIER-BOREL transforms. For further information we refer to the book [BDS] and the papers [Som6], [Som7] and [Som8]. Essential investigations were devoted to distributional boundary values of monogenic functions on manifolds, see also [BDS] and [Som9]. In his Habilitationsschrift F.SOMMEN tackled a new type of problems which are connected with the so-called monogenic operators. For readers who are interested in quantum physics we refer to the definitive paper of F.GÜRSEY and H.C.TZE [GT]. Further physical applications are contained in [LB] and [Spr4]. More recently some papers on RIEMANN-HILBERT boundary problems by XU ZHENYUAN [Xu], [XC] appeared. In the same field the Dissertation by I.STERN is settled in which a wide class of such problems is considered, cf.[Ste].

REFERENCES

[Ahl]: AHLFOHRS,L.:Clifford Numbers and Möbius Transformations in R^n, in: NATO and SERC Workshop on "Clifford Algebras and their Applications in Mathematical Physics" University of Kent, ed. by J.S.R. CHISHOLM and A.K. COMMON, Reidel-Dordrecht 1986

[AL]: AHLFORS,L.;LOUNESTO,P.: Some remarks on Clifford algebras, to appear in Complex variables

[Ale]: ALEKZIDSE,M.A.: A Decomposition Method in Non-orthogonal Functions for the Solution of Boundary Value Problems, Nauka, Moscow (1978), (Russian)

[AHS]: ATIYAH,M.F.;HITCHIN,N.J.;SINGER,I.M.: Selfduality in fourdimensional Riemannian geometry, Proc. Royal Soc., London A 362 (1978), 425-461

[Bal]: BALABAEV,V.E.: Quaternionic analogue of Cauchy-Riemann's system in a four-dimensional complex space and applications, DAN 214 No. 3 (1974), 489-491 (Russian)

[Ba]: BALK,M.B.: Polyanalytic functions, in: LANCKAU,E./ TUTSCHKE,W.: Complex Analysis, Methods, Trends, and Applications, Akademie-Verlag Berlin (1983), 68-84

[Ban]: BANDLE,C.: Isoperimetric Inequalities and Applications, Pitman Publ., Boston-London-Melbourne (1980)

[BJ]: BECHER,P.;JOOS,H.: The Dirac-Kähler Equation and Fermions on the Lattice, Z. Phys. C.-Particles and Fields 15, (1982), 343-365

[B]: BERNSTEIN,S.: Anwendung hyperkomplexer Methoden auf elliptische Aussenraumaufgaben, Diplomarbeit, TU Karl-Marx-Stadt (1987)

[Bi1]: BIZADSE,A.W.: On two-dimensional integrals of Cauchy type, AN Grus. SSR, XVI, (1955), 177-184

[Bi2]: BIZADSE,A.W.: Grundlagen der Theorie der analytischen Funktionen, Akademie-Verlag Berlin (1973)

[Bra]: BRACKX,F.: On (k)-monogenic functions of a quaternion variable, Function Theoretic Methods for Partial Differential Equations, Proceedings Lecture Notes in Mathematics No. 561, Springer-Verlag (1976), p. 138-149

[BCDS]: BRACKX,F.;CONSTALES,D.;DELANGHE,R.;SERRAS,H.: Clifford algebra with Reduce, Suppl. ai rendiconti del circolo matematico di Palermo, Serie II num. 16, (1987), 11-19

[BDS]: BRACKX,F.;DELANGHE,R.;SOMMEN,F.: Clifford analysis, Pitman, Boston-London-Melbourne (1982)

[BP]: BRACKX,F.;PINCKET,W.: The Newtonian potential for a generalized Cauchy-Riemann operator in Euclidean space, Ghent State University, Preprint (1980)

[BT]: BUDINICH,P.;TRAUTMANN,A.: The Spinorial Chessboard, ISBN 3-540-19078-3,(1988)

[Bu]: BURES,J.: Some integral formulas in complex Clifford analysis, Suppl. Rend. Circ. Mat. Palermo, Serie II, Num. 3, (1984), 81-87

[BS]: BURES,J.;SOUCEK,V.: Generalized hypercomplex analysis and its integral formulas, Complex Variables Theory Appl., no. 1, (1985), 53-70

[BM]: BURGESS,G.;MAHAJERIN,E.:The fundamental collocation method applied to the nonlinear Poisson equation in two dimensions, Computers and Structures Vol. 27, No. 6, (1987),763-767

[Che]: CHEVALLEY,C.: The Algebraic Theory of Spinors, Columbia University Press, New York, (1954)

[Cru]: CRUMEYROLLE,A.: Algebres de Clifford et Spineurs, Universite Paul Sabatier, Toulouse, (1974)

[Dea]: DEAVOURS, C.A.: The quaternionic calculus, Am. Math. Monthly 80 (1973), 995-1008

[DL]: DEETER,C.R.;LORD,M.E.: Further theory of operational calculus on discrete analytic functions, J. Math. Anal. Appl. 26, (1969), 92-113

[Del1]: DELANGHE,R.: On Regular-analytic Functions with Values in a Clifford Algebra, Math. Ann. 185, (1970), 91-111

[Del2]: DELANGHE,R.: On the singularities of functions with values in a Clifford algebra, Math. Ann. 196, (1972), 293-319

[Del3]: DELANGHE,R.: Morera's Theorem for functions with values in a Clifford algebra, Simon Stevin 43, (1969-70), 129-140

[Del4]: DELANGHE,R.: On regular points and Liouville's theorem for functions with values in a Clifford algebra, Simon Stevin 44, (1970-71), 55-56

[DB1]: DELANGHE,R.;BRACKX,F.: Hypercomplex function theory and Hilbert modules with reproducing kernel, Proc. London Math. Soc. 37, (1978), 545-576

[DB2]: DELANGHE,R.;BRACKX,F: Regular Solutions at Infinity of a Generalized Cauchy-Riemann Operator, Simon Stevin 53, (1979), No. 1

[DB3]: DELANGHE,R.;BRACKX,F.: Duality in hypercomplex function theory, J. Functional Analysis 37, (1980), 164-181

[DB4]: DELANGHE,R.;BRACKX,F.: Runge's theorem in hypercomplex function theory, J. of Approximation Theory 29, (1980)

[DR]: DENTONI,P.;RIZZA,G.P.: Una nuova classe di funzioni in un algebra reale, Rend. Ist. Mat. Univ. Trieste 4, (1972), 171-181

[Dez]: DEZIN,A.A.: On the solvable extensions of partial differential operators, Outlines Joint Symp. Part. Diff. Equ. Novosibirsk, Academ. Sci. USSR, Sibirian Branch, Moscow, (1963), 65-66

[Du]: DUFFIN,R.J.: Basic properties of discrete analytic functions, Duke Math. J. 23, (1956), 335-363

[DD]: DUFFIN,R.J.;DURIS,C.S.: A convolution product for discrete function theory, Duke Math. J. 31, (1964), 199-220

[Ede]: EDENHOFER,J.: Analytische Funktionen auf Algebren, Dissertation, Technische Universtät München, (1973)

[Ei]: EITZINGER,W.: Bericht zum Numerik-Grundpraktikum (unveröffentlicht), TU Karl-Marx-Stadt, (1988)

[Fra]: FRANKE,R.: Scattered data interpolation: Test of some methods, Math. Comp. 38, 181-200

[Fue1]: FUETER,R.: Analytische Funktionen einer Quaternionenvariablen, Comment. Math. Helv. 4, (1932), 9-20

[Fue2]: FUETER,R.: Reguläre Funktionen einer Quaternionenvariablen, Math. Inst. d. Universität Zürich, (1940)

[Fue3]: FUETER,R.: Funktionentheorie im Hyperkomplexen, Math. Inst. d. Universität Zürich, (1949)

[Fue4]: FUETER,R.: Über die analytische Darstellung der regulären Funktionen einer Quaternionenvariablen, Comm. Math. Helv. 8, (1936), 371-378

[Fue5]: FUETER,R.: Die Funktionentheorie der Differentialgleichungen $\Delta u=0$ und $\Delta\Delta u=0$ mit 4 reellen Variablen, Comm. Math. Helv. 7, (1935), 307-330

[Fab]: FABER,G.: Beweis, dass unter allen homogenen Membranen von gleicher Fläche und gleicher Spannung die kreisförmige den tiefsten Grundton gibt, Sitzungsber. der Bayer. Akad. der Wiss., (1923), 169-172

[Fe]: FERRAND,J.: Fonctions preharmonique et fonctions preholomorphes, Bulletin des Sciences Mathematique, sec. series, vol. 68, (1944), 152-180

[Gai]: GAIER,D.: Vorlesungen über Approximation im Komplexen, Birkhäuser-Verlag Basel-Boston-Stuttgart,(1980)

[Ge]: GEGELIA,T.G.: On the boundedness of singular operators, Soobshz. Akad. Nauk, Gruz. SSR 20:5, (1968), 517-523, (Russian)

[Gir]: GIRAUD,G.: Equations a integrales principales, Ann. Sc. Ecole norm super. 51, fasc. 3 et 4, (1934)

[GJ]: GÖCKELER,M.;JOOS,H.: On KÄHLER's geometric description of Dirac fields, Preprint DESY, (1983), 83-128

[Gol1]: GOLDSCHMIDT,B.: Verallgemeinerte analytische Vektoren im R^n, Thesis, Universität Halle-Wittenberg, (1980)

[Gol2]: GOLDSCHMIDT,B.: A Cauchy Integral Formula for a Class of Elliptic Systems of Partial Differential Equations of First Order in the Space, Math. Nachr. 108, (1982), 167-178

[Gol3]: GOLDSCHMIDT,B.: Properties of Generalized Analytic Vectors in R^n, Math. Nachr. 103, (1981), 245-254

[Gol4]: GOLDSCHMIDT,B.: Existence and representation of solutions of a class of elliptic systems of partial differential equations of first order in the space, Math. Nachr. 108, (1982), 159-166

[Gue1]: GÜRLEBECK,K.: Über Interpolation und Approximation verallgemeinert analytischer Funktionen, Wiss. Inf. 34, TH Karl-Marx-Stadt (1982)

[Gue2]: GÜRLEBECK,K.: Hypercomplex Factorization of the Helmholtz Equation, ZAA 5 (2), (1986), 125-131

[Gue3]: GÜRLEBECK,K.: Approximate solution of stationary Navier-Stokes equations, Math. Nachr., to appear

[Gue4]: GÜRLEBECK,K.: Interpolation and best approximation in spaces of monogenic functions, WZ d. TU Karl-Marx-Stadt 30, (1988), H.1, 38-41

[Gue5]: GÜRLEBECK,K.: Über die optimale Interpolation verallgemeinert analytischer quaternionenwertiger Funktionen und ihre Anwendung zur näherungsweisen Lösung wichtiger räumlicher Randwertaufgaben der Mathematischen Physik, Dissertation, TH Karl-Marx-Stadt, (1984)

[Gue6]: GÜRLEBECK,K.:Grundlagen einer diskreten räumlich verallgemeinerten Funktionentheorie und ihrer Anwendungen, Thesis, TU Karl-Marx-Stadt, (1988)

[GL]:GÜRLEBECK,K.;LÖSCH,U.: Zur numerischen Lösung von Randwertproblemen für die biharmonische Gleichung, WZ d. TU Karl-Marx-Stadt 29 (1987) H.2, 200-203

[GSS]: GÜRLEBECK,K.;SCHÜRER,A.;SPRÖSSIG,W.: Application of Boundary Collocation Methods in Physics and Engineering, WZ d. TH Karl-Marx-Stadt 28,(1986), H.2, 168-174

[GS1]: GÜRLEBECK,K.;SPRÖSSIG,W.: Methods of Quaternionic Analysis for Analytical and Numerical Consideration of Boundary Value Problems, to appear in Seminar Analysis of Academy of Science of the GDR

[GS2]: GÜRLEBECK,K.;SPRÖSSIG,W.: An application of quaternionic analysis to the solution of time-independent Maxwell equations and of Stokes′ equations, Suppl. Rend. Circ. Mat. Palermo, Serie II, num. 14, (1987), 61-76

[GS3]: GÜRLEBECK,K.;SPRÖSSIG,W.: A Generalized Leibniz Rule and a Discrete Quaternionic Analysis, Suppl. Rend. Circ. Mat. Palermo, Serie II, num. 16, (1987), 44-64

[GS4]: GÜRLEBECK,K.;SPRÖSSIG,W.: A Unified Approach to Estimation of Lower Bounds for the First Eigenvalue of Several Elliptic Boundary Value Problems, Math. Nachr. 131, (1987), 183-199

[GST]: GÜRLEBECK,K.;SPRÖSSIG,W.;TASCHE,M.: Numerical Realization of Boundary Collocation Methods, ISNM Vol. 75, Birkhäuser Verlag Basel, (1985), 206-217

[GT]: GÜRSEY,F.;TZE,H.C.: Complex and quaternionic analyticity in chiral and gauge theories, Ann. of Physics 128, (1980), 29-130

[Hab1]: HABETHA,K.: Eine Bemerkung zur Funktionentheorie in Algebren, In: MEISTER,V.E.;WECK,N.;WENDLAND,W.: Lecture Notes in Mathematics Vol. 561, Springer, Berlin-Heidelberg-New York, (1976)

[Hab2]: HABETHA,K.: Function Theory in Algebras, In: LANCKAU,E.;TUTSCHKE,W.: Complex analysis, Akademie-Verlag Berlin, (1983), 225-239

[Hab3]: HABETHA,K.: Eine Definition des Kroneckerindexes im R^{n+1} mit Hilfe der Cliffordanalysis, ZAA 5, (1986), No. 2, 133-137

[Hae]: HAEFELI,H.G.: Hyperkomplexe Differentiale, Comment. Math. Helv. 20, (1947), 382-420

[Ham]: HAMILTON,W.R.: Elements of Quaternions, London, Langmans Green, (1866)

[Hay]: HAYABARA,S.: Operational calculus on the discrete analytic functions, Math. Japon. 11, (1966), 35-65

[Her]: HERSCH,J.: Sur la frequence fondamentale d'une membrane vibrante evoluation par difant et principe de maximum, J. Math. Phy. Appl. 11, (1960), 387-413

[Hes1]: HESTENES,D.: Multivector functions, J. Math. Anal. Appl. 24, (1968), 467-473

[Hes2]: HESTENES,D.: New foundations of classical mechanics, Reidel-Dordrecht, (1987)

[HS]: HESTENES,D.;SOBCZYK,G.: Clifford Algebra to Geometric Calculus - A new language for mathematics and physics, Reidel-Dordrecht, (1985)

[Hil1]: HILE,G.: Hypercomplex function theory applied to partial differential equations, Ph. D. dissertation, Indiana Univ., Bloomington (1972)

[Hil2]: HILE,G.: Representation of solutions of a special class of first-order systems, J. Diff. Eq. 25, (1977), 410-424

[HP]: HILE,G.;PROTTER,M.: Maximum principles for a class of first-order elliptical systems, J. Diff. Eq. 24, (1977), 136-151

[Hö]: HÖRMANDER,L.: The Analysis of Linear Partial Differential Operators I, Distribution Theory and Fourier Analysis, Springer-Verlag, Berlin-Heidelberg-New York-Tokyo, (1983)

[I1]: IFTIMIE,V.: Fonctions hypercomplexes, Bull. Math. Soc. Sci. Math. R.S.R. 9 (57), (1965), 279-332

[I2]: IFTIMIE,V.: Operateurs du type de Moisil-Teodorescu, Bull. Math. Soc. Sci. Math. R.S.R. (58), (1966), 271-305

[Ima]: IMAEDA,K.: A new formulation of classical electrodynamics, Nuovo Cimento 32 A,(1976), 138-162

[J]: JOHN,F.: Plane waves and spherical means applied to partial differential equations, Intersc. Publ., New York, London, (1955)

[KA]: KANTOROWITSCH,L.W.;AKILOW,G.,P.: Funktionalanalysis in normierten Räumen, Akademie-Verlag, Berlin, (1964)

[Ka]: KAWOHL,B.: Estimates for the first eigenvalue of a special elliptic system, Preprint d. Universität Heidelberg, (1987)

[KS]: KAWOHL,B.;SWEERS,G.: Remarks on eigenvalues and eigenfunctions of a special elliptic system, Preprint d. Universität Heidelberg, (1987)

[Ke]: KERSTEN,H.: Über die gleichmässige Approximierbarkeit harmonischer Funktionen in regulären Gebieten einschliesslich des Randes, Dissertation RWTH Aachen,(1977)

[Ko]: KOBELKOV,G.M.: Resolution of the problem of stationary free convection, DAN SSSR 255 No. 2, 277-281, (Russian)

[Kra]: KRAHN,E.: Über eine von Rayleigh formulierte Minimaleigenschaft des Kreises, Math. Ann. 94, (1924), 97-100

[Lad]: LADYZENSKAJA,D.A.: The mathematical theory of viscous impressible flow, 2nd engl. Ed., Gordon and Breach, New York-London-Paris, (1969)

[LU]: LADYZENSKAJA,D.A.;URALCEVA,N.N.: Linear and quasilinear equations of elliptical type, Nauka, Moscow (2. Ed.), (1973), (Russian)

[Len]: LENSE,J.: Kugelfunktionen, Geest & Portig K.G., Leipzig, (1950)

[LP]: LEVINE,H.A.;PROTTER,M.H.: Unrestricted Lower Bounds for Eigenvalues for Classes of Elliptic Equations and Systems of Equations, Math. Meth. in the Appl. Sci. 7, (1985), 210-222

[LM]: LIONS,J.L.;MAGENES,E.: Problemes aux limites non homogenes et applications, vol. 1 et 2, Paris, Dunod, (1968)

[Lou1]: LOUNESTO,P.: Spinor valued regular functions in hypercomplex analysis, Thesis, Helsinki University of Technology, (1979)

[Lou2]: LOUNESTO,P.: Clifford algebras and spinors, NATO & SERC Advanced Research Workshop on Clifford Algebras and their Applications in Mathematical Physics, University of Kent, Canterbury, 15-27 September 1985, Reidel-Dordrecht, (1986), 25-37

[Lou3]: LOUNESTO,P.: Scalar Products of Spinors and an Extension of Brauer-Wall Groups, Found. of Physics, Vol. 11, No. 9/10, (1981), 721-740

[LB]: LOUNESTO,P.;BERGH,P.: Axially Symmetric Vector Fields and their Complex Potentials, Compl. Var. Vol. 2, (1983), 139-150

[LK]: LOUNESTO,P.;KUSTAANHEIMO,P.: Spinor Function Theory, Simon Stevin Vol. 58, No.3, (1984), 193-218

[LMV]: LOUNESTO,P.;MIKKOLA,R.;VIERROS,V.: CLICAL - User Manual, Institute of Mathematics, Helsinki University of Technology, Espoo (1988)

[LS]: LOUNESTO,P.;SPRINGER,A.: Möbius transformations and Clifford algebras in euclidean and anti-euclidean spaces, to appear in Reidel-Dordrecht

[Mal]: MALONEK,H.: Zum Holomorphiebegriff in höheren Dimensionen, Thesis, Halle, Pädagogische Hochschule, (1987)

[Mar]: MARINOV,M.S.: On a class of hyperholomorphic functions, Comptes rendus del' Academie bulgar.des Sciences, T. 41 No. 9, (1988), 15-17

[Mej]: MEJLIHZON,A.S.: On the notion of monogeneous quaternions, DAN SSSR 59, 3, (1948), 431-434, (Russian)

[Mi1]: MICHLIN,S.G.: Multidimensional Singular Integrals and Integral Equations, Gos. Isd. Fiz.-Mat., Moscow, (1962), (Russian)

[Mi2]: MICHLIN,S.G.: Singular Integral Equations, UMN, 3, 25, (1948), 29-112

[Mi3]: MICHLIN,S.G.: Konstanten in einigen Ungleichungen der Analysis, Teubner-Texte zur Mathematik 39, Teubner, Leipzig, (1981)

[Mi4]: MICHLIN,S.G.: Numerische Realisierung von Variationsmethoden, Akademie-Verlag, Berlin, (1969)

[MP]: MICHLIN,S.G.;PRÖSSDORF,S.: Singular Integral Operators, Akademie-Verlag, Berlin, (1986)

[MT]: MOISIL,G.C.;TEODORESCU,N.: Fonctions holomorphes dans l'espace, Matematicae, Cluj, 5, (1931), 142-150

[MN]: MUHAMED-NASER: Hyperholomorphic functions, Sib. Math. Journ. XII, No. 6, (1971), 1327-1340

[Mue1]: MÜLLER,CL.: Spherical harmonics, Lecture Notes in Math., Berlin, (1966)

[Mue2]: MÜLLER,CL.: Neue Verfahren zur Lösung der elliptischen Randwertprobleme der Math. Physik, Vorträge der Rheinisch-Westfäl. Akad. d. Wiss. 288, Westdeutscher Verlag Opladen, (1979), 27-68

[Nau]: NAUMANN,J.: On the interior regularity of weak solutions of the stationary Navier-Stokes equations, Publ. del Departimento di Matematica dele universita di Pisa, No. 156, Giugno, (1986), 1-38

[Obo1]: OBOLASVILI,E.I.: Multidimensional generalized holomorphic vectors, Diff. Urav. 11, 1, (1975), 108-115, (Russian)

[Obo2]: OBOLASVILI,E.I.: Generalized Cauchy-Riemann systems in R^n, Trudy Tbil. Mat. Inst. LVIII, (1978), 168-174, (Russian)

[Pen1]: PENROSE,R.: Twister algebra, J. Math. Phys. 8, (1967), 345-366

[Pen2]: PENROSE,R.: Null hypersurface initial data for classical fields of arbitrary spin and for general relativity, Gen. Relat. and Gravitation 12, (1980), 225-264

[Por]: PORTEOUS,I.: Topological Geometry, Van Nostrand-Reinhold, London, (1969)

[Pro]: PROTTER,M.H.: A lower bound for the fundamental frequency of a convex region, Proc. of the Americ. Math. Soc., vol. 81, No. 1, (1981), 65-70

[Ray]: RAYLEIGH,I.W.S.: The theory of sound, 2nd Edition, London, (1896)

[RT]: REDKOP,D.;THOMPSON,J.L.: Use of fundamental solutions in the collocation method in axisymmetric elastostatics, Comp. Struct. 17, (1983), 485-490

[Rie]: RIESZ,M.: Clifford numbers and spinors, Lecture Series No. 38, Institute for Physical Sciences and Technology, Maryland, (1958)

[Riz]: RIZZA,G.B.: Monogenic functions on the real algebras and conformal mappings, Boll. Unione Mat. Ital., Ser. 4, 12 Suppl. Fac. 3, (1975), 437-450

[Ry1]: RYAN,J.: Complexified Clifford Analysis, Complex Variables 1, (1982), 119-149

[Ry2]: RYAN,J.: Clifford Analysis with Generalized Elliptic and Quasi Elliptic Functions, Appl. Analysis 13, (1982), 151-171

[Ry3]: RYAN,J.: Singularities and Laurent expansions in complex Clifford analysis, Appl. Anal. 16, No. 1, (1983), 33-49

[Saa]: SAAK,E.M.: On the theory of multidimensional elliptic systems of first order, Sov. Math. Dokl. Vol. 16, No. 3, (1975)

[Sar]: SARD,A.: Approximation Based on Nonscalar Observations, J. of Approximation Theory 8, (1973), 315-334

[Sam]: SAMARSKIJ,A.A.: Theory of finite difference methods, Nauka, Moscow, (1977), (Russian)

[SLM]: SAMARSKIJ,A.A.;LASAROV,R.D.;MAKAROV,W.L.: Finite difference methods for differential equations with generalized solutions,Vysshaja Shkola, Moscow, (1987), (Russian)

[Scha]: SCHABAT,B.V.: Introduction in complex analysis, Nauka, Moscow, (1969), (Russian)

[Schou]: SCHOUTEN,J.A.: Ricci-calculus, 2nd Edition, Springer-Verlag, Berlin, (1954)

[S]: SCHÜRER,A.:Konstruktion vollständiger Systeme zur Approximation von Lösungen für Randwertaufgaben der Wärmeleitgleichung, WZ d. TH Karl-Marx-Stadt, 293-298

[Sh1]: SEVTSENKO,V.I.: Multidimensional local and global vector-homomorphisms for a class of quaternionic equations, DAN SSSR 153:2, (1963), 300-302, (Russian)

[Sh2]: SEVTSENKO,V.I.: Hilbert's problem for holomorphic vector, DAN SSSR, 169:6, (1966), 1285-1292, (Russian)

[Sm]: SMIRNOV,W.I.: Lehrgang der höheren Mathematik, Teil V, DVW, Berlin, (1967)

[Smo]: SMOLITZKI,CH.: Estimation of the derivative of fundamental functions, DAN SSSR 24, Nr. 2, (1950)

[Som1]: SOMMEN,F.: Formale randwaarden van monogene functies en elliptische systemen in het vlak, Habilitation, Ghent, (1984)

[Som2]: SOMMEN,F.: A generalized version of the Fourier-Borel transform, Bull. Soc. Roy. Sci., Liege 50, No. 5-8, (1981), 203-218

[Som3]: SOMMEN,F.: Monogenic differential forms and homology theory, Proc. Royal Irish Academy, 84 A, (1984), 87-109

[Som4]: SOMMEN,F.: Some Connections Between Clifford Analysis and Complex Analysis, Complex Variables, Vol. 1, (1982), 97-118

[Som5]: SOMMEN,F.: Spherical Monogenic Functions and Analytic Functionals on the Unit Sphere, Tokyo Journal of Mathematics, Vol. 4, No. 2, (1981), 427-456

[Som6]: SOMMEN,F.: Hypercomplex Fourier and Laplace Transforms I,Illinois Journal of Mathematics, Vol. 26, No. 2, (1982), 332-352

[Som7]: SOMMEN,F.: Hypercomplex Fourier and Laplace Transforms II, Complex Variables, Vol. 1, (1983), 209-238

[Som8]: SOMMEN,F.: Microfunctions with values in a Clifford algebra II, Sci. Papers College Arts Sci. Uni. Tokyo 36, no. 1, (1986), 15-37

[Som9]: SOMMEN,F.: Monogenic functions on surfaces, Zeitschrift für angewandte Mathematik, 361, (1985), 145-161

[Sou1]: SOUCEK,V.: Regularni funkce quaternionve promenne, Thesis, Karls-Universität Prag, (1980)

[Sou2]: SOUCEK,V.: Complex-quaternionic analysis applied to spin-$\frac{1}{2}$-massless fields, Complex Variables, Theory and Applications 1, no. 4, (1983), 327-346

[Sou3]: SOUCEK,V.: Boundary value type and initial value type integral formulas for massless fields, Twistor Newsletters, no. 14, (1982)

[Sou4]: SOUCEK,V.: Complex-quaternionic analysis to spin-massless fields, Preprint

[Sou5]: SOUCEK,V.: H-valued differential forms on H, Suppl. Rend. Circ. Mat. Palermo, Serie II, No. 3, (1984), 293-299

[Sp1]: SPRÖSSIG,W.: Analoga zu funktionentheoretischen Sätzen im R^n, Beiträge zur Analysis 12, (1978), 113-126

[Sp2]: SPRÖSSIG,W.: Über eine mehrdimensionale Operatorrechnung über beschränkten Gebieten des E_n, Thesis, TH Karl-Marx-Stadt, (1979)

[Sp3]: SPRÖSSIG,W.: Über die Regularisierung eines Systems zweidimensionaler singulärer Integralgleichungen, Dissertation, TH Karl-Marx-Stadt, (1974)

[Sp4]: SPRÖSSIG,W.: Anwendung der analytischen Theorie der Quaternionen zur Lösung räumlicher Probleme der linearen Elastizität, ZAMM 59, (1979), 741-743

[Sp5]: SPRÖSSIG,W.: Untere Abschätzungen des 1. Eigenwertes spezieller elliptischer Randwertaufgaben, WZ d. TH Leuna-Merseburg 27, 5, (1985), 556-560

[Sp6]: SPRÖSSIG,W.: Lösungsdarstellungen einer Klasse spezieller Operatorgleichungen, Math. Nachr. 90, (1979), 135-147

[SG]: SPRÖSSIG,W.;GÜRLEBECK,K.: A hypercomplex method of calculating stresses in three-dimensional bodies, Suppl. Rend. Circ. Mat. Palermo, Serie II, num. 6, (1984), 271-284

[SW]: STEIN,E.;WEISS,G.: Generalization of the Cauchy-Riemann equations and representations of the rotation group, Amer. J. of Math. 90, (1968), 163-196

[Ste]: STERN,I.: Dissertation, MLU Halle-Wittenberg, (1989)

[Sud]: SUDBERY,A.: Quaternionic analysis, Math. Proc. Camb. Soc. 85, (1979), 199-225

[Such]: SUCHOMLINOV,G.A.: On the extension of linear functionals in linear normed spaces and linear quaternionic spaces, Mat. sbornik (2),3, (1938), 353-358

[Ta]: TASCHE,M.: Eine einheitliche Herleitung verschiedener Interpolationsformeln mittels der Taylorschen Formel der Operatorrechnung, ZAMM 61, (1981), 379-393

[Tem]: TEMAM,R.: Navier-Stokes Equations, Vol. 2, Studies in Mathematics and its Applications, 2nd Edition, North-Holland Publ. Comp., (1979)

[Te]: TESCHKE,H.: Über die Darstellung harmonischer und metaharmonischer Funktionen in Gebieten mit nichtglatten Rändern, Dissertation, RWTH Aachen, (1978)

[Tu]: TUTSCHKE,W.: Grundlagen der Funktionentheorie, DVW, Berlin, (1967)

[Vah]: VAHLEN,K.TH.: Über Bewegungen und komplexe Zahlen, Math. Ann. 55, (1902), 585-593

[Vek]: VEKUA,I.N.: Generalized analytic functions, Reading, (1962)

[Vla]: VLADIMIROV,W.S.: Gleichungen der Mathematischen Physik, DVW, Berlin, (1972)

[Vin]: VINOGRADOV,V.S.: Investigations on boundary value problems for elliptic systems of first order, Thesis, Moscow, Steklov Inst., (1972), (Russian)

[Val]: VALLI,A.: On the Integral Representation of the Stokes System, Rend. Sem. Mat. Univ. Padova, Vol. 74, (1985)

[Wal]: WALSH,J,L.: Interpolation and approximation by rational functions in the complex domain, Amer. Math. Soc., Providence, R.I., (1969)

[Xu]: XU,ZHENYAN: On Riemann boundary value problem for regular functions with values in a Clifford algebra, Kexue Toughao (Science Bulletin) 32 (13), (1987)

[XC]: XU,ZHENYAN;CHEN,TIN: On Riemann-Hilbert problem for regular functions with values in a Clifford algebra, Kexue Toughao (Science Bulletin) 33 (10), (1988)

[Zei]: ZEIDLER,E.: Vorlesungen über nichtlineare Funktionalanalysis II, -Monotone Operatoren- , Teubner-Texte zur Mathematik Bd. 9, Teubner, Leipzig, (1977)

[Z]: ZEILBERGER,D.: A New Approach to the Theory of Discrete Analytic Functions, J. Math. Anal. Appl. 57, (1977), 350-367

SUBJECT INDEX

NOTATIONS

1 **Inverse and Improperly Posed Problems in Differential Equations**
Proceedings of the Conference on Mathematical and Numerical Methods, held in Halle/Saale (GDR) from May 29 to June 2, 1979
Edited by Gottfried Anger
296 pages, 1979

2 **Fundamentals of Computation Theorie FCT '79**
Proceedings of the Conference on Algebraic, Arithmetic and Categorial Methods in Computation Theory, held in Berlin/Wendisch-Rietz (GDR) September 17–21, 1979
Edited by Lothar Budach
576 pages, 1979

3 **Decidability and Generalized Quantifiers**
By Andreas Baudisch, Detlef Seese, Hans-Peter Tuschik, and Martin Weese
XII, 235 pages, 1980

4 **Zur Elastizitäts- und Thermoelastizitätstheorie**
Potentialtheoretische Methoden in Anwendung auf verschiedene Modelle
Von Lothar Jentsch und Johannes Maul
157 Seiten, 1980

5 **Systems Analysis and Simulation 1980**
Proceedings of the International Symposium held at Berlin (GDR) September 1–5, 1980
Edited by Achim Sydow
496 pages, 1980

6 **Nichtlineare diskrete Optimierung**
von Eberhard Girlich und Michail M. Kowaljow
218 Seiten, 1981

7 **Completeness Problems in the Structural Theory of Automata**
By Jürgen Dassow
148 pages, 1981

8 **Darstellung von Lösungen linearer elliptischer Differentialgleichungen**
Von Günther Wildenhain
92 Seiten, 1981

9 **Introduction to the Spectral Theory of Operators in Spaces with an Indefinite Metric**
By I. S Iohvidov, M. G. Krein, and H. Langer
120 pages, 1982

10 **Nonlinear Time-discrete Systems**
A General Approach by Nonlinear Superposition
By Michael Gössel
115 pages, 1982

11 **Preprimal Algebras**
By Klaus Denecke
162 pages, 1982

12 **Ergodic Theory and Related Topics**
Proceedings of the Conference held in Vitte/Hiddensee (GDR) October 19–23, 1981
Edited by Horst Michel
232 pages, 1982

13 **Singularly Perturbed Differential Equations**
By H. Goering, A. Felgenhauer, G. Lube, H.-G. Roos, and L. Tobiska
176 pages, 1983

14 **Leben wir in einer Volterra-Welt?**
Ein ökologischer Zugang zur angewandten Systemanalyse
Von Manfred Peschel und Werner Mende
233 Seiten, 1983

15 **Markov Decision Problems with Countable State Spaces**
Optimality Criteria – Algorithms – Clustering
By Hans Michael Dietz and Volker Nollau
174 pages, 1983

16 **Stochastische Verfahren zur Suche nach dem Minimum einer Funktion**
von Ryszard Zieliński und Peter Neumann
133 Seiten, 1983

17 **Invertibility and Asymptotics of Toeplitz Matrices**
By Albrecht Böttcher and Bernd Silbermann
200 pages, 1983

18 **Structural Induction on Partial Algebras**
Introduction to Theory and Application of Partial Algebras – Part II
By Horst Reichel
205 pages, 1984

19 **Algebraic Methods for Toeplitz-like Matrices and Operators**
By Georg Heinig and Karla Rost
212 pages, 1984

20 **Frege Conference 1984**
Proceedings of the International Conference held at Schwerin (GDR), September 10–14, 1984
Edited by Gerd Wechsung
408 pages, 1984

21 **Numerical Analysis of Selected Semilinear Differential Equations**
Edited by Christian Großmann, Thomas Riedrich, and Hartmut Schönheinz
203 pages, 1985

22 **Singular Ordinary Differential Operators and Pseudodifferential Equations**
By Johannes Elschner
200 pages, 1985

23 **Lotka-Volterra-Approach to Cooperation and Competition in Dynamic Systems**
Proceedings of the 5th Meeting of UNESCO's Working Group on Systems Theory held on the Wartburg, Eisenach (GDR) at March 5–9, 1984
Edited by Werner Ebeling and Manfred Peschel
271 pages, 1985

24 **Convergence Structures 1984**
Proceedings of the Conference on Convergence held in Bechyně (Czechoslovakia), September 24–28, 1984
Edited by Josef Novák, Werner Gähler, Horst Herrlich, and Jan Mikusiński
254 pages, 1985

25 **Parcella '84**
Proceedings of the II. International Workshop on Parallel Processing by Cellular Automata and Arrays held in Berlin (GDR), September 25–27, 1984
Edited by Wolfgang Händler, Tamás Legendi, and Gottfried Wolf
239 pages, 1985

26 **Multiobjective and Stochastic Optimization Based on Parametric Optimization**
By Jürgen Guddat, Francisco Guerra Vasquez, Klaus Tammer, and Klaus Wendler
175 pages, 1985

27 28 **Systems Analysis and Simulation 1985**
I: Theory and Foundations
II: Applications
Proceedings of the International Symposium held in Berlin (GDR), August 26–31, 1985
Edited by Achim Sydow, Manfred Thoma, and Robert Vichnevetsky
per volume 495 pages, 1985

29 **Parcella '86**
Proceedings of the III. International Workshop on Parallel Processing by Cellular Automata and Arrays held in Berlin (GDR), September 9–11, 1986
Edited by Tamás Legendi, Dennis Parkinson, Roland Vollmar, and Gottfried Wolf
279 pages, 1986 ISBN 3-05-500175-3

30 **Fuzzy Sets Applications, Methodological Approaches, and Results**
Proceedings of the International Workshop on Fuzzy Sets Applications cosponsored by IIASA and convened in cooperation with IAEA held on the Wartburg, Eisenach (GDR), at March 3–8, 1985
Edited by Steffen Bocklisch, Sergei Orlovski, Manfred Peschel, and Yasushi Nishiwaki
233 pages, 1986 ISBN 3-05-500016-16-1

31 **Mathematical Methods of Specification and Synthesis of Software Systems '85**
Proceedings of the International Spring School, held in Wendisch-Rietz (GDR), April 22–26, 1985
Edited by Wolfgang Bibel and Klaus P. Jantke
245 pages, 1986 ISBN 3-05-500018-8

32 **A Model Theoretic Oriented Approach to Partial Algebras**
Introduction to Theory and Application of Partial Algebras – Part I
By Peter Burmeister
319 pages, 1986 ISBN 3-05-500176-1

33 **Finite Difference Method on Irregular Networks**
A Generalized Approach to Elliptic Problems of Second Order
By Bernd Heinrich
206 pages, 1986 ISBN 3-05-500328-4

34 **Modern Algorithms for Large Sparse Eigenvalue Problems**
By Arnd Meyer
125 pages, 1987 ISBN 3-05-500332-2

35 **Parametric Optimization and Related Topics**
Edited by Jürgen Guddat, Hubertus Th. Jongen, Bernd Kummer, and František Nožička
407 pages, 1987 ISBN 3-05-500337-3

36 **Discretization in Differential Equations and Enclosures**
Edited by Ernst Adams, Rainer Ansorge, Christian Großmann, and Hans-Görg Roos
250 pages, 1987 ISBN 3-05-500362-4

37 **Factorization of Measurable Matrix Functions**
By Georgi Semenovich Litvinchuk and Ilya Matveyevich Spitkovskiĭ
372 pages, 1987 ISBN 3-05-500368-3

38 **Parallel Algorithms and Architectures**
Proceedings of the International Workshop on Parallel Algorithms and Architectures held in Suhl (GDR), May 25–30, 1987
Edited by Andreas Albrecht, Hermann Jung, and Kurt Mehlhorn
205 pages, 1987 ISBN 3-05-500397-7

39 **Parametric Integer Optimization**
By Bernd Bank and Reinhard Mandel
138 pages, 1988 ISBN 3-05-500398-5

40 **Computer Analysis of Images and Patterns 1987**
Proceedings of the II. International Conference CAIP '87 on Automatic Image Processing, held in Wismar (GDR), September 2–4, 1987
Edited by Leonid P. Jaroslavskii, Azriel Rosenfeld, and Wolfgang Wilhelmi
282 pages, 1987 ISBN 3-05-500451-5

41 **Multivariable Control**
A Graph-theoretic Approach
By Kurt J. Reinschke
274 pages, 1988 ISBN 3-05-500452-3

42 **Equilibrium Distributions of Branching Processes**
By Alfred Liemant, Klaus Matthes, and Anton Wakolbinger
240 pages, 1988 ISBN 3-05-500453-1

43 **Andreotti-Grauert Theory by Integral Formulas**
By Gennadi Markovič Henkin and Jürgen Leiterer
270 pages, 1988 ISBN 3-05-500510-4

44 **Continual Means and Boundary Value Problems in Function Spaces**

By Efim Mikhailovich Polishchuk

164 pages, 1988 ISBN 3-05-500512-0

45 **Advances in Mathematical Optimization**

Invited papers dedicated to Prof. Dr. Dr. h. c. F. Nožička on occasion of his 70th birthday

Edited by Jürgen Guddat et al.

332 pages, 1988 ISBN 3-05-500543-0

46 47 **System Analysis and Simulation 1988**

Proceedings of the International Symposium held in Berlin (GDR), September 12–16, 1988

I: Theory and Foundations

II: Applications

Edited by Achim Sydow, Spyros G. Tzsafestas, and Robert Vichnevetsky

426 425 pages, 1988 ISBN 3-05-500654-2 ISBN 3-05-500655-0

48 **Parcella '88**

Proceedings of the IV. International Workshop on Parallel Processing by Cellular Automata and Arrays, held in Berlin (GDR), October 17–21, 1988

Edited by Gottfried Wolf, Tamás Legendi, and Udo Schendel

379 pages, 1988 ISBN 3-05-500656-9

49 **Algebraic and Logic Programming**

Proceedings of an International Workshop held in Gaussig (GDR), November 14–18, 1988

Edited by Jan Grabowski, Pierre Lescanne, and Wolfgang Wechler

278 pages, 1988 ISBN 3-05-500655-8

50 **Asymptotics for Elliptic Mixed Boundary Problems**

Pseudo-Differential and Mellin Operators in Spaces with Conormal Singularity

By Stephan Rempel and Bert-Wolfgang Schulze

418 pages, 1989 ISBN 3-05-500676-3

51 **Geobild '89**

Proceedings of the 4th Workshop on Geometrical Problems of Image Processing held in Georgenthal (GDR), March 13–17, 1989

Edited by Albrecht Hübler, Werner Nagel, Brian David Ripley, and Günter Werner

198 pages, 1989 ISBN 3-05-500677-1

52 **Splines in Numerical Analysis**

Contributions to the International Seminar ISAM-89 held in Weissig (GDR), April 24–28, 1989

Edited by Jochen W. Schmidt and Helmut Späth

198 pages, 1989 ISBN 3-05-500679-8

53 **Complex Methods on Partial Differential Equations**

Aspects in Complex Analysis

Edited by Claudio I. Withalm

258 pages, 1989 ISBN 3-05-500680-1

54 **Markov Processes and Control Theory**

Edited by Heinz Langer and Volker Nollau

234 pages, 1989 ISBN 3-05-500754-9

55 **Computer Analysis of Images and Patterns**

Proceedings of the III. International Conference CAIP '89 on Automatic Image Processing held in Leipzig (GDR), September 8–10, 1989

Edited by Klaus Voss, Dmitry Chetverikov, and Gerald Sommer

259 pages, 1989 ISBN 3-05-500755-7

56 **Quaternionic Analysis and Elliptic Boundary Value Problems**

By Klaus Gürlebeck and Wolfgang Sprößig

254 pages, 1989 ISBN 3-05-500756-5

57 **Computers in Science and Higher Education**

Contributions to the Conference BIT '89 held in Berlin (GDR), June 1989

Edited by Jan Grabowski

321 pages, 1989 ISBN 3-05-500774-3

58 **Wissenschaftliches Rechnen mit Ergebnisverifikation**

herausgegeben von Ulrich Kulisch

VI, 262 Seiten, 1989 ISBN 3-05-500816-2

The main aim of the book consists in the statement of a new strategy for solving linear and nonlinear boundary value problems in bounded domains of R^3. The volume gives a summarizing presentation of a complete hypercomplex solution theory including analytical and numerical investigations. Using a special operator calculus and a hypercomplex function theory, the questions of existence, uniqueness, representation formulas, and regularity of solutions of elliptic boundary value problems are studied. Numerical methods which are well-adapted to quaternionic calculus are included too. An insight is given into boundary collocation methods. A discrete quaternionic function theory and its applications to constructing and investigating finite difference methods are published here for the first time.
The book is written for mathematicians, physicists, and engineers in research institutes interested in methods of quaternionic analysis and its application to the numerical solution of boundary value problems.

Hauptinhalt des Buches ist die Vorstellung einer neuen Strategie zur Lösung linearer und nichtlinearer Randwertprobleme in beschränkten Gebieten des R^3. Der Band gibt eine zusammenfassende Darstellung einer in sich geschlossenen Lösungstheorie elliptischer Randwertprobleme, die analytische und numerische Untersuchungen beinhaltet. Unter Benutzung einer speziellen Operatorenrechnung und einer hyperkomplexen Funktionentheorie werden die Fragen nach der Existenz und Eindeutigkeit, nach Darstellungsformeln und nach der Regularität der Lösungen untersucht. Numerische Verfahren, die der hyperkomplexen Betrachtungsweise gut angepaßt sind, werden ebenfalls betrachtet. So wird ein Einblick in die Theorie der Randkollokationsverfahren gegeben. Ein diskretes Modell der Funktionentheorie quaternionenwertiger Funktionen mit Anwendungen auf Konstruktion und analytische Untersuchung von Differenzenverfahren wird hier erstmals veröffentlicht.
Der Band ist geschrieben für Mathematiker, Physiker und Ingenieure in Forschungseinrichtungen, die interessiert sind an Fragestellungen und Methoden der Quaternionenanalysis und an ihrer Anwendung auf die numerische Lösung von Randwertproblemen.

ISBN 3-05-500756-5
ISSN 0138-3019

Bei Fragen zur Produktsicherheit wenden Sie sich bitte an:
If you have any questions regarding product safety,
please contact:

Walter de Gruyter GmbH
Genthiner Straße 13
10785 Berlin
productsafety@degruyterbrill.com